D1348468

Numerical Methods for Engineering Application

Numerical Methods for Engineering Application

JOEL H. FERZIGER

Department of Mechanical Engineering
Stanford University
Stanford, California

A Wiley-Interscience Publication

JOHN WILEY & SONS

New York · Chichester · Brisbane · Toronto

Library of Congress Cataloging in Publication Data:

Ferziger, Joel, H.
Numerical methods for engineering application

"A Wiley-interscience publication."
Bibliography: p.
Includes index.
1. Engineering mathematics. 2. Numerical analysis.
Title.
TA335.F47 515'.02462 81-1260
ISBN 0-471-06336-3 AACR2

Printed in the United States of America

10 9 8 7 6 5 4 3 2 1

To my mother Bessie

and the memory of my late father Moe Ferziger

Preface

Many of the traditional fields of engineering have undergone enormous changes in the past 30 years. In place of the handbook correlation methods that were once commonly used, engineers find themselves going ever more frequently to methods based on fundamental principles and the equations derived from them. A major reason for this is that the rate of technological development has increased so rapidly that it is simply impossible to produce the data that traditional methods would require.

One of the key technological changes is the development of the electronic digital computer whose beginnings are only about 30 years old. The best machines of that day could hardly compete with the cheap and readily available hand calculator of today. We have come a long way and there is no reason to believe that the end is yet in sight. Over the past 30 years the capability of computers has increased by something like two orders of magnitude per decade and, while the cost of machines has gone up, it has not increased at anything like the rate at which their performance has. As a result, the effective cost of computation has decreased by more than an order of magnitude per decade. While we can hardly expect these trends to last forever, there is still no indication of a deviation from the curve of cost versus time and the trend can be expected to continue for at least another decade.

The total cost of doing computing is the sum of the hardware (machine) costs and the software (human) costs. In many applications, hardware costs have become a small part of the total and this is causing a kind of minirevolution in the computer business. In many applications of interest to engineers, however, hardware cost remains dominant. This means that the cost of doing computing will probably continue to decrease relative to the cost of other engineering functions. Consequently, we can anticipate a continuation of the trend toward the computerization of many traditional engineering tasks. This trend covers all areas of engineering, but not all to the same degree.

Not all engineers will use computers heavily in their work, but a large enough number will to render courses in the application of computers to the solution of engineering problems a necessary part of the modern engineering curriculum. Such courses are offered by mathematics and computer science departments and are appropriate for many engineering students. But there is a wide variety of students in engineering, each of whom has a different and legitimate set of needs. It is difficult to meet all of these needs with a single set of courses and there is a tendency toward the teaching of some mathematics and computer science courses by engineering departments.

The material in this book comes from a sequence of courses that the author has offered in the Mechanical Engineering Department at Stanford University and has proven to be of interest to students in related departments. As presently structured, there are four courses in the sequence: the first covers linear algebra, including its numerical aspects and its application to the solution of nonlinear problems; the second covers analytical methods for the solution of partial differential equations; the third deals with numerical analysis, especially those aspects related to the solution of ordinary and partial differential equations; and the fourth covers the application of numerical methods to fluid mechanics. Other courses in numerical methods, including some in the finite element method, are offered by other departments.

This book is based on the third and part of the fourth in the sequence of courses described above, principally the numerical solution of ordinary and partial differential equations. (Notes for the others are available from the author.) It is intended for people who will be using the methods rather than those who will be developing new methods. In practical terms, this means that the emphasis is on intuitive understanding of what makes a good numerical method and what distinguishes it from poorer methods, and not on the analysis of methods. In particular, the analysis of the stability and accuracy of particular methods, which rightly plays a central role in numerical analysis, is given less attention in this work than in standard numerical analysis texts. Correct results will be presented here, but the long analyses necessary to derive them are sometimes referenced rather than given. This is not to say that analysis is not important; it is. The audience to which this book is addressed, however, is one that will generally accept the result without seeing the proof, as long as it can be assured that the proof does exist.

The arrangement of the subjects in this work is a fairly standard one. Interpolation is treated first because it is the cornerstone on which other methods are based. Quadrature, a subject that finds considerable use by engineers, is treated next. This is followed by a chapter on numerical methods for ordinary differential equations. Initial value problems are the focus of the first half of this chapter and boundary value problems are dealt with in the second half. The book closes with a chapter on partial differential equations.

The approach is one in which each subject is built on the subject(s) preceding it. This seems to be a near optimum approach, as it allows the student to see the interconnections between methods that sometimes appear disjoint. It is also simpler to grasp the entire subject when the framework is visible.

The point of view taken is that only "good" methods (defined as those that are efficient and well used) should be presented. Better a few pearls than a beach of sand. For this reason, this book does not contain a compendium of all known methods. It is devoted to a relatively few methods chosen either for the pedagogical reason that they are easy to understand and analyze or because they are commonly used. There are, of course, other good methods that could have been included and were not. The methods described are as good as any.

The author would like to end this preface by giving thanks to the people without whom this work could not have been completed. I learned much of what I know about numerical analysis (especially in how to think about it) from my good friend Dr. Samuel Schechter; Professors Joseph Oliger and Gene Golub were also important contributors in this regard. Help in assembling the examples was very cheerfully provided by Sr. Juan Bardina and is greatly appreciated; and a number of his students and colleagues too numerous to mention by name, also played an important role in suggesting changes and correcting errors in the text. Ms. Ann Ibaraki did service beyond the call of her job in typing and assembling the original version. Later versions were typed by Mss. Ruth Korb, Romen Rey, and Charlean Hampton. Finally, thanks are due to my colleague, friend, and department head Professor Bill Reynolds for his encouragement.

JOEL H. FERZIGER

Stanford University
February 1981

Contents

CHAPTER 1 INTERPOLATION 1

1. Lagrange Interpolation, 2
2. Hermite Interpolation, 12
3. Splines, 12
4. Tension Spline, 20
5. Parametric and Multidimensional Interpolation, 21

Problems, 23

CHAPTER 2 INTEGRATION 24

1. Newton-Cotes Formulas, 25
2. Richardson Extrapolation, 30
3. Romberg Integration, 32
4. Adaptive Quadrature, 37
5. Gauss Quadrature, 41
6. Singularities, 47
7. Concluding Remarks, 48

Problems, 49

CHAPTER 3 ORDINARY DIFFERENTIAL EQUATIONS 50

1. Numerical Differentiation, 51
2. Euler's Method, 57
3. Stability, 60
4. Backward or Implicit Euler Method, 69
5. Accuracy Improvement, 72
6. Predictor-Corrector and Runge–Kutta Methods, 76
7. Multistep Methods, 84
8. The Choice of Method and Automatic Error Control, 92
9. Systems of Equations—Stiffness, 95

10. Systems of Equations—Inherent Instability, 104
11. Boundary Value Problems: I. Shooting, 105
12. Boundary Value Problems: II. Direct Methods, 110
13. Boundary Value Problems: III. Higher Order Methods, 115
14. Boundary Value Problems: IV. Nonuniform Grids, 120
15. Boundary Value Problems: V. Finite Element Methods, 124
16. Boundary Value Problems: VI. Eigenvalue Problems, 126

CHAPTER 4. PARTIAL DIFFERENTIAL EQUATIONS 135

1. Parabolic PDEs: I. Explicit Methods, 138
2. Parabolic PDEs: II. The Crank–Nicolson Method, 147
3. Parabolic PDEs: III. The Dufort–Frankel Method, 152
4. Parabolic PDEs: IV. The Keller Box Method and Higher Order Methods, 155
5. Parabolic PDEs: V. Two and Three Dimensions—Alternating Direction Implicit (ADI) Methods, 158
6. Parabolic PDEs: VI. Other Coordinate Systems and Nonlinearity, 167
7. Elliptic PDEs: I. Finite Differencing, 171
8. Elliptic PDEs: II. Jacobi Iteration Method, 177
9. Elliptic PDEs: III. Gauss-Seidel Method, 184
10. Elliptic PDEs: IV. Line Relaxation Method, 187
11. Elliptic PDEs: V. Successive Overrelaxation (SOR) Method, 189
12. Elliptic PDEs: VI. Alternating Direction Implicit (ADI) Methods, 197
13. Elliptic PDEs: VII. Finite Element Methods, 202
14. Discrete Fourier Transforms, 207
15. The Fast Fourier Transform (FFT) Algorithm, 212
16. Elliptic PDEs: VIII. Fourier Methods, 216
17. Elliptic PDEs: IX. Boundary Integral Methods, 218
18. Hyperbolic PDEs: I. Review of Theory, 221
19. Hyperbolic PDEs: II. Method of Characteristics, 224
20. Hyperbolic PDEs: III. Explicit Methods, 234
21. Hyperbolic PDEs: IV. Implicit Methods, 241
22. Hyperbolic PDEs V. Splitting Methods, 244

APPENDIX A. SOLUTION OF TRIDIAGONAL SYSTEMS 253
APPENDIX B. THE NEWTON–RAPHSON METHOD 255
REFERENCES AND ANNOTATED BIBLIOGRAPHY 260
INDEX 263

Numerical Methods for Engineering Applications

1

Interpolation

Interpolation is the process of "reading between the lines" of a table or of fitting a smooth curve to a limited set of data. We take it up first for a number of reasons, the most obvious of which is that interpolation is frequently used for estimating quantities from tabulated data. A more important reason is that many numerical differentiation and integration procedures are derived by using interpolation to find a smooth approximation and then differentiating or integrating the result.

There are two kinds of interpolation, depending on the type of data provided and the kind of result wanted. In the standard type of interpolation we are given a set of data points and want a curve that passes smoothly *through* them. In least squares interpolation the data generally have some uncertainty associated with them and we want to find a smooth curve that passes sufficiently *near* the data points. In standard interpolation the equation of the approximating curve must have as many parameters as there are data points; in least squares fitting the number of parameters typically is much smaller than the number of data points. We will deal only with the standard case. Least squares fitting is an important topic, especially for experimentalists, but space does not permit its inclusion here.

The basic problem of interpolation may be stated as follows. Given a set of data (x_i, y_i), $i=1,2,\ldots n$, find a smooth curve $f(x)$ that passes through the data. We require the following of the interpolating curve:

1. From the problem statement, we must have

$$f(x_i)=y_i \qquad i=1,2,\ldots n \tag{1.1}$$

2. The function should be easy to evaluate.
3. It should also be easy to integrate and differentiate.
4. It should be linear in the adjustable parameters (to simplify the problem of finding them).

The choice of interpolating function depends on what one means by smoothness and on the function to be approximated. Many functions have been used, the most common of which are polynomials of various kinds because they satisfy criteria (2) and (3) better than any other type of function. Even among polynomial interpolations there are a number of classes. We begin with the simplest—Lagrange interpolation.

1. LAGRANGE INTERPOLATION

In Lagrange interpolation we pass a polynomial of lowest possible degree through the n given data points. Since n parameters are needed, the degree required is $n-1$ so that

$$f(x)=a_{n-1}x^{n-1}+a_{n-2}x^{n-2}+\cdots a_1x+a_0 \tag{1.1.1}$$

The straightforward approach to finding the coefficients is to plug Eq. 1.1.1 into Eq. 1.1. We obtain

$$a_{n-1}x_i^{n-1}+a_{n-2}x_i^{n-2}+\cdots a_1x_i+a_0=y_i \qquad i=1,2,\dots n \tag{1.1.2}$$

which represents a set of n linear algebraic equations in the n unknowns $a_0, a_1,\dots a_{n-1}$, since the x_i and y_i are known. This set can be solved by standard linear equation solvers, but this is not a good way to proceed because (1) it requires a computer if n is larger than 4 or 5, (2) Eqs. 1.1.2 become ill-conditioned for n larger than about 5, and (3) it is better to have a closed form expression in any case. (By ill-conditioned we mean that solution of the system of equations is very sensitive to small changes in the data. When an ill-conditioned system is solved, small errors are magnified and the results may contain large errors. In extreme cases the solution is completely invalid.)

There is another approach. From Eq. 1.1.2 we see that the coefficients $a_0, a_1,\dots a_{n-1}$ must be linear combinations of the y_i. The most general expression that is linear in each of the y_i and a polynomial of degree $n-1$ in x is

$$f(x)=\sum_{k=1}^{n} L_k(x)y_k \tag{1.1.3}$$

where the $L_k(x)$ are polynomials of degree $n-1$. Thus $f(x)$ must have this form.

The problem is to find the $L_k(x)$. It is important to note that they do not depend on the values of y_k. Consequently, we choose the y_k judiciously to simplify the problem. In particular, suppose we let one of the y_k, say y_j, be unity and let all of the others be zero. Then substitution of Eq. 1.1.3 into Eq. 1.1 gives

$$L_j(x_i)=\delta_{ij} \qquad i=1,2,\dots n \tag{1.1.4}$$

where δ_{ij} is the Kronecker symbol:

$$\delta_{ij}=\begin{cases}1 & i=j\\ 0 & i\neq j\end{cases} \tag{1.1.5}$$

In Eq. 1.1.4 the choice of j is arbitrary, so it must hold for *all* values of j. Thus $L_j(x)$ is a polynomial of degree $n-1$ that is zero when $x=x_1, x_2,\ldots x_{j-1}, x_{j+1},\ldots$ or x_n and unity when $x=x_j$. Any polynomial of degree n can be factored into a constant multiple of a product of n factors $(x-x_l)$ where the x_l are the zeros of the polynomial. Since $L_j(x)$ is a polynomial of degree $n-1$ and we know all of its $n-1$ zeros, it must have the form

$$L_j(x)=C_j(x-x_1)(x-x_2)\cdots(x-x_{j-1})(x-x_{j+1})\cdots(x-x_n) \tag{1.1.6}$$

where C_j is a constant. The value of C_j is easily determined from the requirement that $L_j(x_j)=1$. We find

$$C_j=(x_j-x_1)^{-1}(x_j-x_2)^{-1}\cdots(x_j-x_{j-1})^{-1}(x_j-x_{j+1})^{-1}\cdots(x_j-x_n)^{-1} \tag{1.1.7}$$

so that

$$L_j(x)=\frac{(x-x_1)\cdots(x-x_{j-1})(x-x_{j+1})\cdots(x-x_n)}{(x_j-x_1)\cdots(x_j-x_{j-1})(x_j-x_{j+1})\cdots(x_j-x_n)} \tag{1.1.8}$$

For later applications it is convenient to introduce the polynomial of degree n

$$F(x)=(x-x_1)(x-x_2)\cdots(x-x_n) \tag{1.1.9}$$

In terms of $F(x)$, we can write

$$L_j(x)=\frac{C_jF(x)}{x-x_j} \tag{1.1.10}$$

Any numerical method, by definition, produces an approximation to something we wish to compute. In application we rarely need the exact value of a quantity, but there is always a minimum acceptable tolerance. We need to know whether the method used produces the required accuracy. To answer this question for Lagrange interpolation, we suppose that the given data are exactly the values that some smooth function $y(x)$ takes at the points $x_1, x_2,\ldots x_n$. Then $f(x)-y(x)$ is a function that is zero at each of the n data points. Also, $F(x)$ is a polynomial that is zero at these points. Now consider the function

$$g(x)=y(x)-f(x)-AF(x) \tag{1.1.11}$$

where A is a constant. We will choose A so that $g(x)=0$ at some point $x_1<x_0<x_n$. Thus $g(x)$ has at least $(n+1)$ zeros $x_0, x_1, \ldots x_n$. Since g is smooth, it must have a minimum or maximum between each pair of zeros. Therefore, $g'(x)$ has at least n zeros, $g''(x)$ has at least $(n-1)$ zeros,... and, continuing the process, $g^{(n)}(x)$ has at least one zero. Let ξ be this zero, so that $g^{(n)}(\xi)=0$. Since f is a polynomial of degree $(n-1)$, $f^{(n)}=0$. Also, by differentiating Eq. 1.1.9, we find $F^{(n)}=n!$ so that

$$g^{(n)}(\xi)=y^{(n)}(\xi)-An!=0 \tag{1.1.12}$$

Thus, solving for A, we have

$$A=\frac{y^{(n)}(\xi)}{n!} \tag{1.1.13}$$

and

$$y(x)=f(x)+\frac{y^{(n)}(\xi)}{n!}F(x) \tag{1.1.14}$$

where $x_1<\xi<x_n$. The last term is the desired error estimate for Lagrange interpolation. As one might expect, the larger $y^{(n)}(\xi)$, that is, the less smooth the function, the greater the error in the interpolation. Also, we see from the definition of $F(x)$ that the error will be greater for wider spacing between the data points and also will be greater near the end points x_1 and x_n than near the center of the range. This accords with what intuition would tell us.

We now give a number of examples that illustrate the properties of Lagrange interpolation. These examples have been chosen to illustrate some important points and are used later with other methods.

Example 1.1

For the first example we will take e^x for $0<x<1$. This is a smooth function that should be fit well by any interpolation. Using only three points, we can do the calculation by hand and it is useful to do so. With $x_1=0$, $x_2=0.5$, and $x_3=1$, we have

$$f(x)=f(x_1)\frac{(x-x_2)(x-x_3)}{(x_1-x_2)(x_1-x_3)}+f(x_2)\frac{(x-x_1)(x-x_3)}{(x_2-x_1)(x_2-x_3)}$$

$$+f(x_3)\frac{(x-x_1)(x-x_2)}{(x_3-x_1)(x_3-x_2)}$$

At $x=0.25$, we have

$$f(0.25)=1\frac{(-0.25)(-0.75)}{(-0.5)(-1)}+(1.648721)\frac{(0.25)(-0.75)}{(0.5)(-0.5)}$$

$$+(2.718282)\frac{(0.25)(-0.25)}{(1)(0.5)}$$

$$=0.375000+1.236541-0.339785$$

$$=1.271756$$

The exact value of the function is $e^{0.25}=1.2840254$ so that the error is 0.0123 or just about 1%.

It is interesting to compare the error estimate (Eq. 1.1.14) with the actual error. There are a couple of difficulties that arise in doing so. The first is that we do not know what the value of ξ is that makes Eq. 1.1.14 correct. To be safe, we can choose the value that gives the largest error estimate (assuming we can find it!). A more serious problem is that the derivative required is usually very difficult to calculate. In this example, because the function is simple, there is no difficulty. Choosing $\xi=1$ for safety, we find the estimated error to be

$$\varepsilon=\frac{e^1}{3!}F(0.25)=\frac{2.71828}{6}(0.25)(-0.25)(-0.75)=0.0212$$

which is nearly double the actual error. This is actually not too bad; error estimates of this kind sometimes can be much worse than this.

The actual error over the full range of $0<x<1$ is shown in Fig. 1.1. As expected, the error oscillates between the data points (at which it must be zero) and it is evenly distributed over the range. This is fairly typical.

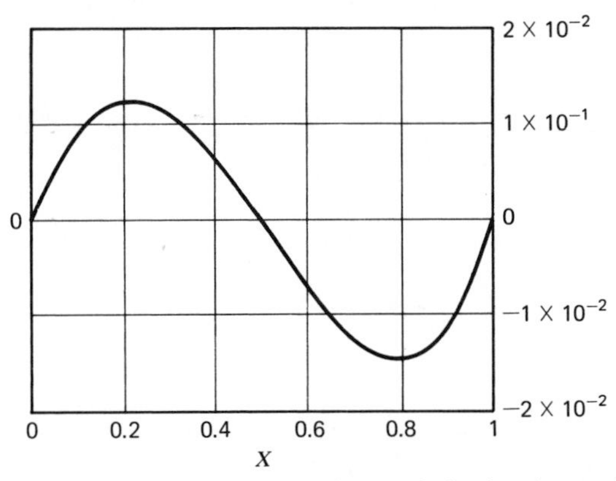

Fig. 1.1. Error in Lagrange interpolation of e^x using three points.

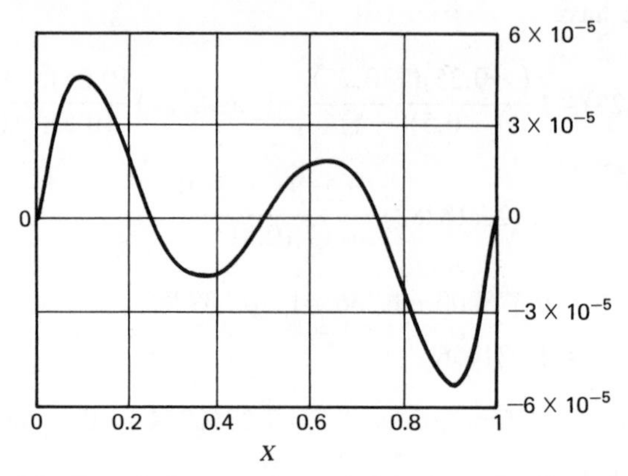

Fig. 1.2. Error in Lagrange interpolation of e^x using five points.

Before going to a larger number of points, let us see how Lagrange interpolation works as an extrapolator. At $x=-0.25$ we find $f(x)=0.83345$ versus an exact value of $e^{-0.25}=0.77880$. The error is now 0.055, more than four times what it was at $x=0.25$ on an absolute basis and even worse in terms of relative error. It is clear that *extrapolation is a much riskier process than interpolation*, a fact of considerable importance in numerical analysis.

Interpolating e^x using five points rather than three produces the results shown in Fig. 1.2. As might be expected, the errors are much smaller. Still, they oscillate, and the largest errors are found near the ends of the interval. This accords with our expectation.

In Fig. 1.3 we have given the maximum error for Lagrange interpolation of e^x versus the number of points used. The dashed line is obtained from the estimate given above. The estimate agrees well with actual results when n is small, but is much too low when n is large because for large n, the error in the interpolation is much smaller than the error produced by roundoff. (See appendix A for a short discussion of roundoff error.) The roundoff error then

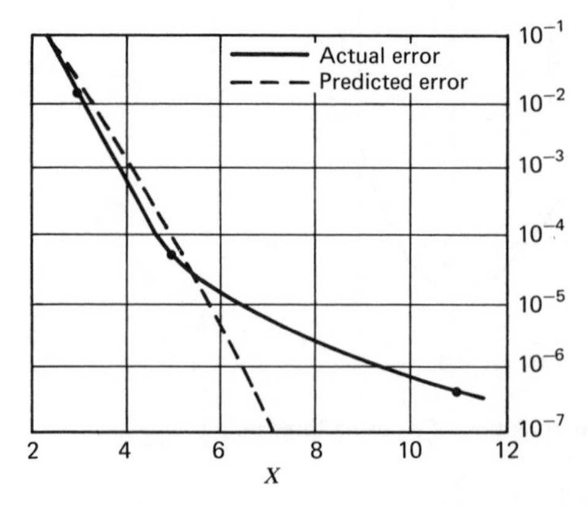

Fig. 1.3. Maximum error in Lagrange interpolation of e^x versus number of points used.

dominates, and the estimate, which does not include roundoff, becomes meaningless. The machine used in this calculation (a Digital Equipment 20/50) has a relative precision of about 10^{-7}; in other words, a number can be represented to seven significant figures. Thus the error cannot be smaller than 10^{-7}. If we had gone to a still higher order, both the number of numerical operations and the roundoff error would increase and the error would actually become larger with increasing n. This brings us to another point that cannot be overemphasized—*one should never use a method whose accuracy is much greater than that of the computer used*. Using such a method increases the cost with no increase in accuracy.

Example 1.2

As a second case, we take sine function $\sin 2\pi x$ on $0<x<1$. The question is one of determining how many points are required to describe a full period of this function. Note that if only two or three points are used, the given values would all be zero (or some other constant if the phase were nonzero) and the interpolation would be meaningless. Figures 1.4 and 1.5 show the results with four and seven points, respectively. The maximum error is much larger than for the exponential function. This is what we might expect intuitively—wiggly functions are much more difficult to fit with polynomials than smooth ones.

The trend of maximum error versus the number of points used shown in Fig. 1.6 is also a little peculiar. The observed error does not follow a smooth curve. The reason is that the placement of data points on the sine curve is quite important. This example shows that a method that is good in one case can be quite poor in another. One of the keys to doing good numerical work is fitting the method to the problem.

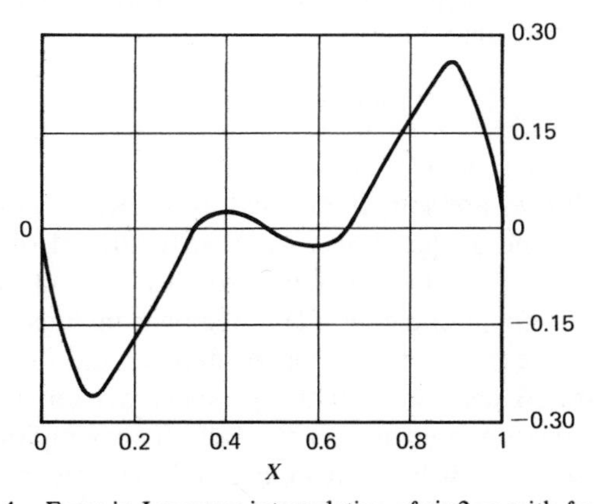

Fig. 1.4. Error in Lagrange interpolation of $\sin 2\pi x$ with four points.

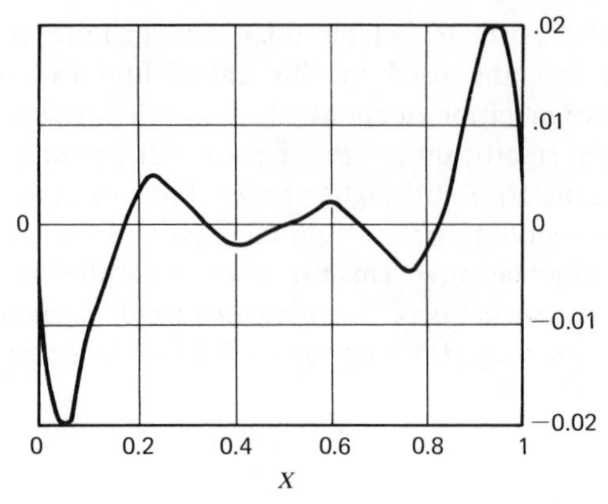

Fig. 1.5. Error in Lagrange interpolation of $\sin 2\pi x$ with seven points.

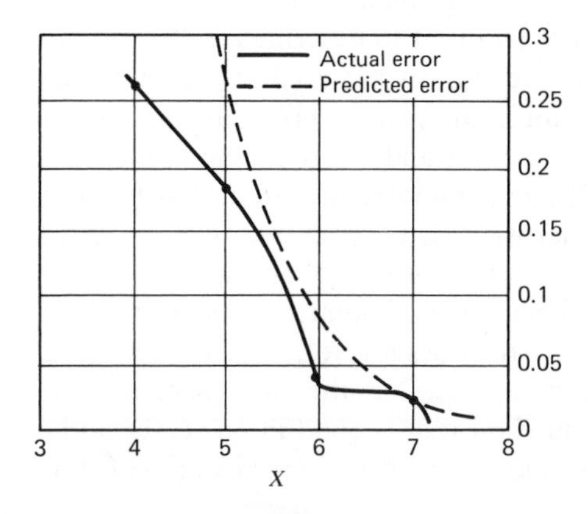

Fig. 1.6. Maximum error in Lagrange interpolation of $\sin 2\pi x$ versus number of points used.

Example 1.3

The third example illustrates another difficult problem for interpolation procedures, the case of a function with concentrated curvature. A good example is the so-called superellipse

$$y=(1-x^m)^{1/m}$$

For $m=2$ the curve is a circle; for large values of m the curve is almost a square. The superellipses are sometimes used as a means of fairing between a circular cross section and a square one. Another feature of these curves is that they become vertical at $x=1$, which is also difficult for a polynomial to represent. The results for a circle ($n=2$) are shown in Figs. 1.7 and 1.8. The error is again much larger than it was for the exponential function and, in this case, it is concentrated near $x=1$. This is expected from the nature of the function. Also, the maximum error does not fall off very rapidly as the number of points is increased as can be seen in Fig. 1.9. The reason is that as data points very close to $x=1$ are included, the polynomial is forced to deal more

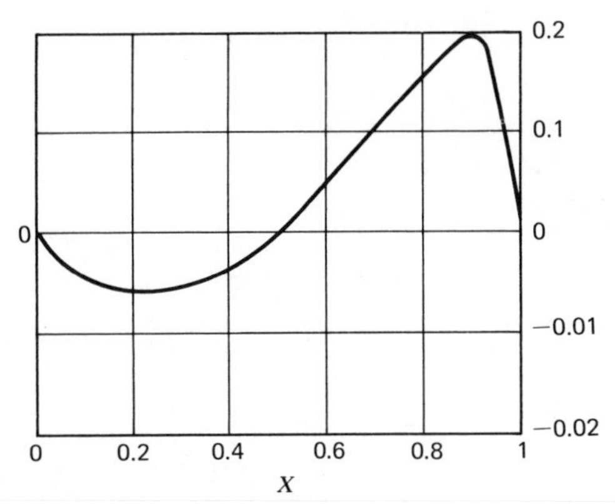

Fig. 1.7. Error in Lagrange interpolation of $(1-x^2)^{1/2}$ using three points.

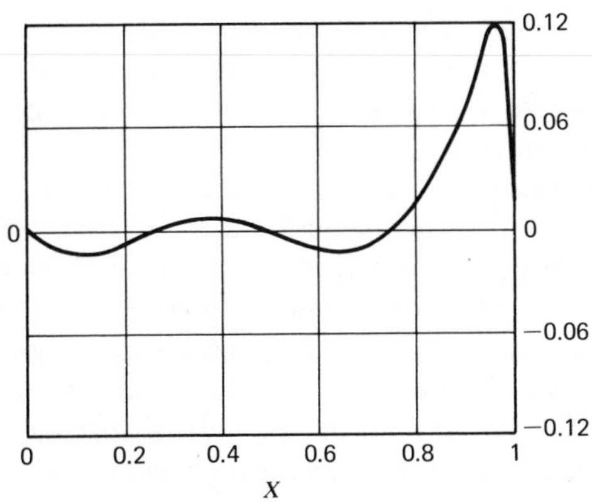

Fig. 1.8. Error in Lagrange interpolation of $(1-x^2)^{1/2}$ using five points.

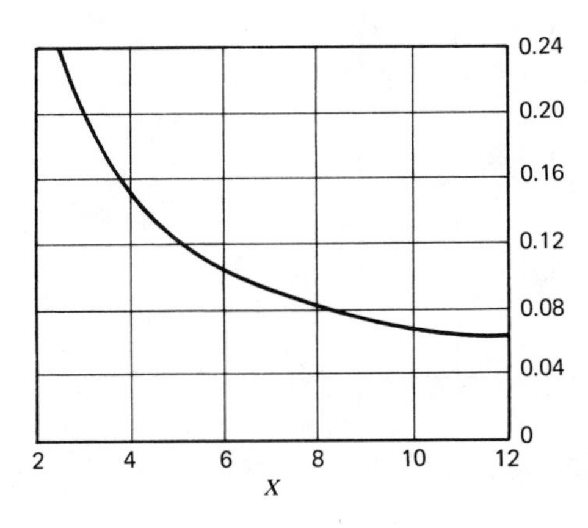

Fig. 1.9. Maximum error in Lagrange interpolation of $(1-x^2)^{1/2}$ versus number of points used.

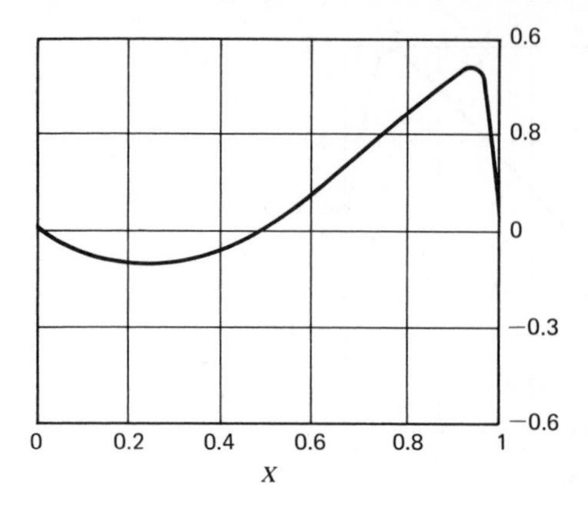

Fig. 1.10. Error in Lagrange interpolation of $(1-x^4)^{1/4}$ using three points.

accurately with the large slope of the curve in this vicinity. As we have seen, a polynomial cannot do that very well and the result is not satisfactory.

The case of $n=4$ is shown in Figs. 1.10 to 1.12. The results are similar to those obtained for $n=2$, but are much more exaggerated. The error is even more concentrated where the curvature is large and it falls off more slowly with an increasing number of points.

It might be anticipated that since the error is concentrated near $x=1$, we might get improvement by using unequal spacing of the mesh points such that more points are in the neighborhood of $x=1$. This is in fact the case, but the improvement is not dramatic. More improvement is obtained by using *piece-wise* Lagrange interpolation. Instead of fitting a single high order polynomial to the entire curve, we fit lower order polynomials to sections of it. This method is quite flexible, since one can use narrow spacing in regions of large curvature or high slope and the results are generally quite satisfactory. The

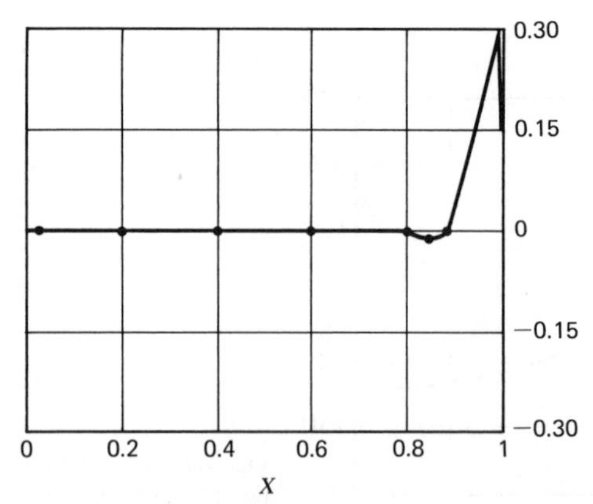

Fig. 1.11. Error in Lagrange interpolation of $(1-x^4)^{1/4}$ using ten points.

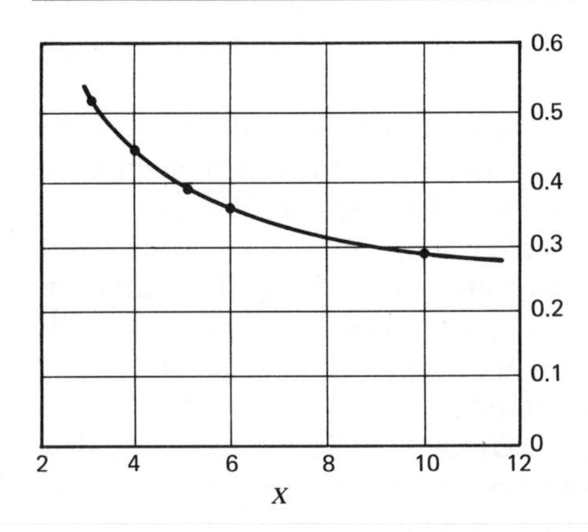

Fig. 1.12. Maximum error in Lagrange interpolation of $(1-x^4)^{1/4}$ versus number of points used.

simplest example of piecewise Lagrange interpolation is the familiar piecewise linear ("connect-the-dots") method. As we point out later, piecewise Lagrange interpolation forms the basis for a number of the most popular numerical methods.

These examples show a number of important characteristics of polynomial interpolation. The interpolated curves tend to oscillate about the exact result. Smooth functions are treated more accurately than oscillatory ones or ones with concentrated curvature. For these reasons, Lagrange interpolation with more than three or four points is rarely used. Piecewise Lagrange interpolation offers some improvement, but suffers from having discontinuous derivatives at the points that join the segments and may cause trouble if the result is to be differentiated. The methods described in the following sections attempt to overcome this.

Higher accuracy can be obtained by having the computer use double (or higher) precision. The combination of high precision arithmetic and accurate algorithms is the best way to achieve highly accurate results, but should be used only when really needed. Double precision sometimes can be used to determine whether a bad result is due to roundoff accumulation or to the algorithm itself.

Lagrange interpolation is rarely used as a method by itself for reasons that are explained below. It does form the basis for a number of other methods, which is why it has been included here. We will not give a FORTRAN procedure for Lagrange interpolation as we will for other methods.

Lagrange interpolation in the form presented above requires a considerable amount of computation. There are a number of alternative but equivalent computational methods (such as divided differences) that reduce the number of numerical operations required to achieve the interpolated result. They are not presented here, as our main interest in Lagrange interpolation is as a basis for other numerical methods.

2. HERMITE INTERPOLATION

There are occasions when we want smoothness beyond that provided by Lagrange interpolation. Such smoothness can be obtained in a number of ways. One of the simplest is to provide not only the value of the function y_i at each point, but also the value of its derivative y_i'. We then have *Hermite* interpolation. Since the data required often are not available, this method is not used frequently as an interpolation technique. It does, however, provide the basis for other important numerical techniques (Gauss quadrature, in particular) and is included here for this reason.

The problem may be stated as follows. Given y_i and y_i' at each of n data points $x_1, x_2, \dots x_n$, find a polynomial of degree $(2n-1)$ that matches all the data. The analysis is very similar to that for Lagrange interpolation, so we will be brief. The polynomial must be linear in all the y_i and y_i', so we have

$$f(x)=\sum_{k=1}^{n} U_k(x)y_k+\sum_{k=1}^{n} V_k(x)y_k' \qquad (1.2.1)$$

where $U_k(x)$ and $V_k(x)$ are polynomials of degree $(2n-1)$ with the following properties:

$$U_k(x_j)=\delta_{jk} \qquad U_k'(x_j)=0$$

$$V_k(x_j)=0 \qquad V_k'(x_j)=\delta_{jk} \qquad (1.2.2)$$

The required polynomials can be constructed by using the properties of the Lagrange polynomial and the results may be expressed in terms of $L_k(x)$. The result is

$$U_k(x)=\left[1-2L_k'(x_k)(x-x_k)\right]L_k^2(x)$$

$$V_k(x)=(x-x_k)L_k^2(x) \qquad (1.2.3)$$

Finally, the error can be estimated by

$$y(x)=f(x)=\frac{y^{(2n)}(\xi)}{(2n)!}F^2(x) \qquad (1.2.4)$$

where ξ again is some point in the interval $x_1<\xi<x_n$.

3. SPLINES

Some disadvantages of Lagrange and Hermite interpolation that have been pointed out are that they can be computationally difficult, they may be inaccurate if the number of data points is large, and the piecewise versions produce discontinuities in derivatives at the joint points. They are useful for deriving numerical methods (as we point out in Chapter 3), but the curves they produce generally are not very satisfactory.

For curve fitting purposes we would like a smooth curve similar to what a drafter might produce. An old drafter's method uses a spline—a thin, flexible rod that can be bent to fit through the given data points. The fit is made by holding the drafting board vertically and placing weights on the spline to make it pass through the data points. The equation from mechanics describing the shape of the spline is

$$EIy^{iv} = F \tag{1.3.1}$$

where E and I are constants that are of no importance here and F is the applied force. Since the force is applied only at discrete points, $F=0$ between the weights and the curve is piecewise cubic. Furthermore, by integrating Eq. 1.3.1 in the vicinity of a weight, we find that the third derivative may be discontinuous at the weights, but the function and its first two derivatives are continuous.

Thus we define a spline interpolating curve by the following criteria:

1. The curve is piecewise cubic, that is, the coefficients of the polynomial are different on each interval (x_i, x_{i+1}).
2. The curve passes through the given data $(x_i, y_i, i=1,2,\dots n)$.
3. The first and second derivatives are continuous at the weight or node points x_i. The end points require special treatment, which we will take up below.

We begin by noting that the criteria given above require that $f''(x)$ be piecewise linear and continuous. The second derivative is thus completely specified by giving its value at the node points. In fact, using the Lagrange formula with $n=1$, we have

$$f_i''(x) = f''(x_i)\frac{x_{i+1}-x}{x_{i+1}-x_i} + f''(x_{i+1})\frac{x-x_i}{x_{i+1}-x_i} \tag{1.3.2}$$

The subscript on f'' indicates that this formula holds only for $x_i \leqslant x \leqslant x_{i+1}$. Equation 1.3.2 is easily integrated twice to give a cubic containing two constants of integration. These may be evaluated from the conditions that the function fit the data, that is, $f_i(x_i)=y_i$ and $f_i(x_{i+1})=y_{i+1}$. After some calculation, we find

$$\begin{aligned} f_i(x) = {} & f''(x_i)\frac{(x_{i+1}-x)^3}{6\Delta_i} + f''(x_{i+1})\frac{(x-x_i)^3}{6\Delta_i} \\ & + \left[\frac{y_i}{\Delta_i} - \frac{\Delta_i}{6}f''(x_i)\right](x_{i+1}-x) \\ & + \left[\frac{y_{i+1}}{\Delta_i} - \frac{\Delta_i}{6}f''(x_{i+1})\right](x-x_i) \end{aligned} \tag{1.3.3}$$

where $\Delta_i = x_{i+1} - x_i$. If we can find $f''(x_i)$, the formula will be complete. We have satisfied all the criteria enumerated above except continuity of the first derivative, so we must use this condition to find the $f''(x_i)$. Continuity of f' is enforced by carrying out the following operations:

1. Differentiate $f_i(x)$ and set $x = x_{i+1}$; that is, evaluate the derivative of f at the right-hand limit of interval i.
2. Differentiate $f_{i+1}(x)$ and set $x = x_{i+1}$; that is, evaluate the derivative of f at the left-hand limit of interval $i+1$.
3. Equate the two results obtained.

The result of these operations is the following set of equations:

$$\frac{\Delta_{i-1}}{6} f''(x_{i-1}) + \frac{(\Delta_{i-1} + \Delta_i)}{3} f''(x_i) + \frac{\Delta_i}{6} f''(x_{i+1}) = \frac{y_{i+1} - y_i}{\Delta_i} - \frac{y_i - y_{i-1}}{\Delta_{i-1}} \tag{1.3.4}$$

This is a set of linear algebraic equations for $f''(x_i)$. Since the set of equations is tridiagonal, that is, the ith equation contains only $f''(x_{i-1})$, $f''(x_i)$, and $f''(x_{i+1})$, it is easily solved (see Appendix B). The remaining problem is that for $i = 1$ or n one cannot apply continuity of f', so we only have $n-2$ equations for n unknowns, and two further conditions are needed. These require some assumptions and/or approximations in the end intervals. Some possibilities are the following:

1. *Periodic spline.* This assumption is useful when the data are part of a repeating periodic curve and represents the fact that the ends meet. We have

$$f''(x_1) = f''(x_{n-1}) \qquad f''(x_2) = f''(x_n) \tag{1.3.5}$$

2. *Parabolic runout.* We set

$$f''(x_1) = f''(x_2) \qquad f''(x_n) = f''(x_{n-1}) \tag{1.3.6}$$

which makes f'' constant on the end intervals and f quadratic there.

3. If the end of the spline is free, no moment will be exerted and

$$f''(x_1) = f''(x_n) = 0 \tag{1.3.7}$$

This gives the so-called natural spline—the curve that minimizes the total curvature of the spline. Physically, it corresponds to pinning the ends—simple supports in beam theory.

4. *Cantilever.* A combination of (2) and (3) above gives the condition

$$f''(x_1) = \lambda f''(x_2) \qquad f''(x_n) = \lambda f''(x_{n-1}) \qquad 0 \leqslant \lambda \leqslant 1 \tag{1.3.8}$$

which represents a cantilevered beam. This leaves the problem of picking λ. $\lambda=0$ corresponds to (3) and $\lambda=1$ to (2).

5. Still another choice is given by Forsythe, Malcolm, and Moler (1977), who suggest passing a Lagrange cubic through the first four data points and then matching the third derivative of this cubic to that of the spline at the end point. This makes the end condition take the form

$$-\Delta_1 f''(x_1)+\Delta_1 f''(x_2)=\alpha_1 y_1+\alpha_2 y_2+\alpha_3 y_3+\alpha_4 y_4 \tag{1.3.9}$$

where the α_i are the coefficients of the difference formula. For a uniform mesh, it becomes $\alpha_1=-\alpha_4=(1/6\Delta)$ and $\alpha_2=-\alpha_3=(-1/2\Delta)$. A similar treatment is used at the other end.

For equally spaced intervals Eq. 1.3.4 becomes

$$\tfrac{1}{6}[f''(x_{i-1})+4f''(x_i)+f''(x_{i+1})]=\frac{y_{i+1}-2y_i+y_{i-1}}{\Delta^2} \tag{1.3.10}$$

One can show that for this case the error is given by

$$f(x)-y(x)\simeq\frac{\Delta^4}{96}y^{iv}_{\max} \tag{1.3.11}$$

This is similar to the error in the cubic Lagrange, but the advantages and disadvantages of the spline are best illustrated by examples.

Example 1.4

This example will serve to introduce a method of investigating numerical methods that is used a number of times later in the text. By taking the simplest nontrivial case, it is possible to gain insight into a method. In the case of spline interpolation the simplest case that illustrates the important features has four data points. This gives one segment in which the function is actually cubic and two end intervals. For further simplicity we will assume that the data points are equally spaced so that Eq. 1.3.10 can be used. For $i=2$, we have

$$f''(x_1)+4f''(x_2)+f''(x_3)=\frac{6}{\Delta^2}(y_1-2y_2+y_3) \tag{A}$$

If we abbreviate $f''(x_i)=f_i''$ and use the boundary condition Eq. 1.3.8, this becomes

$$(4+\lambda)f_2''+f_3''=\frac{6}{\Delta^2}(y_1-2y_2+y_3) \tag{B}$$

Similarly, we find for $i=3$ that

$$f_2''+(4+\lambda)f_3''=\frac{6}{\Delta^2}(y_2-2y_3+y_4) \tag{C}$$

These two equations are readily solved for f_2'' and f_3''.

To make things still simpler, we will compute the function at the midpoint $x=(x_2+x_3)/2$ for which, Eq. 1.3.3 gives

$$f\left(\frac{x_2+x_3}{2}\right)=\frac{1}{2}(y_2+y_3)-\frac{\Delta^2}{16}(f_2''+f_3'') \tag{D}$$

The first term on the right-hand side is the value produced by linear interpolation between the two central data points. The last term is evidently a correction that can be evaluated from Eqs. B and C and gives

$$f\left(\frac{x_2+x_3}{2}\right)=\left(\frac{1}{2}+\frac{3}{8(5+\lambda)}\right)(y_2+y_3)-\frac{3}{8(5+\lambda)}(y_1+y_4) \tag{E}$$

The interpolated value does not depend strongly on the value of λ (as we would hope).

This result can be applied to the functions used in the examples in the preceding section. For $y=e^x$, we find that Eq. E with $\lambda=0$ underestimates $e^{-0.5}$ by about 5×10^{-3} and with $\lambda=1$ the interpolated value is low by 5×10^{-4}. Clearly, both results are quite good, which again shows that the exponential is easily interpolated.

Interpolation of $y=\sin 2\pi x$ using the spline will in fact give $y(0.5)=0$, but this is due to the symmetry of the function and not to the accuracy of the method. It is more interesting to consider $\sin \pi x$ instead. Then $\lambda=0$ gives 0.9959 and $\lambda=1$ gives 0.9743, which indicates that the pinned end works best for this function. This is the case for a wide variety of functions.

Finally, we apply the result to $(1-x^4)^{1/4}$. We find $f(0.5)=1.0306$ for $\lambda=0$ and $f(0.5)=1.0424$ for $\lambda=1$. Both of these are far above the exact value 0.9840 and are in fact above the maximum value of 1.0000 that the function takes on the interval $0\leqslant x\leqslant 1$. This, too, is typical. The extreme concentration of curvature of this function puts a large bending moment on the end of the spline, which is transmitted throughout the curve as "porpoising" or undesirable oscillation.

These observations are supported further by the next example.

Example 1.5

Since the spline is frequently used as an interpolation tool, we have given a computer program for it. All of the programs in this book have the form of

```
      PROGRAM MAIN
C-----THIS PROGRAM ACCEPTS AS INPUT THE COORDINATES OF THE
C-----DATA POINTS (ASSUMED TO BE IN ORDER) AND COMPUTES
C-----THE SPLINE INTERPOLATION AT THE DESIRED POINTS.
C-----SUBROUTINES SPLINE AND SPEVAL ARE CALLED BY THIS PROGRAM.
      DIMENSION X(20),Y(20),FDP(20)
C-----THE NUMBER OF POINTS IS REQUESTED FIRST.
      WRITE (5,100)
  100 FORMAT ( ' GIVE THE NUMBER OF DATA POINTS')
      READ (5,110) N
  110 FORMAT ( I )
C-----NOW THE INPUT DATA ARE REQUESTED.
      DO 1 I=1,N
      WRITE (5,120) I
  120 FORMAT ( ' TYPE X(',I2,')' )
      READ (5,130) X(I)
  130 FORMAT ( F )
      WRITE (5,140) I
  140 FORMAT ( ' TYPE Y(',I2,')' )
      READ (5,130) Y(I)
    1 CONTINUE
C-----HAVING ALL THE DATA, WE CALL SUBROUTINE SPLINE
      CALL SPLINE (N,X,Y,FDP)
C-----READ THE VALUE(S) OF X FOR WHICH THE INTERPOLATION
C-----IS DESIRED.  ANY NUMBER CAN BE USED.
   10 WRITE (5,150)
  150 FORMAT ( ' TYPE THE VALUE OF X FOR WHICH YOU WANT THE FUNCTION')
      READ (5,130) XX
C-----WE ARE NOW READY TO EVALUATE THE SPLINE.
      CALL SPEVAL (N,X,Y,FDP,XX,F)
C---- WRITE THE RESULT
      WRITE (5,160) XX,F
  160 FORMAT ( ' X= ',F12.6,2X,'INTERPOLATED VALUE = ',F12.6)
C-----THE FOLLOWING ASKS WHETHER FURTHER VALUES ARE WANTED
      WRITE (5,170)
  170 FORMAT ( ' IF YOU WANT ANOTHER INTERPOLATED VALUE, TYPE 1')
      READ (5,110) IFF
      IF (IFF.EQ.1) GO TO 10
      STOP
      END
```

subroutines in standard FORTRAN and "driver" routines, which will be used for obtaining the numerical results we need to illustrate points about the method. The subroutines are written as simply as possible, follow the notation of the text, and are accompanied by comments. They are not the most efficient codes of their kind, but are nearly so. Finally, they do not incorporate all of the possible devices for avoiding trouble, but these are rarely needed and the trouble is usually fairly easily spotted. The driver routines are written in interactive FORTRAN because the interactive ("time share") mode is well suited for learning about numerical methods. Both the subroutines and the drivers may need modification for different machines, but these should be minimal.

We proceed to look at some of the results obtained. Since many of the results are similar to those for the Lagrange polynomial, fewer graphic results will be presented.

Figure 1.13 shows the results for the exponential. For this case both methods are accurate and the Lagrange method is actually somewhat superior. Another

```
      SUBROUTINE SPLINE (N,X,Y,FDP)
C-----THIS SUBROUTINE COMPUTES THE SECOND DERIVATIVES NEEDED
C-----IN CUBIC SPLINE INTERPOLATION.  THE INPUT DATA ARE:
C-----N = NUMBER OF DATA POINTS
C-----X = ARRAY CONTAINING THE VALUES OF THE INDEPENDENT VARIABLE
C-----     (ASSUMED TO BE IN ASCENDING ORDER)
C-----Y = ARRAY CONTAINING THE VALUES OF THE FUNCTION AT THE
C-----     DATA POINTS GIVEN IN THE X ARRAY
C-----THE OUTPUT IS THE ARRAY FDP WHICH CONTAINS THE SECOND
C-----DERIVATIVES OF THE INTERPOLATING CUBIC SPLINE.
      DIMENSION X(20),Y(20),A(20),B(20),C(20),R(20),FDP(20)
C-----COMPUTE THE COEFFICIENTS AND THE RHS OF THE EQUATIONS.
C-----THIS ROUTINE USES THE CANTILEVER CONDITION.  THE PARAMETER
C-----ALAMDA (LAMBDA) IS SET TO 1. BUT THIS CAN BE USER-MODIFIED.
C-----A,B,C ARE THE THREE DIAGONALS OF THE TRIDIAGONAL SYSTEM;
C----- R IS THE RIGHT HAND SIDE.  THESE ARE NOW ASSEMBLED.
      ALAMDA = 1.
      NM2 = N - 2
      NM1 = N - 1
      C(1) = X(2) - X(1)
      DO 1 I=2,NM1
      C(I) = X(I+1) - X(I)
      A(I) = C(I-1)
      B(I) = 2.*(A(I) + C(I))
      R(I) = 6.*((Y(I+1) - Y(I))/C(I) - (Y(I) - Y(I-1))/C(I-1))
    1 CONTINUE
      B(2) = B(2) + ALAMDA * C(1)
      B(NM1) = B(NM1) + ALAMDA * C(NM1)
C-----AT THIS POINT WE COULD CALL A TRIDIAGONAL SOLVER SUBROUTINE
C-----BUT THE NOTATION IS CLUMSY SO WE WILL SOLVE DIRECTLY.  THE
C-----NEXT SECTION SOLVES THE SYSTEM WE HAVE JUST SET UP
      DO 2 I=3,NM1
      T = A(I)/B(I-1)
      B(I) = B(I) - T * C(I-1)
      R(I) = R(I) - T * R(I-1)
    2 CONTINUE
      FDP(NM1) = R(NM1)/B(NM1)
      DO 3 I=2,NM2
      NMI = N - I
      FDP(NMI) = (R(NMI) - C(NMI)*FDP(NMI+1))/B(NMI)
    3 CONTINUE
      FDP(1) = ALAMDA * FDP(2)
      FDP(N) = ALAMDA * FDP(NM1)
C-----WE NOW HAVE THE DESIRED DERIVATIVES SO WE RETURN TO THE
C-----MAIN PROGRAM.
      RETURN
      END
```

```
      SUBROUTINE SPEVAL (N,X,Y,FDP,XX,F)
C-----THIS SUBROUTINE EVALUATES THE CUBIC SPLINE GIVEN
C-----THE DERIVATIVE COMPUTED BY SUBROUTINE SPLINE.
C-----THE INPUT PARAMETERS N,X,Y,FDP HAVE THE SAME
C-----MEANING AS IN SPLINE.
C-----XX = VALUE OF INDEPENDENT VARIABLE FOR WHICH
C-----     AN INTERPOLATED VALUE IS REQUESTED
C-----F =  THE INTERPOLATED RESULT
      DIMENSION X(20),Y(20),FDP(20)
C-----THE FIRST JOB IS TO FIND THE PROPER INTERVAL.
      NM1 = N - 1
      DO 1 I=1,NM1
      IF (XX.LE. X(I+1)) GO TO 10
    1 CONTINUE
C-----NOW EVALUATE THE CUBIC
   10 DXM = XX - X(I)
      DXP = X(I+1) - XX
      DEL = X(I+1) - X(I)
      F = FDP(I)*DXP*(DXP*DXP/DEL - DEL)/6.
     1   +FDP(I+1)*DXM*(DXM*DXM/DEL - DEL)/6.
     2   +Y(I)*DXP/DEL + Y(I+1)*DXM/DEL
      RETURN
      END
```

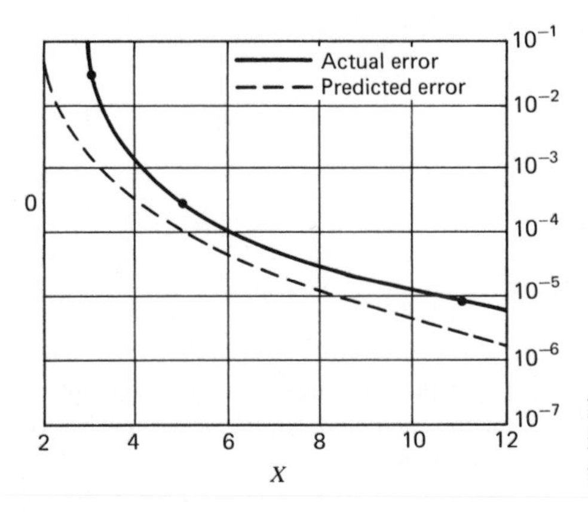

Fig. 1.13. Maximum error in cubic spline interpolation of e^x versus number of points used.

choice of spline end conditions might produce better results, but since our experience above has shown that different end conditions are best for different functions, this is not a worthwhile exercise.

Figure 1.14 shows that the spline gives a smaller error than Lagrange interpolation for the sine function.

For the circle, the spline, like the Lagrange, produces the largest error where the slope becomes large near $x=1$. Maximum errors are shown in Fig. 1.15. The Lagrange method produces a slightly smaller error, but neither method is very good. In both cases the problem can be reduced by using fewer points for small x where the curve is relatively flat and more closely spaced points near $x=1$. When this is done, the spline will be found to be superior; the Lagrange polynomial will oscillate about the exact curve, while the spline will fit it quite well. Similar findings hold for the superellipse with $n=4$.

Because the spline is capable of producing accuracy similar to that of the Lagrange polynomial when the curve is easy to fit and does much better than the Lagrange polynomial in the difficult cases in which the function oscillates, it is usually the preferred method of fitting a function. This is especially true

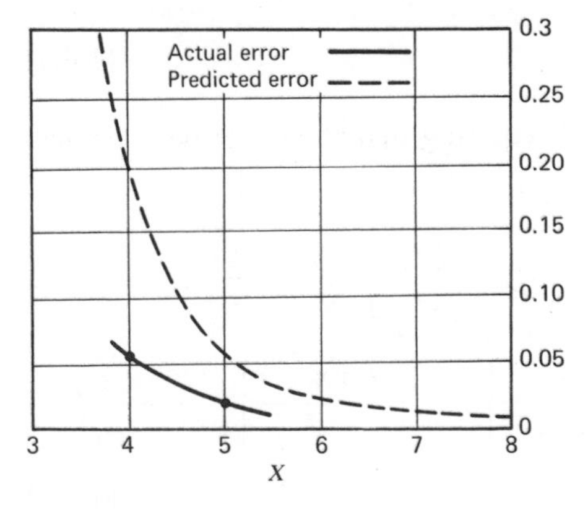

Fig. 1.14. Maximum error in cubic spline interpolation of $\sin 2\pi x$ versus number of points used.

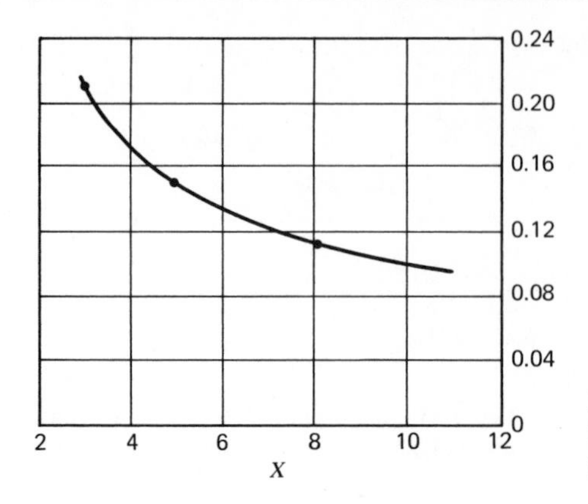

Fig. 1.15. Maximum error in cubic spline interpolation of $(1-x^2)^{1/2}$ versus number of points used.

when the number of data points is large. In this case expense of computing the Lagrange polynomial becomes very large and the tendency to oscillate about the correct result becomes more pronounced. The spline suffers from neither problem and is generally well behaved when the number of points is large.

The next two sections address two remaining problems—dealing with functions with concentrated curvature and dealing with functions with infinite slope.

4. TENSION SPLINE

In most cases the spline gives a smooth fit to data. In some severe cases, however, it wiggles too much. A cure for this is to put the spline in tension—that is, to pull on the ends of the "thin strip." Clearly, if no tension is exerted, we will get the ordinary spline described in the previous section; but if the tension is made extremely high, a linear spline (piecewise linear function) will result.

The effect of the tension is to replace Eq. 1.3.2 by

$$f_i'' - \sigma^2 f_i = \left[f''(x_i) - \sigma^2 y_i \right] \frac{(x_{i+1}-x)}{(x_{i+1}-x_i)} + \left[f''(x_{i+1}) - \sigma^2 y_{i+1} \right] \frac{(x-x_i)}{(x_{i+1}-x_i)} \tag{1.4.1}$$

On integrating this equation twice and requiring that the curve pass through the proper node points, we obtain

$$\begin{aligned} f(x) = {} & \frac{f''(x_i)}{\sigma^2} \frac{\sinh[\sigma(x_{i+1}-x)]}{\sinh[\sigma(x_{i+1}-x_i)]} + \left[y_i - \frac{f''(x_i)}{\sigma^2} \right] \frac{(x_{i+1}-x)}{(x_{i+1}-x_i)} \\ & + \frac{f''(x_{i+1})}{\sigma^2} \frac{\sinh[\sigma(x-x_i)]}{\sinh[\sigma(x_{i+1}-x_i)]} + \left[y_{i+1} - \frac{f''(x_{i+1})}{\sigma^2} \right] \frac{(x-x_i)}{(x_{i+1}-x_i)} \\ & x_i < x < x_{i+1} \end{aligned} \tag{1.4.2}$$

Then on applying continuity of the first derivative, we find the equation for the $f''(x_i)$ to be

$$\left[\frac{1}{\Delta_{i-1}}-\frac{\sigma}{\sinh\sigma\Delta_{i-1}}\right]\frac{f''(x_{i-1})}{\sigma^2}+\left[\frac{\sigma\cosh\sigma\Delta_{i-1}}{\sinh\sigma\Delta_{i-1}}-\frac{1}{\Delta_{i-1}}+\frac{\sigma\cosh\sigma\Delta_i}{\sinh\sigma\Delta_i}-\frac{1}{\Delta_i}\right]\frac{f''(x_i)}{\sigma^2}+\left[\frac{1}{\Delta_i}-\frac{\sigma}{\sinh\sigma\Delta_i}\right]\frac{f''(x_{i+1})}{\sigma^2}=\frac{y_{i+1}-y_i}{\Delta_i}-\frac{y_i-y_{i-1}}{\Delta_{i-1}} \tag{1.4.3}$$

Again we have a tridiagonal system of equations for the second derivatives. The coefficients are more difficult to evaluate, as is the resulting spline, but excellent results have been obtained with the use of tension splines. Generally it is recommended that if the natural spline is not satisfactory, one should try the tension spline with the tension parameter set equal to unity as a first guess. If the results still wiggle too much, the tension can be increased.

Endpoint conditions for the tension spline may be modeled after those for the natural spline. Any of the conditions given in the preceding section may be used, but one should replace $f''(x_i)$ by $f''(x_i)-\sigma^2 y_i$.

Finally, we note that the concept of a spline—a curve that is made of pieces that are individually extremely smooth and are connected by compatibility conditions at the nodes—has been extended to a number of other functions. A linear spline is nothing more or less than a piecewise linear (connect-the-dots) fit that is commonly used. Quadratic splines have been used on occasion; they have the property that they require only one end condition and lead to a bidiagonal set of equations that can be solved by a unidirectional elimination procedure similar to the backward elimination part of the Gauss elimination algorithm. In a few applications in which functions vary rapidly, exponential splines have been used. These are computationally more difficult (if a constant in the exponent is allowed to be one of the parameters, the problem is nonlinear).

5. PARAMETRIC AND MULTIDIMENSIONAL INTERPOLATION

As the examples have shown, none of the methods discussed so far is capable of dealing with infinite slopes. Although one can avoid the problem in simple cases by interchanging the roles of the independent and dependent variables, this will not work when the curves have closed loops. This problem commonly arises when a function of two independent variables $f(x, y)$, is to be described by giving the contours $f=\text{constant}$. Prime examples of this type occur in geographical and weather mapping. They require multidimensional interpolation to locate the contours and parametric interpolation to draw them smoothly. Both of these topics are described only briefly and we begin with the latter.

Parametric Interpolation

Suppose we wish to draw a smooth curve through a set of data that forms a closed loop. Part of the problem is that the function is multivalued; for each x there are at least two possible values of y, and vice versa. It is very hard to create a routine that will not jump from one branch of the curve to another. One way out of this difficulty is parametric interpolation. We regard both x and y as dependent on a third variable, which we will call s. The choice of the parameter s is rather free, as we require only that s increase monotonically along the curve. A simple choice is the arc length along the curve. Of course the actual arc length is not known, but it is sufficient to use the distances between data points to supply the values at the nodes. The first point is assigned $s=s_1=0$; at the second $s=s_2=[(x_2-x_1)^2+(y_2-y_1)^2]^{1/2}$, at the third $s=s_3=s_2+[(x_3-x_2)^2+(y_3-y_2)^2]^{1/2}$, and so on. Then the parametric curves $x(s)$ and $y(s)$ are treated by the usual spline method. For a closed curve, periodic end conditions are appropriate; for other cases, other end conditions can be applied. The desired interpolation then is obtained by computing both x and y, the coordinates of a point on the curve, for a given value of s. If the value of y is needed for a particular x, we first find the s corresponding to the given x and then compute y. Finding s requires solving the cubic (either by formula or iteratively), and the extra computation involved is the chief drawback of the method. Nevertheless, the method is good enough to be in common use.

Example 1.6

To see how the method works, suppose we are given a number of points on a circle of radius R. In this case parametric interpolation is almost equivalent to spline fitting $x(s)=\cos(s/R)$ and $y(s)=\sin(s/R)$. (The representation would be exact if s were the exact arc length $R\theta$.) We already know how the spline deals with these functions so there is no need to give detailed numerical results here.

Multidimensional Interpolation

The simplest methods in multidimensional interpolation are extensions of one-dimensional methods. For example, if a function of x and y is given on a regular array of points—that is, on the points of a rectangular grid—one can do the following. For each line $x=x_i$ on which the data are given, compute the second derivatives (which we call $f_i''(y_j)$) of the spline fit in y by the method of the preceding section. These second derivatives then may be considered functions of x and may be interpolated by spline fitting in x. To evaluate the interpolated value at a given (x, y), we first determine in which box it is placed. Given that $x_i < x < x_{i+1}$ and $y_j < y < y_{j+1}$, we can apply the spline fits to $f''(x_i, y_j)$ and $f''(x_{i+1}, y_j)$ to find $f''(x, y_j)$. In the same manner, from $f''(x_i, y_{j+1})$ and $f''(x_{i+1}, y_{j+1})$ we can find $f''(x, y_{j+1})$. Finally, using $f''(x, y_j)$ and $f''(x, y_{j+1})$ we can compute $f(x, y)$.

Problems

1. Another set of functions that are frequently used as test cases are $y_n = x^n$ on $0 \leqslant x \leqslant 1$. The lower order cases are very smooth, but the higher order ones become increasingly sharply curved. Use linear and quadratic Lagrange interpolation on these functions for $n = 3$, 6, and 10, respectively. Find the maximum error and the point at which it occurs. Compare the result obtained with the estimate given in the text.
2. In Chapter 3 we use interpolation formulas to derive finite difference approximations to derivatives. This is done by differentiating the interpolation formula and evaluating it at the point at which the derivative is required.

    a. Given the value of $f(x_1)$ and $f(x_2)$, use linear Lagrange interpolation to find $f'(x_1)$ and $f'(x_2)$.

    b. Given $f(x_1)$, $f(x_2)$, and $f(x_3)$, use quadratic Lagrange interpolation to find f' and f'' at each of the three points.

3. Write a subroutine for Lagrange interpolation with any number of points up to 20.
4. Use the program in Problem 3 to determine how the Lagrange interpolation performs on the sinusoidal function with a greater number of points, say 10, 15, and 20.
5. A quadratic spline is a continuous curve that has a continuous first derivative, but not necessarily a continuous second derivative. Derive a set of equations for the first derivative of the spline at the node points. What are appropriate end condition(s) for this interpolation?
6. Using four points, apply spline interpolation to the function $f(x) = x^6$ on $0 \leqslant x \leqslant 1$. Which value of λ in Eq. 1.3.8 gives the best result for $x = 0.5$?
7. Compute the spline of $\sin 2\pi x$ interpolation with 10, 15, and 20 points, using the program given (or your own). Plot the error versus x.
8. Apply parametric interpolation to the ellipse $x^2 + (y^2/2) = 1$. Use 8, 12, 16, and 20 more or less evenly spaced points.

2

Integration

The computation of integrals is frequently required in engineering, and it is necessary to distinguish the various cases that might arise. In the normal case one needs to calculate the integral of a function that can be evaluated at any arbitrary point; the function is smooth and finite everywhere on the range on integration, and the limits are finite. Occasionally we are faced with a problem in which one or more of the conditions above are violated. If the function blows up at some point on the range on integration, or at least one of the limits is infinite, we say that the integral is singular. Special methods are required for the evaluation of such integrals. They will be taken up in Section 6 of this chapter. Another possibility is one in which the values of the function are given at a fixed set of points and there may be uncertainty in the values that are provided. Such cases usually arise in conjunction with experimental data and are best treated by fitting a least squares curve through the data and integrating the resulting curve. Since we have not treated least squares curve fitting in the preceding chapter, this case will not be treated here.

We look first at the nonsingular case. All of the quadrature formulas that are considered approximate the integral by a weighted sum of the values of the integrand at points on the interval of integration—that is, they amount to an approximation of the form

$$\int_a^b f(x)\,dx \cong \sum_{i=1}^{n} w_i f(x_i) \tag{2.1}$$

where $a \leqslant x_1 < x_2 < \cdots < x_n \leqslant b$. The numbers w_i are called the *weights* and the points x_i at which the function is to be evaluated are usually called the *abscissas*; both depend on n as well as i. A quadrature formula is simply some particular choice for the weights and abscissas. The simplest approach is to use evenly spaced abscissas and choose the weights to give the best approximation. This method has the advantage that it can be systematically improved by successive halving of the interval size (see below). A more sophisticated approach allows the abscissas to be chosen in a way that increases the accuracy. In either case the parameters will depend on the criterion used to measure the accuracy of the method.

Any quadrature formula should approach the exact integral as the number of points becomes very large. An almost obvious approach to obtaining accuracy is to estimate the integral using the quadrature formula with n points, repeat the procedure with twice as many points, and then compare the two results. If they agree within the desired tolerance, we can accept the last result as accurate. If they do not agree, we double the number of points again and continue until convergence is achieved. It is easiest to carry out this type of calculation when the abscissas are evenly spaced; if need be, one can always subdivide the integral and treat each of the parts with evenly spaced abscissas. For formulas with evenly spaced abscissas, the error in the approximation is almost always proportional to an integral power of the interval between successive abscissas. Thus, with $n+1$ evenly spaced points, the interval $h=(b-a)/n$. If, as h is decreased and n is increased, the error goes to zero as h^m, we have an mth order approximation.

Alternatively, we can require that a quadrature formula integrate polynomials of any degree up to and including m exactly; m is then a measure of the quality of a formula.

The first quadrature formulas we consider use Lagrange interpolation on an evenly spaced set of points. We derive the formulas and some of their properties and discuss how one chooses a formula of this class. When high accuracy is desired, a further trick that not only decreases the size of the interval but simultaneously increases the order of the approximation may prove very valuable. The concept behind this method is introduced in Section 2 and is applied to the convergence of quadrature formulas in Section 3.

Another approach is to allow the computer to estimate the error in the result from analytically based formulas. This can be done for each interval and one can then improve the accuracy only in those intervals in which the accuracy is not yet satisfactory. Such adaptive methods are discussed in Section 4.

In some cases the evaluation of the integrand is costly and it is important to use a quadrature method that yields maximum accuracy for a given number of evaluations of the function and a different criterion of goodness is desirable. The method of choice then becomes Gauss quadrature. This method and some of its relatives are discussed in Section 1.5.

The final section of this chapter considers methods for dealing with singular integrals, particularly integrals whose integrands blow up somewhere in the range of integration.

1. NEWTON–COTES FORMULAS

One of the easiest ways to obtain useful quadrature formulas is to use Lagrange interpolation on an evenly spaced mesh and integrate the result. In this manner one obtains the Newton–Cotes formulas, the first two of which are familiar to most readers.

We begin by dividing the range of integration into n intervals. Including the ends, $n+1$ points are needed. Thus

$$x_j = a + jh \qquad h = \frac{b-a}{n} \qquad j = 0, 1, 2, \dots n \tag{2.1.1}$$

We can pass the Lagrange interpolating polynomial of degree n through the points $(x_j, f(x_j))$, $j = 0, 1, 2, \dots n$

$$P(x) = \sum_{k=0}^{n} L_k(x) f(x_k) \tag{2.1.2}$$

This polynomial then can be integrated in a straightforward manner to give the desired result

$$\int_a^b f(x)\,dx \simeq \int_a^b P(x)\,dx = \sum_{k=0}^{n} f(x_k) \int_a^b L_k(x)\,dx = (b-a) \sum_{k=0}^{n} C_k^n f(x_k) \tag{2.1.3}$$

where the C_k^n are pure numbers defined by

$$C_k^n = (b-a)^{-1} \int_a^b L_k(x)\,dx \tag{2.1.4}$$

These are called Cotes numbers, which are tabulated in Table 2.1. A few of the properties of Cotes numbers are easily discovered. Since the Lagrange polynomial obviously gives an exact fit to a constant, $f(x) = 1$ is integrated exactly by the Newton–Cotes formulas of any order. Thus

$$\sum_{k=0}^{n} C_k^n = 1 \tag{2.1.5}$$

Furthermore, we can reverse the direction of integration; that is we could have

Table 2.1. Newton–Cotes Coefficients

n	N	NC_0^n	NC_1^n	NC_2^n	NC_3^n	NC_4^n	NC_5^n	NC_6^n	Error
1	2	1	1						$8.3\ 10^{-2}\Delta^3 f^{\mathrm{II}}$
2	6	1	4	1					$3.5\ 10^{-4}\Delta^5 f^{\mathrm{IV}}$
3	8	1	3	3	1				$1.6\ 10^{-4}\Delta^5 f^{\mathrm{IV}}$
4	90	7	32	12	32	7			$5.2\ 10^{-7}\Delta^7 f^{\mathrm{VI}}$
5	288	19	75	50	50	75	19		$3.6\ 10^{-7}\Delta^7 f^{\mathrm{VI}}$
6	840	41	216	27	272	27	216	41	$6.4\ 10^{-10}\Delta^9 f^{\mathrm{VIII}}$

integrated from b to a instead of from a to b. Consequently, Cotes numbers possess the symmetry property

$$C_k^n = C_{n-k}^n \tag{2.1.6}$$

Aside from the job of computing the Cotes numbers, which is a straightforward but tedious task, the job is now complete.

The first two special cases are of interest. For $n=1$ we obtain the well-known *trapezoid rule*

$$\int_a^b f(x)\,dx = \tfrac{1}{2}[f(a)+f(b)](b-a) \tag{2.1.7}$$

For $n=2$ the Cotes formula reduces to *Simpson's rule*, the well-known formula obtained by passing a parabola through three points

$$\int_a^b f(x)\,dx = \tfrac{1}{6}\left[f(a)+4f\left(\frac{a+b}{2}\right)+f(b)\right](b-a) \tag{2.1.8}$$

The higher order Cotes formulas are less well known, but Cotes numbers for them are listed in Table 2.1.

In actual practice one rarely computes an integral by applying a single Cotes formula to the entire interval. Rather, the interval is broken into subintervals, which may or may not be equal in size, and the quadrature formula is applied to each interval. If the interval (a, b) is broken into n equal intervals of width h each, and the trapezoid rule is applied to each subinterval, we obtain the quadrature formula

$$\int_a^b f(x)\,dx = h\left\{\sum_{k=0}^{n} f(x_k) - \tfrac{1}{2}[f(x_0)+f(x_n)]\right\} \qquad x_k = a + \frac{k}{n}(b-a) \tag{2.1.9}$$

Simpson's rule may be applied in the same manner, but the number of subintervals must be even, since the formula is applied to pairs of intervals. One finds

$$\int_a^b f(x)\,dx = \frac{h}{3}[f(x_0)+4f(x_1)+2f(x_2)+\cdots 2f(x_{n-2})+4f(x_{n-1})+f(x_n)] \tag{2.1.10}$$

Next we consider the question of the accuracy of these formulas. Some idea as to the error produced can be obtained from the error estimate for Lagrange interpolation. The problem with this approach is that the value of ξ in Eq. 1.1.12 depends on x, so it is difficult to obtain an accurate estimate of the error. In fact accurate estimation of the error is moderately difficult and the interested reader is referred to one of the more advanced numerical analysis texts for the derivation. The results are given in the last column of Table 2.1. Note that the error estimate given is for a single interval $\Delta = b - a$. If a single formula is used over many subintervals, the formula should be applied to each subinterval and the results summed.

Table 2.1 shows that each even approximation is a significant improvement over the odd one preceding it, but the odd approximations are not much better than their even predecessors. For this reason the odd approximations (other than the trapezoid rule) are almost never used.

Another question that is frequently asked is: Which approximation is best for a particular application? Generally, a reasonable rule of thumb is that one should use the approximation for which the numerical coefficient given in the last column of Table 2.1 is closest to the desired relative error. Alternative approaches that avoid much of this difficulty are the Romberg method presented in Section 3 and the adaptive method presented in Section 4.

The formulas given above are called *closed* Newton–Cotes formulas because they use the values of the function at the end points in the approximation. There are also *open* Newton–Cotes formulas, which do not use the end points as abscissas. These formulas are not as accurate as the closed formulas and are rarely used. The first of these formulas is worth noting, however, as it is used later:

$$\int_a^b f(x)\,dx = f\left(\frac{a+b}{2}\right)(b-a) \tag{2.1.11}$$

and is known as the *midpoint rule*. It is obtained by assuming that the function is constant.

Example 2.1

The function e^x, used in the interpolation examples in Chapter 1, is very smooth and should be easy to compute. The exact result is:

$$I = \int_0^1 e^x\,dx = e - 1 = 1.718281828$$

First we will apply the trapezoid rule with various numbers of points. Since the error should scale like the square of the interval size, we will also compute ε/h^2 where ε is the error and h the interval size. With $n=2$ we have $h=\frac{1}{2}$ and

$$I \simeq \tfrac{1}{2}\left[\tfrac{1}{2}(e^0 + e^1) + e^{-0.5}\right] = 1.7539$$

which larger than the exact result by 0.036. For other values of n results are given in Table 2.2. For a function as smooth as this one the error estimate works very well, even for relatively large interval sizes, and the error is nearly proportional to h^2.

Table 2.2. Trapezoid Integration of e^x

Number of Intervals $=n$	Error$=\varepsilon$	$\varepsilon n^2=\varepsilon/h^2$
1	0.141	0.141
2	0.036	0.144
3	0.016	0.144
4	0.0089	0.143
5	0.0057	0.143

Example 2.2

Trying the same integral using the Newton–Cotes formulas applied to the entire interval gives the results shown in Table 2.3. These errors follow the estimates given in Table 2.1 very closely. Since the number of function evaluations in Simpson's rule is the same as that for two-interval trapezoid integration, a comparison of $n=2$ results in Tables 2.2 and 2.3 shows the clear superiority of Simpson's rule. In the same manner, comparison of corresponding entries for larger n shows that when high accuracy is desired, *use of higher order formulas is more efficient than use of more intervals*. This is a result of considerable applicability in numerical analysis.

Table 2.3. Newton–Cotes Integration of e^x

n	ε
1	0.141
2	5.79×10^{-4}
3	2.58×10^{-4}
4	8.6×10^{-7}
5	2.8×10^{-7}

Example 2.3

The full wave sine function is not a good choice for an example of integration. The exact result

$$\int_0^1 \sin 2\pi x\,dx=0$$

will be produced by many quadrature formulas as a consequence of the symmetry of the function, not the quality of the method. We consider, therefore, the integral

$$I=\int_0^1(1-x^2)^{1/2}\,dx=\frac{\pi}{4}=0.785398$$

Table 2.4. Trapezoid Integration of $(1-x^2)^{1/2}$

n	Value	ε	ε/h^2
1	0.5000	0.2854	0.285
2	0.6830	0.1024	0.410
3	0.7294	0.0560	0.504
4	0.7489	0.0365	0.584
5	0.7593	0.0261	0.653

Table 2.5. Newton–Cotes Integration of $(1-x^2)^{1/2}$

n	I	ε
1	0.5000	0.2854
2	0.7440	0.0414
3	0.7581	0.0273
4	0.7727	0.0127
5	0.7754	0.0100
6	0.7792	0.0062

which involves another of the functions used in the interpolation examples in Chapter 1. The results are shown in Table 2.4. The error decreases less rapidly than it does for the smooth function. In fact the error does not become proportional to h^2 until the interval size becomes very small because all derivatives of the integrand become very large near the end point $x=1$.

For the same reason, Newton–Cotes integration of this function converges to the correct result very slowly. The results are given in Table 2.5. The decrease of the error is much slower than it was for e^x, but greater improvement is still obtained in going from an odd to an even approximation than vice versa.

2. RICHARDSON EXTRAPOLATION

An ideal integration subroutine is one that allows the user to provide the function and interval of integration and produces the result within an accuracy specified by the user. Of course, this should also be accomplished with a minimum of effort. In principle Newton–Cotes quadrature can produce any desired accuracy by repeatedly halving the interval until the results converges. This is a fairly costly method, however, and improved approaches have been developed. Romberg integration is one such procedure. It is based on the concept of Richardson extrapolation (also called deferred approach to the limit), which is a useful numerical technique in its own right. A second method is described in Section 4.

As with many good ideas, the concept is simple. Suppose we require some quantity g, not necessarily an integral, which we cannot evaluate exactly. Suppose also that we have some method of approximating it. Most approximations depend on a parameter, say h, which can be made small, and we will denote the approximation $g(h)$. Examples of such approximations are the quadrature formulas of the preceding section in which g is the integral and h is the size of the subinterval. It often happens that the approximation can be represented by a Taylor series in the small parameter h, so that

$$g(h)=g+c_1h+c_2h^2+\cdots \tag{2.2.1}$$

where $c_1, c_2, \ldots$ are constants. Some of the terms may be missing. For example, the series may contain only the even terms, so that it is actually a Taylor series in h^2. All of the terms except the first represent errors that we would prefer to eliminate.

Now suppose that we have calculated $g(h)$ for some particular value of h. In the preceding section it was suggested that we repeat the calculation with h replaced by $h/2$. Then we would obtain the value

$$g\left(\frac{h}{2}\right)=g+\tfrac{1}{2}c_1h+\tfrac{1}{4}c_2h^2+\cdots \tag{2.2.2}$$

In making this calculation we have roughly doubled the amount of calculation and halved the error. The idea of Richardson extrapolation is that by combining $g(h)$ and $g(h/2)$ we can obtain a still better approximation. More precisely, what we do is subtract Eq. 2.2.1 from twice Eq. 2.2.2 to get

$$g_1(h)=2g\left(\frac{h}{2}\right)-g(h)=g+c_2'h^2+c_3'h^3+\cdots \tag{2.2.3}$$

which, if h is small, is considerably more accurate than either of the two values from which it was derived. Furthermore, the procedure can be continued indefinitely. Thus having g_1 we can form

$$g_2(h)=\tfrac{1}{3}\left[4g_1\left(\frac{h}{2}\right)-g_1(h)\right]=g+c_3''h^3+\cdots \tag{2.2.4}$$

which is still more accurate. Proceeding in this way, at the nth stage of this process we compute

$$g_n(h)=\frac{2^ng_{n-1}(h/2)-g_{n-1}(h)}{2^n-1}=g+0(h^{n+1}) \tag{2.2.5}$$

If it is known a priori that the approximation $g(h)$ contains only even powers, one can use just the even n operations in the sequence. (No harm will come from using the odd operations other than a waste of computational effort.) We also note that it is not necessary to halve the interval each time. Other fractions can be used, but halving makes maximum use of the previous results and is much easier if the process is to be continued an indefinite number of times.

Example 2.4

A simple and interesting illustration of these ideas is the calculation of 2π by computing the perimeter of regular polygons inscribed in a unit circle. A little trigonometry suffices to show that the perimeter of an n-sided inscribed polygon is

$$P_n = 2n \sin\frac{\pi}{n} = 2\pi - \frac{\pi^3}{3n^2} + 0\left(\frac{1}{n^4}\right) \tag{2.2.6}$$

This is a case in which only even terms occur. One might object that the value of π is needed to calculate $\sin n/\pi$, but this is not the case because one can use the trigonometric formula for the sine of the half angle. The results of applying Richardson extrapolation to this problem are shown in Table 2.6. The first column gives the values computed using Eq. 2.2.6 and the succeeding columns give the extrapolated values. It is amazing that we obtain eight place accuracy (the limit on the calculator used by the author) with only $n=16$, whereas the original formula with $n=512$ yields only five place accuracy. Richardson extrapolation has many applications in numerical analysis, one of which is given in the next section.

3. ROMBERG INTEGRATION

Romberg integration is nothing more than a combination of the methods presented in the preceding two sections. The idea is to take a relatively inaccurate quadrature method and improve it by using Richardson extrapolation. Thus we can start with as simple a method as the trapezoid rule.

First we estimate the integral using the trapezoid rule with a single interval. The calculation is then repeated with 2 intervals, 4 intervals, 8 intervals, and so on, the results of which can be called I_0^1, I_0^2, I_0^4, and so on. In each of these calculations half of the needed function values have already been used in the preceding calculation and need not be recalculated. Thus the whole computation can be done for little more than the cost of the last trapezoid rule evaluation. Since the trapezoid rule is second order accurate, we can calculate improved values of the integral by Richardson extrapolation Thus we calculate

$$I_1^n = \frac{4I_0^n - I_0^{n/2}}{3} \tag{2.3.1}$$

The results so obtained turn out to be precisely what we would get using

Table 2.6. Richardson Extrapolation Calculation of 2π ($2\pi = 6.283185308$)

n	P_n^0	$P_{n+1}^2 = -\dfrac{P_n^0 - 4P_{2n}^0}{3}$	$P_{n+1}^2 = -\dfrac{P_n^1 - 16P_{2n}^1}{15}$	$P_{n+1}^3 = -\dfrac{P_n^2 - 64P_{2n}^2}{63}$	$P_{n+1}^4 = -\dfrac{P_n^3 - 256P_{2n}^3}{255}$
1	0				
2	4	5.3333...			
4	5.65685	6.209139	6.267526		
8	6.122935	6.278295	6.2829056	6.2831496	
16	6.242890	6.282875	6.2831808	6.2831852	6.2831853
32	6.27309				
64	6.28066				
128	6.28255				
256	6.283028				
512	6.283146				

Table 2.7. A Schematic of Romberg Integration

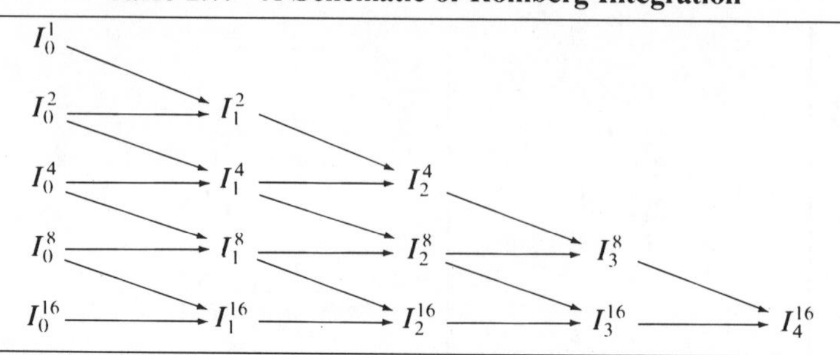

Simpson's rule with n intervals. With considerable difficulty (the interested reader is again referred to the standard numerical analysis texts) one can show that the error for the trapezoid rule is a Taylor series in h^2. So the error in Eq. 2.3.1 is fourth order (we know this because it is Simpson's rule).

Since the error is fourth order, we may apply Richardson extrapolation

$$I_2^n = \tfrac{1}{15}\left(16 I_1^n - I_1^{n/2}\right) \tag{2.3.2}$$

This is sixth order accurate but is not a Newton–Cotes formula. The process may be continued indefinitely using the general formula

$$I_l^n = \frac{4^l I_{l-1}^n - I_{l-1}^{n/2}}{4^l - 1} \tag{2.3.3}$$

The resulting values may be put in tabular form as in Table 2.7.

The order in which the calculations are actually done is I_0^1, I_0^2, I_1^2, I_0^4, I_1^4, I_2^4, I_0^8, and so on. When the maximum extrapolation that can be done with a given number of points is reached, that is, when I_n^{2n} has been calculated, it is compared with I_{n-1}^{2n}. If the agreement is satisfactory, the calculation is terminated. If not, the trapezoid integral with $2n$ points is computed and extrapolated; the process is repeated until the desired result is obtained.

In principle, arbitrary accuracy can be achieved. In fact, however, accuracy is limited by the computational accuracy of the computer used and the function evaluating algorithm. One should not demand accuracy greater than an order of magnitude poorer than either the machine or the algorithm can deliver.

Example 2.5

The function e^x is easy to integrate and the efficiency of Romberg integration is shown in Table 2.8. The correct figures are indicated by underscores and it is clear that the Romberg method is very effective and efficient for the computation of the integrals of smooth functions.

Table 2.8. Romberg Integration of e^x (Exact = 1.718281828)

n	I_0^n	I_1^n	I_2^n	I_3^n
1	1.85140914			
2	1.753931093	1.718861153		
4	1.727221905	1.718318824	1.718828269	
8	1.720518591	1.718284153	1.718281842	1.718281829

Note: The number of evaluations of the integrand is the number of intervals plus one.

Example 2.6

As would be expected from the previous results, the function $(1-x^2)^{1/2}$ is difficult to integrate with the Romberg method. There is another class of functions that are difficult to integrate and, since they occur more often in practice, we will look at one of them. These are functions in which the largest contribution to the result comes from a relatively small portion of the range of integration. A good example of such an integral is

$$I=\int_0^1 e^{-a^2x^2}\,dx=\left(\frac{\sqrt{\pi}}{2a}\right)\operatorname{erf}(a)$$

where erf(a) is the error function that is tabulated in a number of sources. For small a the integrand is very smooth and any method will do well and the behavior of the Romberg method for this function is similar to that for e^x. For large a the integrand is highly peaked near $x=0$ and the function is difficult to integrate.

We solve this problem with the aid of a computer program. Program MAIN simply handles the input-output functions and the work is actually done in subroutine ROMBRG. A routine for evaluating the function must be supplied separately. In this case function evaluation is very simple; in other cases the function evaluation may be very elaborate. Generally we expect that the evaluation of the integrand is the most time-consuming part of computing an integral and the number of evaluations required, therefore, is used conventionally as a measure of the quality of a method. Note also that the subroutine assumes that one parameter (a) is needed to define the function. Obviously this could be generalized.

The results for $a=1$ are shown in Table 2.9. A very severe test of the method is provided by taking $a=10$ for which the integrand is peaked near the origin. The results are shown in Table 2.10. A large number of evaluations is required. The accuracy obtained with 129 points is in fact the best we can do in single precision because roundoff error is beginning to become important. This is indicated by the minor improvement in going from 129 points to 257 points.

Table 2.9. Romberg Integration of e^{-x^2}

Number of Evaluations	Error
3	3×10^{-4}
5	9×10^{-6}
9	1×10^{-7}
17	$<10^{-10}$

Table 2.10. Romberg Integration of e^{-100x^2} (Exact=0.088622683)

n	Number of Evaluations	Error
1	3	+0.576
2	5	−0.045
3	9	−0.011
4	17	−0.012
5	33	9.5×10^{-4}
6	65	8.2×10^{-6}
7	129	4.4×10^{-7}
8	257	3.5×10^{-7}

```
      PROGRAM MAIN
C-----THIS PROGRAM USES SUBROUTINE ROMBRG TO COMPUTE THE
C-----INTEGRAL OF EXP-(C*X)**2.  IT BEGINS BY REQUESTING
C-----THE VALUES OF C, DESIRED ACCURACY, AND THE LIMITS.
C-----THE FUNCTION TO BE INTEGRATED MUST BE SUPPLIED IN A
C-----SEPARATE FUNCTION SUBPROGRAM.
      EXTERNAL FUN
      WRITE (5,50)
   50 FORMAT ( ' THIS PROGRAM COMPUTES THE INTEGRAL OF
     1 EXP-(C*X)**2 BETWEEN LIMITS IT REQUESTS')
   10 WRITE (5,100)
  100 FORMAT ( ' GIVE THE VALUE OF C')
      READ (5,110) C
  110 FORMAT ( F)
      WRITE (5,120)
  120 FORMAT ( ' GIVE DESIRED ACCURACY (IN F FORMAT)')
      READ (5,110) ERR
      WRITE (5,130)
  130 FORMAT ( ' NOW GIVE THE LOWER LIMIT')
      READ (5,110) A
      WRITE (5,140)
  140 FORMAT ( ' FINALLY, GIVE THE UPPER LIMIT')
      READ (5,110) B
C-----HAVING ALL OF THE NEEDED DATA, WE CALL ROMBRG.
      CALL ROMBRG (FUN,A,B,C,ERR,XINT)
C-----ALL THAT'S LEFT IS TO WRITE THE RESULT.
      WRITE (5,150) C,A,B,XINT
  150 FORMAT ( ' INTEGRAL OF EXP-(',F6.3,'X)**2 FROM X=',F6.3,
     1 ' TO X=',F6.3,' IS',F12.6)
C-----ASK FOR ANOTHER SET OF PARAMETERS
      WRITE (5,160)
  160 FORMAT ( ' IF YOU WANT TO COMPUTE ANOTHER INTEGRAL, TYPE 1')
      READ (5,170) IFF
  170 FORMAT ( I)
      IF (IFF.EQ.1) GO TO 10
      STOP
      END
```

```
      FUNCTION FUN(X,C)
      FUN = EXP(-(C*C*X*X))
      RETURN
      END

      SUBROUTINE ROMBRG(FUN,A,B,C,ERR,RES)
C-----THIS PROGRAM COMPUTES INTEGRALS OF USER SUPPLIED
C-----FUNCTIONS BY THE ROMBERG METHOD.  THE INPUTS ARE:
C-----FUN = THE FUNCTION TO BE INTEGRATED
C-----A = LOWER LIMIT
C-----B = UPPER LIMIT
C-----C = PARAMETER NEEDED TO DEFINE FUNCTION (COULD
C-----      BE CHANGED TO AN ARRAY IF NECESSARY.)
C-----ERR = DESIRED ACCURACY, SHOULD NOT BE LESS THAN
C-----         MACHINE PRECISION
C-----THE OUTPUT IS RES, THE RESULT.
      EXTERNAL FUN
C-----THE ARRAY OF APPROXIMATIONS IS CALLED Z.
      DIMENSION Z(10,10)
C-----INITIALIZE THE INDEX AND COMPUTE THE FIRST APPROXIMATION.
      I = 1
      DEL = B - A
      Z(1,1) = .5 * DEL * (FUN(A,C) + FUN(B,C))
C-----THE MAIN LOOP.  THE FIRST PART COMPUTES THE INTEGRAL
C-----USING A 2J+1 POINT TRAPEZOID RULE.  THE METHOD MAKES
C-----MAXIMAL USE OF THE VALUES ALREADY COMPUTED.
   10 J = 2**(I-1)
      DEL = DEL/2.
      I = I + 1
      Z(I,1) = .5 * Z(I-1,1)
      DO 1 K=1,J
      X = A + (2.*K - 1) * DEL
      Z(I,1) = Z(I,1) + DEL * FUN(X,C)
    1 CONTINUE
C-----NOW WE NEED TO DO THE RICHARDSON EXTRAPOLATION.
      DO 2 K=2,I
      Z(I,K) = (4.**(K-1)*Z(I,K-1) - Z(I-1,K-1)) / (4.**(K-1) -1.)
    2 CONTINUE
C-----ERROR CONTROL TIME
      DIFF = ABS (Z(I,I) - Z(I,I-1))
      IF (DIFF. LT. ERR) GO TO 20
C-----THE MAXIMUM NUMBER OF ITERATIONS ALLOWED IS 10.
      IF (I. LT. 10) GO TO 10
      WRITE (5,100)
  100 FORMAT ( ' MORE THAN 10 ITERATIONS REQUIRED, CHECK PARAMETERS')
      STOP
   20 RES = Z(I,I)
      RETURN
      END
```

4. ADAPTIVE QUADRATURE

Although the Romberg method is capable of any desired accuracy (within the inherent limits of the computer used), it may not be the most efficient method.

Function evaluations usually represent the bulk of the cost of carrying out a quadrature, so it is desirable to minimize their number. This and the following sections give methods designed to do this.

One reason why the Romberg method requires so many function evaluations is that the points are evenly spaced throughout the interval of integration. Intuition tells us that we ought to be able to use fewer points in regions in which the function is either small or slowly varying. So we should be able to reduce the number of function evaluations by breaking the interval into a number of subintervals. Then we can compute each subintegral and estimate the error in each of the subintegrals. This enables us to increase the accuracy on only those subintervals for which the accuracy is not within the desired tolerance. There are a number of ways in which this can be done and a number of methods of this kind (called *adaptive* quadrature) have been proposed. As they bear a close resemblance to each other (and to Romberg as well), we will give only one such method here.

Suppose we wish to compute the integral

$$I=\int_a^b f(x)\,dx \tag{2.4.1}$$

with accuracy ε. We might do this by breaking the interval (a,b) into a number of subintervals and using the trapezoid rule with three points (two intervals) to estimate each subintegral. If h is the size of an interval, the error in this approximation is proportional to h^2. Suppose we get another estimate of the integral by breaking h into two subintervals of size $h/2$. Due to the second order accuracy of the trapezoid rule, the error in this improved approximation, say δ_2, is approximately one quarter of the error in the original approximation δ_1. The difference between the two approximations is $\delta_1-\delta_2 \simeq 3\delta_2$, or three times the error in the more accurate approximation. Thus $(\delta_1-\delta_2)/3$ is an estimate of the error and it can be computed from information in the quadrature method itself. If this error is less than $\varepsilon h/(b-a)$, we may accept the present estimate of the integral. If it is not, we halve the subinterval and continue until the error is within the desired accuracy.

Example 2.7

The function e^{-100x^2} used in Example 2.6 provides an ideal illustration of what adaptive quadrature can do. We will use trapezoid integration and demand an accuracy of 0.01 for this illustration. For purposes of this example let $T(a,b)$ denote the trapezoid estimate of

$$\int_a^b e^{-100x^2}\,dx$$

and $I(a,b)$ denote the exact integral.

At the first step we compute

$$T(0,1)=0.50000$$

Next we find

$$T(0,0.5)=0.25000$$

$$T(0.5,1)=0.00000$$

The sum of these gives 0.25000 as a new estimate of $I(0,1)$. The estimated error is $(0.25-0.50)/3 \simeq -0.083$, which is greater than the requested error, so we continue.

Next we deal with the interval (0,0.5). We compute

$$T(0,0.25)=0.12524$$

$$T(0.25,0.5)=0.00024$$

which gives a new estimate of $I(0,0.5)$ as 0.12548 with an error estimate of approximately 0.04. This is more than $\frac{1}{2}$ (0.01) so we continue. We compute

$$T(0,0.125)=0.07560$$

$$T(0.125,0.25)=0.01322$$

so that $I(0,0.25) \simeq 0.08882$ and $\varepsilon \simeq 0.012 > \frac{1}{4}$ (0.01). So we must continue.

The entire process of computing this integral is illustrated in Table 2.11. Several points about adaptive quadrature are illustrated by this example. The method does indeed concentrate the abscissas in the region in which the largest contributions to the integral originate. For this example 23 function evaluations were used. All but two of these are in the first half of the range, and 17 are in the first one-eighth of the range. The relative error in some of the asterisked values, for example, $T(0.25,375)$ may be quite large but they are accepted because their absolute contribution is small, which makes the absolute error small.

For a case such as this one (in which the major contributions to the integral come from a small part of the range) the criterion for acceptance of a given value is quite conservative and the result produced actually is considerably more accurate than the requested error. Here the actual error is approximately 0.0002 versus the required accuracy of 0.01. These results compare very favorably with those produced by the Romberg method.

There is a tradeoff in adaptive quadrature just as there was in some of the methods given earlier. By using a more accurate method such as Simpson's rule as the basic method, fewer subdivisions of the integral are necessary. On the other hand, the function will be evaluated a greater number of times than necessary in some regions and there will be a waste of computation. For

Table 2.11. Integration of e^{-100x^2} by Adaptive Quadrature

0.50000
0 1

0.25000 0.00000
0 1

0.12524 0.00024 0.00000* 0.00000*
0 0.25 0.5 0.75 1

0.07560 0.01322 0.00012* 0.00000*
0 0.125 0.25 0.375 0.5

0.05239 0.02770 0.00748 0.00099
0 0.0625 0.125 0.1875 0.25

0.02980 0.02474 0.01706 0.00976 0.00464 0.00182 0.00060* 0.00016*
0 0.03125 0.0625 0.09375 0.125 0.15625 0.1875 0.21875 0.25

0.01471 0.01156 0.00749* 0.00400* 0.00176* 0.00064*
0.01544 0.01336 0.00953* 0.00561* 0.00272* 0.00109*
0 0.03125 0.0625 0.09375 0.125 0.15625 0.1875
0.015625 0.046875 0.078125 0.109375 0.140625 0.171875

0.00769* 0.00724* 0.00649* 0.00554*
0.00779* 0.00751* 0.00690* 0.00603*
0 0.015625 0.03125 0.046875 0.0625
0.0078125 0.0234375 0.0390625 0.0546875

*Denotes a value that is part of the final estimate. The total of asterisked values is 0.088623.

general purposes Newton–Cotes methods with $n=4$ to $n=8$ have been found to be most useful.

No subroutine for adaptive quadrature is given here, as the method requires a fair amount of logic in order to avoid duplication of function evaluations. For an excellent subroutine QUANC8 (Quadrature Adaptive Newton–Cotes 8 Point) given by Forsythe, Malcolm, and Moler (1977) is recommended.

Other adaptive procedures can be based on any of a number of quadrature formulas.

5. GAUSS QUADRATURE

We now ask for a method that is optimum in the sense of maximum accuracy for a given number of function evaluations. The price that we pay for this property is the lack of a simple method for systematic error reduction.

All of the methods introduced so far use evenly spaced abscissas at least on subintervals. With the abscissas picked a priori we cannot obtain methods of significantly greater accuracy than those already given. We are thus led to consider the possibility of allowing the abscissas to be adjustable parameters chosen to make the resulting formula as accurate as possible. In this instance it is best to use the ability to integrate polynomials of the highest possible order exactly as our criterion of goodness. Since the general formula Eq. 2.1 contains n weights and n abscissas, all of which are now regarded as adjustable, there are a total of $2n$ parameters. If all of them are adjustable, we should be able to integrate a polynomial of degree $2n-1$ exactly using an n point formula. The problem is then one of finding the parameters that accomplish this.

We begin by noting that with the simple transformation

$$\xi=\frac{2x-(a+b)}{b-a} \tag{2.5.1}$$

any integral between two finite limits can be transformed into an integral between the limits -1 and $+1$, so it is sufficient to treat this case. Since we seek to integrate polynomials of degree $2n-1$ using n abscissas, it seems logical to start with an interpolation formula that fits such a formula to n points; in other words, we use the Hermite interpolation formula of Eq. 1.2.1. Integrating that formula we obtain

$$\int_{-1}^{1} w(x)f(x)\,dx=\sum_{k=1}^{n} w_k f(x_k)+\sum_{k=1}^{n} v_k f'(x_k) \tag{2.5.2}$$

where $w(x)$ is a known weight function, which we have inserted for later convenience, and w_k and v_k are given by

$$w_k=\int_{-1}^{1} w(x)U_k(x)\,dx \tag{2.5.3a}$$

$$v_k=\int_{-1}^{1} w(x)V_k(x)\,dx \tag{2.5.3b}$$

and U_k and V_k are given by Eq. 1.2.3. For Eq. 2.5.2 to be the desired quadrature formula it is necessary that all of the v_k be zero. This can be assured by selecting the abscissas properly. For the present it suffices to take only the case in which $w(x)=1$. Then with the explicit expression in Eq. 1.2.3 for V_k we have

$$v_k=\int_{-1}^{1}(x-x_k)L_k^2(x)\,dx \qquad k=1,2,\ldots n \tag{2.5.4}$$

as the quantities we need to force to zero. Using Eq. 1.1.9, we can write this as

$$v_k=\int_{-1}^{1} F(x)L_k(x)\,dx \qquad k=1,2,\ldots n \tag{2.5.5}$$

Recall that $F(x)$ is a polynomial of degree n and the $L_k(x)$ are polynomials of degree $n-1$. Furthermore, if the x_k are all different, the polynomials $L_k(x)$ are linearly independent. Since there can be no more than n linearly independent polynomials of degree $n-1$, the $L_k(x)$ form a complete set (i.e., there are no further polynomials of that degree that are linearly independent of the $L_k(x)$). Hence Eq. 2.5.5 can be interpreted as stating that $F(x)$ is a polynomial of degree n that is orthogonal to any polynomial of degree $n-1$. This polynomial is unique (aside from a multiplicative constant) and is well known in applied mathematics; $F(x)$ must be the Legendre polynomial of degree n, $P_n(x)$. The Legendre polynomials, which have the property that

$$\int_{-1}^{1} P_n(x)P_m(x)\,dx=\delta_{nm} \tag{2.5.6}$$

have been well studied and their properties may be found in many textbooks. We have shown thus that the abscissas must be zeros of the Legendre polynomial of degree n and it is possible to show that they all lie between -1 and $+1$. The weights are then computed by substituting Eq. 1.2.3 for U_k into the integral in Eq. 2.5.3a defining w_k. The resulting integral can be simplified somewhat, but it can only be evaluated numerically. The values will be given below.

The result of this calculation is the quadrature formula

$$\int_{-1}^{1} f(x)\,dx=\sum_{k=1}^{n} w_k f(x_k) \tag{2.5.7}$$

The values of the abscissas x_k and weights w_k for some values of n up to 24 are given in Table 2.12, which has been reproduced from the National Bureau of Standards' *Handbook of Mathematical Functions*.

The error in Gauss quadrature procedure with n abscissas is shown to be

$$\varepsilon \simeq \frac{2^{2n+1}(n!)^4}{(2n+1)(2n!)^3} f^{(2n)}(\xi) \tag{2.5.8}$$

Table 2.12. Abscissas and Weight Factors for Gaussian Integration $\left(\int_{-1}^{+1} f(x)\,dx = \sum_{i=1}^{n} w_i f(x_i)\right)^a$

$$\int_{-1}^{+1} f(x)\,dx \approx \sum_{i=1}^{n} w_i f(x_i)$$

Abscissas $= \pm x_i$ (Zeros of Legendre Polynomials) — Weight Factors $= w_i$

n	$\pm x_i$	w_i
$n=2$	0.57735 02691 89626	1.00000 00000 00000
$n=3$	0.00000 00000 00000	0.88888 88888 88889
	0.77459 66692 41483	0.55555 55555 55556
$n=4$	0.33998 10435 84856	0.65214 51548 62546
	0.86113 63115 94053	0.34785 48451 37454
$n=5$	0.00000 00000 00000	0.56888 88888 88889
	0.53846 93101 05683	0.47862 86704 99366
	0.90617 98459 38664	0.23692 68850 56189
$n=6$	0.23861 91860 83197	0.46791 39345 72691
	0.66120 93864 66265	0.36076 15730 48139
	0.93246 95142 03152	0.17132 44923 79170
$n=7$	0.00000 00000 00000	0.41795 91836 73469
	0.40584 51513 77397	0.38183 00505 05119
	0.74153 11855 99394	0.27970 53914 89277
	0.94910 79123 42759	0.12948 49661 68870
$n=8$	0.18343 46424 95650	0.36268 37833 78362
	0.52553 24099 16329	0.31370 66458 77887
	0.79666 64774 13627	0.22238 10344 53374
	0.96028 98564 97536	0.10122 85362 90376
$n=9$	0.00000 00000 00000	0.33023 93550 01260
	0.32425 34234 03809	0.31234 70770 40003
	0.61337 14327 00590	0.26061 06964 02935
	0.83603 11073 26636	0.18064 81606 94857
	0.96816 02395 07626	0.08127 43883 61574
$n=10$	0.14887 43389 81631	0.29552 42247 14753
	0.43339 53941 29247	0.26926 67193 09996
	0.67940 95682 99024	0.21908 63625 15982
	0.86506 33666 88985	0.14945 13491 50581
	0.97390 65285 17172	0.06667 13443 08688
$n=12$	0.12523 34085 11469	0.24914 70458 13403
	0.36783 14989 98180	0.23349 25365 38355
	0.58731 79542 86617	0.20316 74267 23066
	0.76990 26741 94305	0.16007 83285 43346
	0.90411 72563 70475	0.10693 93259 95318
	0.98156 06342 46719	0.04717 53363 86512
$n=16$	0.09501 25098 37637 440185	0.18945 06104 55068 496285
	0.28160 35507 79258 913230	0.18260 34150 44923 588867
	0.45801 67776 57227 386342	0.16915 65193 95002 538189
	0.61787 62444 02643 748447	0.14959 59888 16576 732081
	0.75540 44083 55003 033895	0.12462 89712 55533 872052
	0.86563 12023 87831 743880	0.09515 85116 82492 784810
	0.94457 50230 73232 576078	0.06225 35239 38647 892863
	0.98940 09349 91649 932596	0.02715 24594 11754 094852
$n=20$	0.07652 65211 33497 333755	0.15275 33871 30725 850698
	0.22778 58511 41645 078080	0.14917 29864 72603 746788
	0.37370 60887 15419 560673	0.14209 61093 18382 051329
	0.51086 70019 50827 098004	0.13168 86384 49176 626898
	0.63605 36807 26515 025453	0.11819 45319 61518 417312
	0.74633 19064 60150 792614	0.10193 01198 17240 435037
	0.83911 69718 22218 823395	0.08327 67415 76704 748725
	0.91223 44282 51325 905868	0.06267 20483 34109 063570
	0.96397 19272 77913 791268	0.04060 14298 00386 941331
	0.99312 85991 85094 924786	0.01761 40071 39152 118312
$n=24$	0.06405 68928 62605 626085	0.12793 81953 46752 156974
	0.19111 88674 73616 309159	0.12583 74563 46828 296121
	0.31504 26796 96163 374387	0.12167 04729 27803 391204
	0.43379 35076 26045 138487	0.11550 56680 53725 601353
	0.54542 14713 88839 535658	0.10744 42701 15965 634783
	0.64809 36519 36975 569252	0.09761 86521 04113 888270
	0.74012 41915 78554 364244	0.08619 01615 31953 275917
	0.82000 19859 73902 921954	0.07334 64814 11080 305734
	0.88641 55270 04401 034213	0.05929 85849 15436 780746
	0.93827 45520 02732 758524	0.04427 74388 17419 806169
	0.97472 85559 71309 498198	0.02853 13886 28933 663181
	0.99518 72199 97021 360180	0.01234 12297 99987 199547

[a] From Abramowitz and Stegun, *Handbook of Mathematical Functions*,[11] National Bureau of Standards, Washington, D.C.

after a laborious calculation. This again shows that a polynomial of degree $2n-1$ is integrated exactly.

As stated earlier, Gauss quadrature gives the best accuracy (in the sense of correctly integrating polynomials of the highest possible order) for a given number of function evaluations. Examination of Table 2.12, however, shows that the abscissas for the various orders are all different. Thus if one wishes to improve the accuracy by going to a higher order method, the calculation must be done from scratch. Consequently, if one desires a particular degree of accuracy and cost is not a factor, Romberg or adaptive integration remain the methods of choice.

There are also a number of variations on the theme of Gauss quadrature. We describe briefly two of the better known techniques of this kind, but there are others. The first one deals with the case of integrals in which one of the limits is infinite. Such integrals can always be transformed so that the range of integration runs from zero to infinity and the quadrature formula takes the form

$$\int_0^\infty e^{-x} f(x)\,dx = \sum_{k=1}^{n} w_k f(x_k) \tag{2.5.9}$$

The exponential weight factor is included for convenience. The abscissas then turn out to be the zeros of the nth polynomial of the set defined by the orthogonality property

$$\int_0^\infty L_m(x) L_n(x) e^{-x}\,dx = \delta_{mn} \tag{2.5.10}$$

which are known as the Laguerre polynomials. The abscissas and weights for this formula are given in Table 2.13.

Finally, when both limits are infinite, we have the formula

$$\int_{-\infty}^\infty e^{-x^2} f(x)\,dx = \sum_{k=1}^{n} w_k f(x_k) \tag{2.5.11}$$

In this case the abscissas are the zeros of the nth Hermite polynomial. These are defined by the orthogonality property

$$\int_{-\infty}^\infty H_m(x) H_n(x) e^{-x^2}\,dx = \delta_{mn} \tag{2.5.12}$$

The abscissas and weights for this formula are given in Table 2.14.

Example 2.8

Gauss quadrature is sufficiently accurate that the error in computing

$$\int_0^1 e^x\,dx$$

Table 2.13. Abscissas and Weight Factors for Laguerre Integration[a]

$$\int_0^\infty e^{-x} f(x)\,dx \approx \sum_{i=1}^{n} w_i f(x_i) \qquad \int_0^\infty g(x)\,dx \approx \sum_{i=1}^{n} w_i e^{x_i} g(x_i)$$

Abscissas $= x_i$ (Zeros of Laguerre Polynomials) Weight Factors $= w_i$

x_i	w_i	$w_i e^{x_i}$
$n=2$		
0.58578 64376 27	(-1)8.53553 390593	1.53332 603312
3.41421 35623 73	(-1)1.46446 609407	4.45095 733505
$n=3$		
0.41577 45567 83	(-1)7.11093 009929	1.07769 285927
2.29428 03602 79	(-1)2.78517 733569	2.76214 296190
6.28994 50829 37	(-2)1.03892 565016	5.60109 462543
$n=4$		
0.32254 76896 19	(-1)6.03154 104342	0.83273 91238 38
1.74576 11011 58	(-1)3.57418 692438	2.04810 243845
4.53662 02969 21	(-2)3.88879 085150	3.63114 630582
9.39507 09123 01	(-4)5.39294 705561	6.48714 508441
$n=5$		
0.26356 03197 18	(-1)5.21755 610583	0.67909 40422 08
1.41340 30591 07	(-1)3.98666 811083	1.63848 787360
3.59642 57710 41	(-2)7.59424 496817	2.76944 324237
7.08581 00058 59	(-3)3.61175 867992	4.31565 690092
12.64080 08442 76	(-5)2.33699 723858	7.21918 635435
$n=6$		
0.22284 66041 79	(-1)4.58964 673950	0.57353 55074 23
1.18893 21016 73	(-1)4.17000 830772	1.36925 259071
2.99273 63260 59	(-1)1.13373 382074	2.26068 459338
5.77514 35691 05	(-2)1.03991 974531	3.35052 458236
9.83746 74183 83	(-4)2.61017 202815	4.88682 680021
15.98287 39806 02	(-7)8.98547 906430	7.84901 594560
$n=7$		
0.19304 36765 60	(-1)4.09318 951701	0.49647 75975 40
1.02666 48953 39	(-1)4.21831 277862	1.17764 306086
2.56787 67449 51	(-1)1.47126 348658	1.91824 978166
4.90035 30845 26	(-2)2.06335 144687	2.77184 863623
8.18215 34445 63	(-3)1.07401 014328	3.84124 912249
12.73418 02917 98	(-5)1.58654 643486	5.38067 820792
19.39572 78622 63	(-8)3.17031 547900	8.40543 248683
$n=8$		
0.17027 96323 05	(-1)3.69188 589342	0.43772 34104 93
0.90370 17767 99	(-1)4.18786 780814	1.03386 934767
2.25108 66298 66	(-1)1.75794 986637	1.66970 976566
4.26670 01702 88	(-2)3.33434 922612	2.37692 470176
7.04590 54023 93	(-3)2.79453 623523	3.20854 091335
10.75851 60101 81	(-5)9.07650 877336	4.26857 551083
15.74067 86412 78	(-7)8.48574 671627	5.81808 336867
22.86313 17368 89	(-9)1.04800 117487	8.90622 621529
$n=9$		
0.15232 22277 32	(- 1)3.36126 421798	0.39143 11243 16
0.80722 00227 42	(- 1)4.11213 980424	0.92180 50285 29
2.00513 51556 19	(- 1)1.99287 525371	1.48012 790994
3.78347 39733 31	(- 2)4.74605 627657	2.08677 080755
6.20495 67778 77	(- 3)5.59962 661079	2.77292 138971
9.37298 52516 88	(- 4)3.05249 767093	3.59162 606809
13.46623 69110 92	(- 6)6.59212 302608	4.64876 600214
18.83359 77889 92	(- 8)4.11076 933035	6.21227 541975
26.37407 18909 27	(-11)3.29087 403035	9.36321 823771
$n=10$		
0.13779 34705 40	(- 1)3.08441 115765	0.35400 97386 07
0.72945 45495 03	(- 1)4.01119 929155	0.83190 23010 44
1.80834 29017 40	(- 1)2.18068 287612	1.33028 856175
3.40143 36978 55	(- 2)6.20874 560987	1.86306 390311
5.55249 61400 64	(- 3)9.50151 697518	2.45025 555808
8.33015 27467 64	(- 4)7.53008 388588	3.12276 415514
11.84378 58379 00	(- 5)2.82592 334960	3.93415 269556
16.27925 78313 78	(- 7)4.24931 398496	4.99241 487219
21.99658 58119 81	(- 9)1.83956 482398	6.57220 248513
29.92069 70122 74	(-13)9.91182 721961	9.78469 584037
$n=12$		
0.11572 21173 58	(- 1)2.64731 371055	0.29720 96360 44
0.61175 74845 15	(- 1)3.77759 275873	0.69646 29804 31
1.51261 02697 76	(- 1)2.44082 011320	1.10778 139462
2.83375 13377 44	(- 2)9.04492 222117	1.53846 423904
4.59922 76394 18	(- 2)2.01023 811546	1.99832 760627
6.84452 54531 15	(- 3)2.66397 354187	2.50074 576910
9.62131 68424 57	(- 4)2.03231 592663	3.06532 151828
13.00605 49933 06	(- 6)8.36505 585682	3.72328 911078
17.11685 51874 62	(- 7)1.66849 387654	4.52981 402998
22.15109 03793 97	(- 9)1.34239 103052	5.59725 846184
28.48796 72509 84	(-12)3.06160 163504	7.21299 546093
37.09912 10444 67	(-16)8.14807 746743	10.54383 74619
$n=15$		
0.09330 78120 17	(- 1)2.18234 885940	0.23957 81703 11
0.49269 17403 02	(- 1)3.42210 177923	0.56010 08427 93
1.21559 54120 71	(- 1)2.63027 577942	0.88700 82629 19
2.26994 95262 04	(- 1)1.26425 818106	1.22366 440215
3.66762 27217 51	(- 2)4.02068 649210	1.57444 872163
5.42533 66274 14	(- 3)8.56387 780361	1.94475 197653
7.56591 62266 13	(- 3)1.21243 614721	2.34150 205664
10.12022 85680 19	(- 4)1.11674 392344	2.77404 192683
13.13028 24821 76	(- 6)6.45992 676202	3.25564 334640
16.65440 77083 30	(- 7)2.22631 690710	3.80631 171423
20.77647 88994 49	(- 9)4.22743 038498	4.45847 775384
25.62389 42267 29	(-11)3.92189 726704	5.27001 778443
31.40751 91697 54	(-13)1.45651 526407	6.35956 346973
38.53068 33064 86	(-16)1.48302 705111	8.03178 763212
48.02608 55726 86	(-20)1.60059 490621	11.52777 21009

[a] From Abramowitz and Stegun, *Handbook of Mathematical Functions*, National Bureau of Standards, Washington, D.C.

Table 2.14. Abscissas and Weight Factors for Hermite Integration[a]

$$\int_{-\infty}^{\infty} e^{-x^2} f(x)\,dx \approx \sum_{i=1}^{n} w_i f(x_i)$$

Abscissas $= \pm x_i$ (Zeros of Hermite Polynomials)

$\pm x_i$	w_i	$w_i e^{x_i^2}$
$n=2$		
0.70710 67811 86548	(-1)8.86226 92545 28	1.46114 11826 611
$n=3$		
0.00000 00000 00000	(0)1.18163 59006 04	1.18163 59006 037
1.22474 48713 91589	(-1)2.95408 97515 09	1.32393 11752 136
$n=4$		
0.52464 76232 75290	(-1)8.04914 09000 55	1.05996 44828 950
1.65068 01238 85785	(-2)8.13128 35447 25	1.24022 58176 958
$n=5$		
0.00000 00000 00000	(-1)9.45308 72048 29	0.94530 87204 829
0.95857 24646 13819	(-1)3.93619 32315 22	0.98658 09967 514
2.02018 28704 56086	(-2)1.99532 42059 05	1.18148 86255 360
$n=6$		
0.43607 74119 27617	(-1)7.24629 59522 44	0.87640 13344 362
1.33584 90740 13697	(-1)1.57067 32032 29	0.93558 05576 312
2.35060 49736 74492	(-3)4.53000 99055 09	1.13690 83326 745
$n=7$		
0.00000 00000 00000	(-1)8.10264 61755 68	0.81026 46175 568
0.81628 78828 58965	(-1)4.25607 25261 01	0.82868 73032 836
1.67355 16287 67471	(-2)5.45155 82819 13	0.89718 46002 252
2.65196 13568 35233	(-4)9.71781 24509 95	1.10133 07296 103
$n=8$		
0.38118 69902 07322	(-1)6.61147 01255 82	0.76454 41286 517
1.15719 37124 46780	(-1)2.07802 32581 49	0.79289 00483 864
1.98165 67566 95843	(-2)1.70779 83007 41	0.86675 26065 634
2.93063 74202 57244	(-4)1.99604 07221 14	1.07193 01442 480
$n=9$		
0.00000 00000 00000	(-1)7.20235 21560 61	0.72023 52156 061
0.72355 10187 52838	(-1)4.32651 55900 26	0.73030 24527 451
1.46855 32892 16668	(-2)8.84745 27394 38	0.76460 81250 946
2.26658 05845 31843	(-3)4.94362 42755 37	0.84175 27014 787
3.19099 32017 81528	(-5)3.96069 77263 26	1.04700 35809 767

$$\int_{-\infty}^{\infty} g(x)\,dx \approx \sum_{i=1}^{n} w_i e^{x_i^2} g(x_i)$$

Weight Factors $= w_i$

$\pm x_i$	w_i	$w_i e^{x_i^2}$
n 10		
0.34290 13272 23705	(- 1)6.10862 63373 53	0.68708 18539 513
1.03661 08297 89514	(- 1)2.40138 61108 23	0.70329 63231 049
1.75668 36492 99882	(- 2)3.38743 94455 48	0.74144 19319 436
2.53273 16742 32790	(- 3)1.34364 57467 81	0.82066 61264 048
3.43615 91188 37738	(- 6)7.64043 28552 33	1.02545 16913 657
n 12		
0.31424 03762 54359	(- 1)5.70135 23626 25	0.62930 78743 695
0.94778 83912 40164	(- 1)2.60492 31026 42	0.63962 12320 203
1.59768 26351 52605	(- 2)5.16079 85615 88	0.66266 27732 669
2.27950 70805 01060	(- 3)3.90539 05846 29	0.70522 03661 122
3.02063 70251 20890	(- 5)8.57368 70435 88	0.78664 39394 633
3.88972 48978 69782	(- 7)2.65855 16843 56	0.98969 90470 923
$n=16$		
0.27348 10461 3815	(- 1)5.07929 47901 66	0.54737 52050 378
0.82295 14491 4466	(- 1)2.80647 45852 85	0.55244 19573 675
1.38025 85391 9888	(- 2)8.38100 41398 99	0.56321 78290 882
1.95178 79909 1625	(- 2)1.28803 11535 51	0.58124 72754 009
2.54620 21578 4748	(- 4)9.32284 00862 42	0.60973 69582 560
3.17699 91619 7996	(- 5)2.71186 00925 38	0.65575 56728 761
3.86944 79048 6012	(- 7)2.32098 08448 65	0.73824 56222 777
4.68873 89393 0582	(-10)2.65480 74740 11	0.93687 44928 841
$n=20$		
0.24534 07083 009	(- 1)4.62243 66960 06	0.49092 15006 667
0.73747 37285 454	(- 1)2.86675 50536 28	0.49384 33852 721
1.23407 62153 953	(- 1)1.09017 20602 00	0.49992 08713 363
1.73853 77121 166	(- 2)2.48105 20887 46	0.50967 90271 175
2.25497 40020 893	(- 3)3.24377 33422 38	0.52408 03509 486
2.78880 60584 281	(- 4)2.28338 63601 63	0.54485 17423 644
3.34785 45673 832	(- 6)7.80255 64785 32	0.57526 24428 525
3.94476 40401 156	(- 7)1.08606 93707 69	0.62227 86961 914
4.60368 24495 507	(-10)4.39934 09922 73	0.70433 29611 769
5.38748 08900 112	(-13)2.22939 36455 34	0.89859 19614 532

[a] From Abramowitz and Stegun, *Handbook of Mathematical Functions*, National Bureau of Standards, Washington D.C.

is less than 10^{-8} with just four points. To provide a fair test we will try to integrate the function of Example 2.6.

$$I(a) = \int_0^1 e^{-a^2x^2}$$

With small a, say $a=1$, four points are sufficient to achieve an accuracy of 10^{-8}. The results given in Table 2.15 show only the even approximations, but for most functions, including this one, the error reduction tends to be orderly.

Comparing the results in Tables 2.9 and 2.15, we see that the Gauss quadrature does produce a smaller error for a given number of points or, conversely, requires fewer evaluations for a given accuracy. A comparison of

Table 2.15. Integration of e^{-x^2} with Gauss Quadrature

n	Error
2	-2.3×10^{-4}
4	-3.4×10^{-7}
6	10^{-8}

Table 2.16. Integration of e^{-100x^2} with Gauss Quadrature (Exact=0.88622693)

n	Absolute Error	Relative Error
2	-8.3×10^{-2}	-0.935
4	$+1.8\times10^{-2}$	-0.212
6	-2.0×10^{-3}	-2.2×10^{-2}
8	$+1.5\times10^{-4}$	$+1.7\times10^{-3}$
10	-1.7×10^{-5}	-1.9×10^{-4}
12	-3.0×10^{-6}	$+3.4\times10^{-5}$
14	-4.6×10^{-7}	-5.2×10^{-6}
16	$+4.9\times10^{-8}$	-5.6×10^{-7}
18	-5.0×10^{-9}	$+5.1\times10^{-8}$

Tables 2.10 and 2.16 show that for a more difficult function the advantage of Gauss is even greater. Adaptive methods are intermediate between Romberg and Gauss. The only disadvantage of Gauss quadrature, as we have said earlier, is the difficulty of accuracy improvement.

6. SINGULARITIES

Occasionally we are faced with the problem of computing an integral whose integrand blows up at some point in the range of integration. Naturally we assume that the singularity is gentle enough for the integral to exist. Such integrals can be computed by the methods already described, provided that a little attention is given to handling the integral near the singular point. Some of the better methods for accomplishing this are now given.

a. Integration by Parts

In many singular integrals it is possible to factor the integrand into the product of two components: one that contains the singularity but is easily integrated analytically, and a second that is not singular and can be differentiated. In such a case integration by parts will often convert the integral into one that is not singular. This is best illustrated by an example:

$$\int_0^\pi \frac{\cos x}{\sqrt{x}}\,dx = 2\sqrt{x}\cos x\Big|_0^\pi + 2\int_0^\pi \sqrt{x}\sin x\,dx \tag{2.6.1}$$

The integrated part is evaluated easily and the remaining integral is not singular. Although it is not the case here, sometimes it is desirable to repeat the integration by parts to obtain a still smoother integral.

b. Singularity Subtraction

A related but slightly different method is to factor the integral into the sum of two parts: one that contains the singularity but is integrable analytically, and a second that is nonsingular but requires numerical quadrature. This method can also be applied to the integral that was described above:

$$\int_0^{\pi} \frac{\cos x}{\sqrt{x}}\,dx = \int_0^{\pi} \frac{dx}{\sqrt{x}} + \int_0^{\pi} \frac{\cos x - 1}{\sqrt{x}}\,dx \tag{2.6.2}$$

Again the first part is easily evaluated and the second is nonsingular. The choice between these two methods is one of convenience and of deciding which gives the easier remaining integral.

7. CONCLUDING REMARKS

Although we have presented only a few quadrature methods, we have concentrated on those that are the most effective and, therefore, the most widely used. A look at any standard numerical analysis text will show that there are a great many more methods. For the most part these have no significant advantages over the methods presented here, except in special cases.

The one case in which it may pay to consider a method other than the ones given here is the case of multiple or iterated integrals. For these there are a number of good methods that can be found in the texts.

Problems

1. Another example of a peaky function is the Lorentz profile $(1+x^2/a^2)^{-1}$. For large a it is well behaved but for small a it is peaked near the origin. Integrate this function from 0 to 1 using (a) Newton–Cotes methods of various orders and (b) Simpson's rule with variable numbers of points. Plot the error versus the number of function evaluations with $a=0.1$ and 1.
2. Integrate the function of Problem 1 using Romberg quadrature. Use the subroutine in the text or one available at your own computer center. Again, plot the error versus the number of function evaluations.
3. Repeat Problem 1 with the superellipse $(1-x^4)^{1/4}$.
4. Repeat Problem 2 with the superellipse of Problem 3.
5. Occasionally integral equations occur in physics and engineering. These are equations of the type

$$\int_0^1 K(x, y) f(y)\,dy + h(x) f(x) = g(x)$$

where $K(x, y)$, $h(x)$, and $g(x)$ are given functions. Propose a method of solving this equation numerically. Suppose that you were asked to solve this equation with an accuracy of ε. How would you approach this problem?

6. Numerically solve the integral equation

$$g(\theta)-\frac{1-\varepsilon}{4\varepsilon}\int_{\theta_0}^{2\pi-\theta_0} g(\theta')\sin\frac{|\theta-\theta'|}{2}\,d\theta'=\frac{\sigma T^4}{2}\left[1+\cos\frac{(\theta+\theta_0)}{2}+\cos\frac{(\theta-\theta_0)}{2}\right]$$

Note that the exact solution is

$$g(\theta)=\frac{\varepsilon\sigma T^4\sin\dfrac{\theta_0}{2}\cos\dfrac{\sqrt{\varepsilon}}{2}(\pi-\theta)}{\sin\dfrac{\theta_0}{2}\cos\dfrac{\sqrt{\varepsilon}}{2}(\pi-\theta_0)+\cos\dfrac{\theta_0}{2}\sin\dfrac{\sqrt{\varepsilon}}{2}(\pi-\theta_0)}$$

(This problem arises in radiative heat transfer in a cylinder with a slot.)

7. Numerically compute

$$\int_0^{0.999}\tan\frac{\pi}{2}x\,dx$$

Be careful near the upper limit!

3

Ordinary Differential Equations

One of the major scientific and engineering applications of numerical methods is the solution of differential equations, ordinary and partial. Partial differential equations soivers are based largely on those for ordinary differential equations, so the latter case is discussed first. As in the preceding chapters, we begin by classifying the problems. The following categories provide a useful means of classification for ordinary differential equations:

First order	Higher order
Single equation	System of equations
Linear	Nonlinear
Homogeneous	Inhomogeneous
Initial value	Boundary value

We note first that nearly any equation of an order higher than first can be reduced to a system of first order equations. Thus the nth order equation

$$y^{(n)}=f(x, y, y', \dots y^{(n-1)}) \tag{3.1}$$

can be reduced to a system of first order equations by defining

$$\begin{aligned} y_1 &= y \\ y_j &= y^{(j-1)} \qquad j=1,2,\dots n-1 \end{aligned} \tag{3.2}$$

so that Eq. 3.1 becomes

$$\begin{aligned} y_j' &= y_{j+1} \qquad j=1,2,\dots n-1 \\ y_n' &= f(x, y_1, y_2, \dots y_n) \end{aligned} \tag{3.3}$$

Thus the first two lines in the above table are essentially equivalent. Furthermore, with the exception of one important difficulty (stiffness), the methods used for systems of equations are similar to those used for single equations and it is sufficient, at least initially, to restrict our attention to the case of a single first order equation.

The distinction between linear and nonlinear equations is not important as long as the equations may be written in the form of Eq. 3.1—in other words, as long as it is possible to isolate the highest derivative. The same methods may be applied in both linear and nonlinear equations. The major difference between them is that nonlinear problems generally require considerably more computation and some attention should be given to this.

Similarly, the distinction between homogeneous and inhomogeneous problems is not important except for the case of linear homogeneous boundary value problems, which become eigenvalue problems. These are discussed after boundary value problems in Section 16 of this chapter.

The distinction between initial and boundary value problems is important. In the initial value case there is a definite starting point and one can "march" the solution with increasing value of the independent variable. For boundary value problems this is not true and the problem is more difficult. Some methods for solving boundary value problems are based on those for initial value problems, so we begin by considering the initial value problem.

The plan of this chapter is first to study methods of estimating derivatives numerically and then to apply these to the development of schemes for solving a single first order ordinary differential equation. The properties of these methods are examined with a view toward finding the key features required for efficient solution. Then we examine systems of equations and look at the difficult problem of stiffness in particular. The chapter ends with a look at methods for solving boundary value problems. Many of the results obtained are applied to the solution of partial differential equations in Chapter 4.

1. NUMERICAL DIFFERENTIATION

Before proceeding to the solution of differential equations, we look at how one estimates derivatives numerically. The problem is that since differentiation is defined only for smooth functions and in any computation we can have only a finite number of discrete values of a function, we need to estimate the derivative from a limited number of values. For the present we assume that the data are the exact values of a smooth function at the data points and further that we need the derivatives only at the data points.

There are at least three approaches to this problem. In the first method we use interpolation to fit a smooth curve through the data points and differentiate it to get the results we want. A second approach uses Taylor series and produces substantially the same results at the cost of somewhat more work, but yields information about the error. Finally, numerical quadrature methods can be used directly to provide schemes for solving ordinary differential equations. This approach is closely related to the use of interpolation.

a. Interpolation

The simplest possibility is to use Lagrange interpolation to provide estimates of derivatives. For later applications expressions for the second derivative as well as the first are developed. The simplest case is, of course, linear Lagrange interpolation for which we have the approximation

$$f(x) \simeq \frac{x_i - x}{x_i - x_{i-1}} f(x_{i-1}) + \frac{x - x_{i-1}}{x_i - x_{i-1}} f(x_i) \tag{3.1.1}$$

Differentiating this equation, we find that the derivative is constant on each interval so that we get the same value at both points:

$$f'(x_{i-1}) \simeq D_+ f \equiv \frac{(f_i - f_{i-1})}{h_i} \tag{3.1.2}$$

$$f'(x_i) \simeq D_- f \equiv \frac{(f_i - f_{i-1})}{h_i} \tag{3.1.3}$$

These are known as the forward and backward difference formulas, respectively, and D_+ and D_- are the forward and backward difference operators. Any approximation of a derivative in terms of values at a discrete set of points is called a *finite difference approximation*. We have used the abbreviations $f_i = f(x_i)$ and $h_i = x_i - x_{i-1}$ to avoid complicating this and later equations. These are useful formulas but since the derivative approximation is different according to whether the forward or backward formula is chosen, it is not clear which formula ought to be used. They are equally accurate and the choice depends on considerations to be discussed later.

Improvement on these formulas can be obtained by using the quadratic Lagrange polynomial

$$f(x) = f(x_{i-1}) \frac{(x - x_i)(x - x_{i+1})}{(x_{i-1} - x_i)(x_{i-1} - x_{i+1})} + f(x_i) \frac{(x - x_{i-1})(x - x_{i+1})}{(x_i - x_{i-1})(x_i - x_{i+1})}$$

$$+ f(x_{i+1}) \frac{(x - x_{i-1})(x - x_i)}{(x_{i+1} - x_{i-1})(x_{i+1} - x_i)} \tag{3.1.4}$$

Differentiating this equation at the midpoint x_i of the interval, we obtain

$$f'(x_i) \simeq -f_{i-1}\left[\left(\frac{h_{i+1}}{h_i}\right)(h_i + h_{i+1})^{-1}\right] + f_i[h_i^{-1} - h_{i+1}^{-1}]$$

$$+ f_{i+1}\left[\left(\frac{h_i}{h_{i+1}}\right)(h_i + h_{i+1})^{-1}\right] \tag{3.1.5}$$

For equally spaced intervals $h_i = h_{i-1} = h$ this reduces to

$$f'(x_i) \simeq \frac{f_{i+1} - f_{i-1}}{2h} = \tfrac{1}{2}(D_+ + D_-)f \tag{3.1.6}$$

which is the average of the forward and backward difference approximations. Equations 3.1.5 and 3.1.6 are called *central difference formulas.* We also give the results at x_{i-1} and x_{i+1} for the case of equally spaced intervals

$$2hf'(x_{i-1}) \simeq -3f_{i-1} + 4f_i - f_{i+1} \tag{3.1.7a}$$

$$2hf'(x_{i+1}) \simeq f_{i-1} - 4f_i + 3f_{i+1} \tag{3.1.7b}$$

These are forward and backward difference formulas. We can also differentiate Eq. 3.1.4 twice to obtain a formula for the second derivative. The result is a constant on the interval (x_{i-1}, x_{i+1}):

$$f''(x_i) \simeq 2h_i^{-1}(h_i + h_{i+1})^{-1} f_{i-1} - \left(\frac{2}{h_i h_{i+1}}\right) f_i + 2h_{1+1}^{-1}(h_i + h_{i+1})^{-1} f_{i+1} \tag{3.1.8}$$

which for equal intervals becomes

$$f'' = \frac{f_{i-1} - 2f_i + f_{i+1}}{h^2} = D_+ D_- f = D_- D_+ f \tag{3.1.9}$$

This formula may be applied to the estimation of the second derivative at any of the three points x_{i-1}, x_i, and x_{i+1}. Depending on the point to which it is applied, it is called the *forward*, *central*, or *backward difference formula.* As we show later, the formulas obtained from the quadratic interpolation formula are more accurate than those derived from the linear approximation.

This procedure obviously can be extended to higher order interpolation methods. Using the cubic polynomials, we can obtain forward and backward difference formulas; the formulas obtained at the center two of the four points do not fit into any of the categories defined above. The quartic Lagrange polynomial yields central, forward, and backward formulas as well as two formulas (at the second and fourth of the five points) that do not fit these categories. Some of these formulas are shown in Table 3.1.

The use of Hermite interpolation as a means of estimating first derivatives does not make sense because it uses first derivatives as data, but it can be used to approximate higher order differential equations directly.

The use of splines as sources of finite difference approximations has received some attention in recent years. The only spline treated in detail in Chapter 1 is the cubic for which a method of computing the second derivative is given by Eq. 1.3.4. It is interesting that for equally spaced intervals the right-hand side

Table 3.1. Coefficients in Difference Formulas

Derivative	f_{i-2}	f_{i-1}	f_i	f_{i+1}	f_{i+2}	Error
Forward formulas						
hf_i'			-1	1		$0(h)$
			-3	4	-1	$0(h^2)$
h^2f_i''			1	-2	1	$0(h)$
Backward formulas						
hf_i'		-1	1			$0(h)$
	1	-4	3			$0(h^2)$
h^2f_i''	1	-2	1			$0(h)$
Central formulas						
$2hf_i'$		-1	0	1		$0(h^2)$
$12hf_i'$	1	-8	0	8	-1	$0(h^4)$
h^2f''		1	-2	1		$0(h^2)$
$12h^2f''$	-1	16	-30	16	-1	$0(h^4)$

of Eq. 1.3.4 is simply the central difference formula derived above. Thus the effect of the spline is to distribute the result in Eq. 3.1.9 over the central point and its two nearest neighbors with weights $\frac{1}{6}$, $\frac{4}{6}$, and $\frac{1}{6}$. Despite the fact that use of the spline requires the solution of a set of simultaneous algebraic equations, the increase in computational effort is small because the system is tridiagonal. The results are only as accurate as the formulas given above, but the error is opposite in sign and it is possible to use this knowledge to obtain a simple formula that is better than either Eq. 3.1.9 or the spline result; we will do this below.

b. Taylor Series

Another simple method of obtaining finite difference approximations to derivatives is by using Taylor series. Using the well-known expansion

$$f(x \pm h) = f(x) \pm hf'(x) + \frac{h^2}{2} f''(x) \pm \cdots \tag{3.1.10}$$

and forming linear combinations of the values of the function at various mesh points x, $x \pm h$, $x \pm 2h$, and so on, we can obtain the desired results. For example, if we seek a formula for the first derivative, we must form a linear combination in which the terms involving $f(x)$ drop out but the terms involving $f'(x)$ do not. Also, by including the values of the function at enough points, we can eliminate the coefficients of some of the higher derivatives. The coefficients can be found by solving a set of linear equations. An alternative and simpler method is to take guidance from the results found by interpolation. Thus using Eq. 3.1.10 with $x=x_i$ to compute the right-hand side of Eq. 3.1.2, we have

$$\frac{f(x_i)-f(x_{i-1})}{h_i}=f'(x_{i-1})-\frac{h_i}{2}f''(x_{i-1})+\cdots \tag{3.1.11a}$$

Using Eq. 3.1.10 at $x=x_{i-1}$ to compute the right-hand side of Eq. 3.1.3, we have

$$\frac{f(x_i)-f(x_{i-1})}{h_i}=f'(x_i)+\frac{h_i}{2}f''(x_i)+\cdots \tag{3.1.11b}$$

In this manner we have shown again that Eq. 3.1.2 and Eq. 3.1.3 are approximations to the first derivative, but we have also shown that the approximations are in error by $\pm hf''/2$ for small values of h. We call such an approximation *first order accurate*; both forward and backward differences are first order estimates of the first derivative.

This procedure can be applied to any of the results found above. It is not important to do this calculation for every case, but the central difference formula for equally spaced intervals is of considerable importance:

$$\frac{f(x_{i+1})-f(x_{i-1})}{2h}\simeq f'(x_i)+\frac{h^2}{6}f'''(x_i)+\cdots \tag{3.1.12}$$

This is a second order accurate approximation for $f'(x_i)$. Note, however, that as an estimate for $f'(x_{i-1})$ or $f'(x_{i+1})$ it becomes identical to the forward or backward formulas with h replaced by $2h$ and is therefore only first order accurate.

These results can be applied to estimating the error in solutions to differential equations. They also provide a means of finding more accurate formulas by means of Richardson extrapolation. We cannot apply the formulas for the derivative using $h/2$ as the mesh size because there is no mesh point at $x_i \pm h/2$, but we can get the estimate with $2h$ as the step size and apply Richardson extrapolation to the values obtained with h and $2h$. (In some cases useful results can be obtained by using intermediate points as if they were mesh points. As long as the final result does not reference the intermediate points, there is no problem.) As examples of the use of Richardson extrapolation in the estimation of derivatives, we note that its application to the first order forward and backward difference formulas in Eqs. 3.1.2 and 3.1.3 produces the second order accurate results of Eqs. 3.1.7a and 3.1.7b. In the same manner, the application of Richardson extrapolation to the central difference formulas in Eqs. 3.1.6 and 3.1.9, which are second order accurate, produces the fourth order accurate formulas given in Table 3.1. Thus the combination of Taylor series and Richardson extrapolation is a powerful tool for constructing accurate difference schemes.

The error is usually smaller for a central difference approximation than for one that is not centered. Also, because there is some error cancellation due to symmetry, an equally spaced mesh generally produces more accurate difference formulas than an irregular one. Finally, for a given type of difference formula, higher accuracy requires including more points in the difference formula. The reason is that the inclusion of more terms in the difference formula introduces more coefficients, which can be chosen to cancel more terms of the Taylor series expansions. Similarly, the calculation of higher order derivatives requires the inclusion of more points in the difference formula. High order multipoint schemes are useful, but in boundary value problems it is difficult to match them to the boundary conditions.

One way to avoid this problem has become popular recently. The derivation of this method can be based on the idea of the spline. If we regard Eq. 1.3.10 with equal spacing as a difference formula (i.e., a means of computing f''), then we can use Taylor series to show that it is second order accurate and the error is precisely equal and opposite to that of Eq. 3.1.10. Consequently, the average of these two equations

$$\frac{h^2}{12}[f''(x_{i-1})+10f''(x_i)+f''(x_{i+1})]=f(x_{i-1})-2f(x_i)+f(x_{i+1}) \tag{3.1.13}$$

is both fourth order accurate and compact in the sense that it references only three mesh points. The fact that it requires the solution of a tridiagonal set of equations is, in many cases, only a slight inconvenience. One can also derive a similar scheme for the first derivative:

$$\frac{h}{6}[f'(x_{i-1})+4f'(x_i)+f'(x_{i+1})]=f(x_{i+1})-f(x_{i-1}) \tag{3.1.14}$$

c. Numerical Integration

It is hardly surprising that numerical quadrature methods can be used as the basis of techniques for the solution of ordinary differential equations. Since the quadrature methods developed in Chapter 2 were based on interpolation methods, however, we expect that any method derived from a quadrature scheme can also be found by the approaches considered above. It is, nonetheless, interesting to see to what some of the better quadrature methods lead in the way of differential equations solvers.

Suppose that we wish to solve the differential equation

$$y'=f(x,y) \tag{3.1.15}$$

and that we have the value of the function $y(x)$ at all points up to x_n. If we can find a procedure for determining $y(x_{n+1})=y(x_n+h)$, we can successively generate $y(x_{n+2})$, $y(x_{n+3}),\ldots$ in the same manner and thus trace out the solution to the differential equation. The first two methods under consideration are based on the midpoint rule, Eq. 2.1.11. Applying the midpoint rule to the interval (x_n, x_{n+1}) we have

$$y_{n+1}-y_n=hf(x_{n+1/2}, y_{n+1/2}) \tag{3.1.16}$$

We use y_n and $y(x_n)$ to denote the computed and exact solutions at x_n and, as before, $h=x_{n+1}-x_n$. This method would be satisfactory if we had a means of evaluating the function $f(x_{n+1/2}, y_{n+1/2})$. Since doing so requires a further approximation, we look instead at the result obtained by applying the midpoint rule to the interval (x_{n-1}, x_{n+1})

$$y_{n+1}-y_{n-1}=2hf(x_n, y_n) \tag{3.1.17}$$

This is known as the *leapfrog method* and is a fairly popular technique.

Application of the trapezoid rule to the differential equation 3.1.15 produces

$$y_{n+1}-y_n=\frac{h}{2}\left[f(x_n, y_n)+f(x_{n+1}, y_{n+1})\right]$$

This may also be obtained by approximating the value of $f(x_{n+1/2}, y_{n+1/2})$ in the midpoint rule formula of Eq. 3.1.15 by the average of the values at n and $n+1$. This method, which has been used very commonly, goes by various names. As an ordinary differential equation solver, it is usually known simply as the *trapezoid rule*.

As a final example we apply Simpson's rule to Eq. 3.1.15. The nature of Simpson's rule requires using the interval (x_{n-1}, x_{n+1}) for this application and we have

$$y_{n+1}-y_{n-1}=\frac{h}{3}\left[f(x_{n-1}, y_{n-1})+4f(x_n, y_n)+f(x_{n+1}, y_{n+1})\right]$$

which is very accurate but is not used often because it is unstable.

2. EULER'S METHOD

The simplest method for solving ordinary differential equations is based on the forward difference formula 3.1.2. This method, called *Euler's method*, is not very good and has been superceded by better methods. The reasons for studying it are that it is simple and one can learn a great deal about ordinary differential equation solvers in general from a thorough study of it. When applied to the differential equation of Eq. 3.1.15, Euler's method gives

$$y_{n+1}-y_n=hf(x_n, y_n) \tag{3.2.1}$$

We will present a number of properties of this method. If y_n is known, y_{n+1} can be computed just by evaluating f; a method having this property is called *explicit*. As we did earlier, we use y_n to represent the computed value of the function at x_n, and $y(x_n)$ to represent the exact value. For convenience we assume that $h=x_{n+1}-x_n$ is independent of n. Comparing Eq. 3.2.1 with the exact Taylor series

$$y(x_{n+1})=y(x_n)+hy'(x_n)+\left(\frac{h^2}{2}\right)y''(x_n)+\cdots \tag{3.2.2}$$

we see that if $y_n=y(x_n)$, that is, if the value of the function at x_n is exact, then

$$y(x_{n+1})-y_{n+1}\simeq\left(\frac{h^2}{2}\right)y_n'' \tag{3.2.3}$$

Thus the error introduced by a single step is proportional to h^2y''. The nomenclature of numerical analysis calls this a first method because the approximation to the derivative is first order accurate as shown in the preceding section. Also, if we wish to compute from $x=0$ to $x=1$, for example, the number of steps of size h will be $1/h$. Thus, ignoring possible error amplification (see below), the error in $y(1)$ will be the product of the number of steps and the error per step, and is proportional to h.

To understand the global behavior of this method, a graphic presentation will probably serve better than mathematical analysis. The solutions of the exact differential equation in Eq. 3.1.15 form a one-parameter family of curves in the $x-y$ plane (the constant of integration can be regarded as the parameter). Suppose we are computing in the direction of increasing x, and consider the case shown in Fig. 3.1*a*. The curves represent a family of solutions of the differential equation. Starting from a point $[x_0, y(x_0)]$ on the desired solution curve, we use Eq. 3.2.1 to compute y_1, an approximation to $y(x_1)$. Since the slope at the initial point is smaller than the average slope of the curve segment $[x_0, y(x_0)]$, $[x_1, y(x_1)]$, y_1 lies below $y(x_1)$. The slope of the solution curve passing through (x_1, y_1) is lower than the slope at $[x_1, y(x_1)]$ and this causes an even larger error at the next step. In fact the error increases exponentially, which will be the case whenever the family of solution curves diverges in the direction of computation.

In the case shown in Fig. 3.1*b* the solution curves converge in the direction of the computation. As in the previous case, (x_1, y_1) lies below the solution curves but now the slope of the solution curve passing through (x_1, y_1) is greater than the slope of the exact solution curve. Consequently, the error due to being off the solution curve is opposite to the error caused by the finite

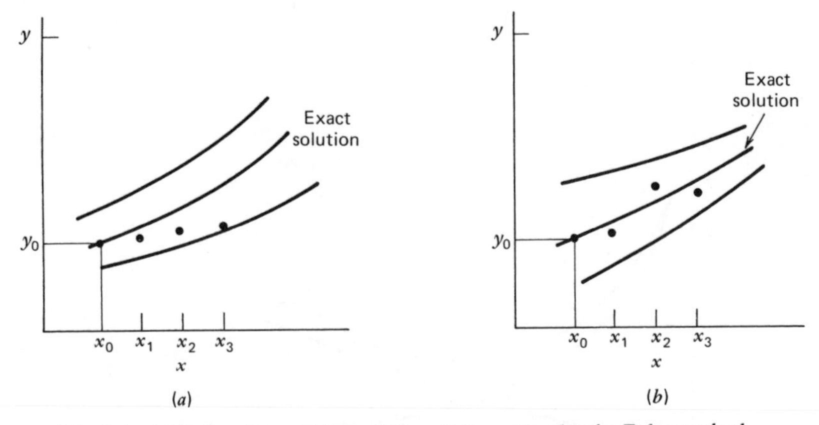

Fig. 3.1. Solution of an ordinary differential equation by the Euler method.

difference approximation. As a result, the rate of error growth is less disastrous.

What distinguishes the two cases from each other is that in the first case $dy/dx=f$ increases with increasing y for fixed x, and in the second it decreases. Thus the behavior is determined by the sign of $\partial f/\partial y$. If $\partial f/\partial y>0$, the solution tends to behave in an unstable manner; but if $\partial f/\partial y<0$, reasonably good results can be expected.

One can obtain an estimate for the global error, in other words, the error after many steps. Not surprisingly it depends greatly on $\partial f/\partial y$. One finds

$$|y(x_n)-y_n|\leqslant(e^{Lnh}-1)\frac{hN}{2L}+|\varepsilon_0||1+hL| \tag{3.2.4}$$

where ε_0 is the initial error

$$\varepsilon_0=y(0)-y_0$$

and L and N are upper bounds on $|\partial f/\partial y|$ and $|y''|$, respectively

$$\frac{\partial f}{\partial y}<L \qquad |y''|<N$$

Eq. 3.2.4 tends to be a very generous estimate and may not be very useful. Its importance is that it shows that it is possible for the error to grow exponentially.

We see from Eq. 3.2.4 that one means of reducing the error is to make the step size h smaller. For Euler's method this results in a decrease in the error that is approximately linear with h. As h is reduced, however, we must increase the number of steps. This results in an increase in both cost and roundoff error. From Fig. 3.2 we see that there is in fact an optimum step size that produces minimum error.

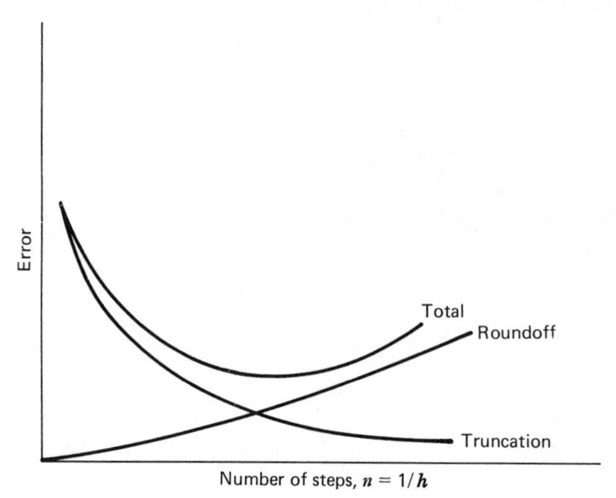

Fig. 3.2. Error versus step size. The curves are illustrative and do not apply to any particular equation.

3. STABILITY

Just as some ordinary differential equations can be solved exactly, it is frequently possible to find exact solutions to difference equations. Generally, if the differential equation has a known exact solution, then a finite difference approximation corresponding to it can probably be solved analytically as well. Since we are looking at single first order differential equations of the type given by Eq. 3.1.15, the ideal choice for a model equation on which we can test this idea is the simplest nontrivial such equation*

$$y' = \alpha y \tag{3.3.1}$$

---

*Any equation of the form of Eq. 3.1.15 can be expanded locally in Taylor series:

$$y' = f(x_0, y_0) + (x - x_0)\frac{\partial f}{\partial x}(x_0, y_0) + (y - y_0)\frac{\partial f}{\partial y}(x_0, y_0) + \cdots$$

The omitted terms contain higher powers of $(x - x_0)$ and/or $(y - y_0)$ and are assumed small. Lumping all of the terms evaluated at the point (x_0, y_0), we can put this equation in the form

$$y' = \alpha y + (\beta_1 + \beta_2 x)$$

Thus the solution can be regarded locally as containing two parts: a particular solution to the inhomogeneous equation and a solution to the homogeneous equation. Both parts of the solution are equally well approximated and the stability characteristics are governed by the homogeneous part of the equation. Thus it is sufficient for our purposes to deal with the homogeneous equation, Eq. 3.3.1.

The exact solution to this equation is well known

$$y(x)=y_0e^{\alpha x} \tag{3.3.2}$$

If we were to solve the differential equation of Eq. 3.3.1 by Euler's method, the finite difference approximation would be

$$y_{n+1}-y_n=\alpha h y_n \tag{3.3.3}$$

This is easily rearranged to give

$$y_{n+1}=y_n(1+\alpha h) \tag{3.3.4}$$

Using this result inductively, we find that the finite difference solution is

$$y_n=y_0(1+\alpha h)^n \tag{3.3.5}$$

Thus the solution to the difference equation is also exponential. In fact Eq. 3.3.5 is the same as Eq. 3.3.2 with the substitution of $(1+\alpha h)$ for $e^{\alpha h}$. Since

$$e^{\alpha h}=1+\alpha h+\frac{(\alpha h)^2}{2}+\cdots \tag{3.3.6}$$

we see that Euler's method produces the first two terms of the Taylor series for the exponential. This is another way of showing that the method is first order accurate.

We can learn something even more important from this solution. Suppose we let the constant α in Eq. 3.3.1 be complex. We do this because higher order systems of differential equations may have oscillatory solutions; this can be simulated by a single equation with a complex constant. If the real part of α is positive, both the exact solution (Eq. 3.3.2) and the approximate solution (Eq. 3.3.5) grow with increasing x. For large values of αh the approximate solution is badly in error, but it is growing.

If the real part of α is negative, however, there can be a more serious difficulty. The exact solution decays with increasing x, but the approximate solution may not decay. In fact if $|1+\alpha h|$ is greater than 1, the numerical solution will grow with increasing x. In the complex αh plane (see Fig. 3.3) the region in which $|1+\alpha h|<1$ is a disc of unit radius centered at the point -1. For any value of αh in the left-half plane outside this circle Euler's method will *yield an increasing solution when it should produce a decaying one*. This is clearly a situation that must be avoided at all costs and is one of the greatest dangers in numerical differential equation solvers—*instability*.

The question of stability of numerical methods is extremely important for both ordinary and partial differential equations. Unfortunately there are a number of different definitions of stability in common use and the subject can

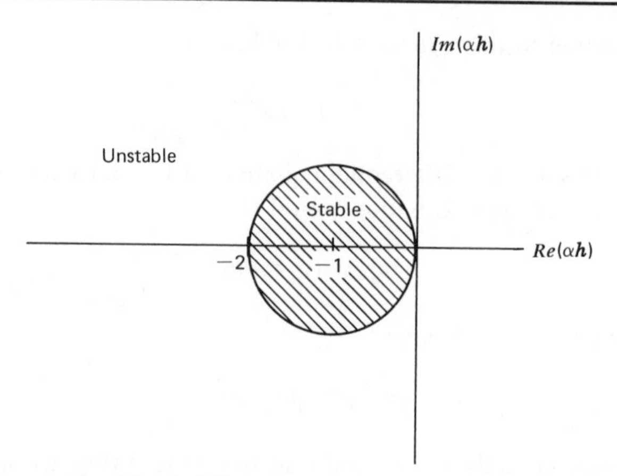

Fig. 3.3. Stability region for Euler's method.

become confusing. The type of stability we have defined above is called A-*stability* in the numerical analysis literature and is the definition most commonly applied by users of numerical methods. Simply put, a method is stable if it produces a bounded solution when it is supposed to and it is unstable if it does not. A-stability is equivalent to this for Eq. 3.3.1. Euler's method is said to be *conditionally stable* because it is stable for some but not all values of αh. A method that is stable for all values of this parameter is said to be *unconditionally stable*; one that is not stable for any value is called *unconditionally unstable*.

One might think that basing our analysis of stability on an equation as simple as Eq. 3.3.1 is not meaningful. As shown in the footnote on p. 60, however, any equation can be linearized locally; essentially α represents the local value of y'/y. Since the stability condition for Eq. 3.3.1 always takes the form of restriction on αh, we surmise that a method will be stable if hy'/y nowhere exceeds this bound. This turns out to be correct; that is, if this condition is obeyed, the solution will not grow when it should not. Note, however, that the question of stability has nothing whatever to do with the accuracy of the solution obtained. It is quite possible for a method to be stable and highly inaccurate. It is also possible for a method to be accurate and unstable. We demonstrate how this case can arise later in this chapter.

When Euler's method is unstable $(1-\alpha h)$ is greater than unity in magnitude and, if real, negative. Thus when Euler's method is unstable it will usually produce a result that alternates in sign at every step. A growing oscillatory solution is an indicator that something is amiss.

Example 3.1

As a first example, let us take equation 3.3.1 with $\alpha=1$, $y'=-y$. The behavior of the solution is shown in Fig. 3.4 for small h. Both the exact and

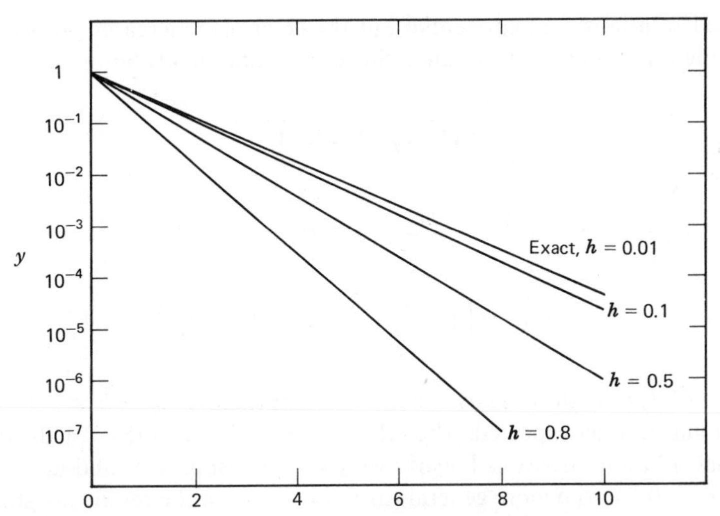

Fig. 3.4. Solution of $y' = -y$ by the Eulers method using various step sizes.

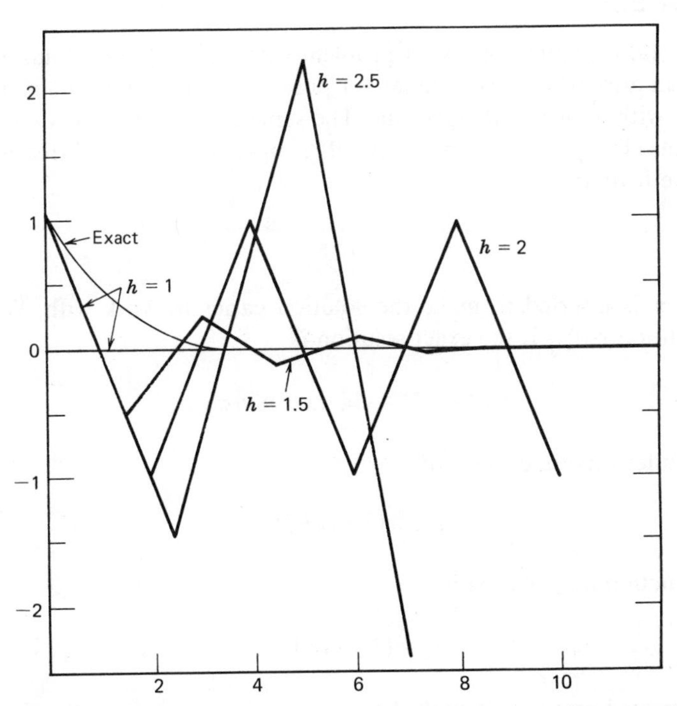

Fig. 3.5. Solution of $y' = -y$ by the Eulers method using various step sizes.

computed solutions are exponential but the error also increases exponentially. In fact the error is easily computed. Since the computed solution is

$$y_n=(1-h)^n \simeq \left(e^{-h}-\frac{h^2}{2}\right)^n$$

$$\simeq e^{-nh}-\frac{nh^2}{2}e^{-(n-1)h}$$

$$=e^{-nh}\left(1-\frac{nh^2}{2}e^h\right) \simeq e^{-x}\left(1-\frac{hx}{2}\right)$$

since $e^h \simeq 1$ for small h. Thus the error is approximately $-hxe^{-x}/2$ and is proportional to h as expected. The relative error, that is, the error divided by the actual solution, increases linearly with x. The results are valid up to about $h=0.1$ ($\alpha h=0.1$ in the more general case). For larger h the results are shown in Fig. 3.5; the computed result behaves reasonably as long as h is less than about 0.5. At $h=1$ the solution goes to zero at the first step and stays there. For $1<h<2$ the solution oscillates with decaying amplitude. Finally, for $h>2$ the onset of instability is obvious.

Example 3.2

As a second example, we take a problem with an oscillating solution. This situation can arise when α is complex (or pure imaginary) and usually arises in connection with a system of equations. The same problem is done two ways to illustrate this. The complex form is actually simpler for analytical treatment, so we will take it first:

$$y'=i\pi y$$

The factor π is inserted to make the solution easier to work with. With the initial condition $y(0)=1$, the exact solution is

$$y(x)=e^{i\pi x}=\cos\pi x+i\sin\pi x$$

Applying Euler's method, we find

$$y_{n+1}=(1+i\pi h)y_n$$

and by induction the solution is

$$y_n=(1+i\pi h)^n$$

This is compared more easily with the exact result if we write $(1+i\pi h)=Ae^{i\theta}$.

The amplitude is

$$A=(1+\pi^2h^2)^{1/2}$$

while the phase angle is

$$\theta=\tan^{-1}\pi h$$

Thus Euler's method has resulted in the replacement of the exact solution $e^{i\pi x}=e^{i\pi hn}$ by $A^n e^{in\theta}$. The amplitude, which is unity in the exact solution, has become $A^n=(1+\pi^2h^2)^{n/2}$, which is greater than unity. Thus a solution that remains bounded has been replaced by a growing one and we have to call the method *unstable* for any h. This is not surprising. Fig. 3.3 shows clearly that the entire imaginary axis lies outside the domain of stability of the Euler method. Despite this, the growth of the function is fairly weak if h is small. In fact

$$(1+\pi^2h^2)^{1/2}\simeq 1+\frac{\pi^2h^2}{2}$$

The amplitude as a function of step size is shown in Fig. 3.6.

We turn to look at the phase of the solution. Euler's method replaces the phase of the exact solution $n\pi h$ by $n\tan^{-1}\pi h$. For small πh the Taylor series

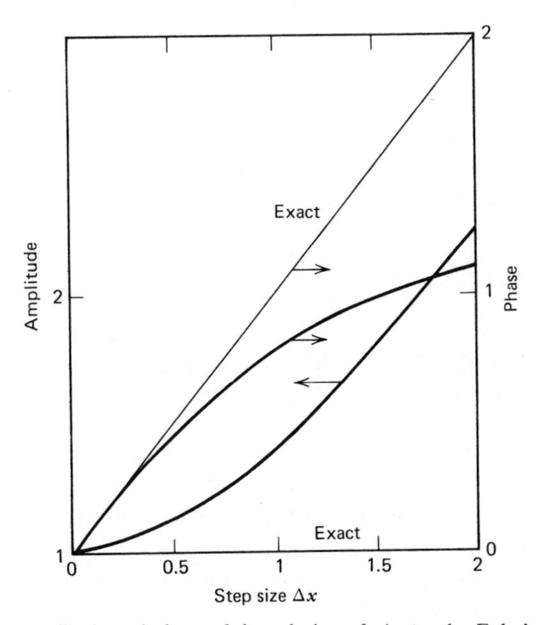

Fig. 3.6. Amplitude and phase of the solution of $y'=i\pi y$ by Euler's method.

expansion

$$\tan^{-1}\pi h \simeq \pi h - \left(\frac{\pi h}{3}\right)^3$$

shows that the error is quite small. For larger values of πh the error becomes much more serious. The phase as a function of step size is also shown in Fig. 3.6.

This example illustrates a number of points that arise when the solutions are oscillatory. The order of a method may not be a good indicator of its quality. Because there are both amplitude and phase errors, the difference between the exact and computed solutions may behave very erratically. There may be regions in which the error is quite small and others in which it is quite large. Consequently, when one is interested in problems with oscillatory solutions, *it is more appropriate to discuss numerical methods in terms of amplitude and phase errors rather than the apparently simpler truncation error.*

Example 3.3

Here we treat the problem of the preceding example in terms of real equations and compute the solution numerically. If we write

$$y = y_1 + iy_2$$

then the differential equation of Example 3.2 can be replaced by the pair of equations

$$y_1' = -\pi y_2$$

$$y_2' = \pi y_1$$

and the initial conditions are $y_1(0) = 1$, $y_2(0) = 0$.

To apply Euler's method we simply write

$$y_{1,n} = y_{1,n} - \pi h y_{2,n}$$

$$y_{2,n} = y_{2,n} + \pi h y_{1,n}$$

This problem can be solved with the program given below. In fact the subroutine will solve the system of equations

$$y_i' = f(x, y, y_2, \dots y_n) \qquad i = 1, 2, \dots n$$

for any number of equations up to 10 (this limit comes from the dimension and is easily increased).

The results obtained are, of course, identical to those obtained earlier, so they need not be given here.

```
      PROGRAM ORDIFF
C-----THIS PROGRAM SOLVES THE SYSTEM OF TWO EQUATIONS OF EXAMPLE 3.3
C-----USING THE EXPLICIT EULER METHOD.  A SUBPROGRAM TO EVALUATE
C-----THE FUNCTIONS IS GIVEN SEPARATELY.
      DIMENSION Y(2)
      EXTERNAL F
C-----FIRST, COLLECT THE DATA NEEDED.
      WRITE (5,100)
  100 FORMAT ( ' TYPE THE INITIAL VALUE OF THE INDEPENDENT VARIABLE')
      READ (5,110) X0
  110 FORMAT ( F)
      WRITE (5,120)
  120 FORMAT ( ' NOW TYPE THE FINAL VALUE')
      READ (5,110) XF
      WRITE (5,130)
  130 FORMAT ( ' TYPE THE INITIAL VALUE OF DEPENDENT VARIABLE 1')
      READ (5,110) Y10
      WRITE (5,140)
  140 FORMAT ( ' TYPE THE INITIAL VALUE OF DEPENDENT VARIABLE 2')
      READ (5,110) Y20
   10 WRITE (5,150)
  150 FORMAT ( ' GIVE THE STEPSIZE')
      READ (5,110) DX
      N = IFIX ((XF-X0)/DX)
      WRITE (5,160)
  160 FORMAT ( ' THE RESULT WILL BE PRINTED EVERY K STEPS, TYPE K')
      READ (5,170) K
  170 FORMAT ( I)
C---- WRITE THE HEADING
      WRITE (5,180)
  180 FORMAT ( 5X,1HX,12X,4HY(1),9X,4HY(2))
C-----SET UP THE PARAMETERS FOR THE SUBROUTINE CALL.
      NCALL = N/K
      NEQ = 2
      X = X0
      Y(1) = Y10
      Y(2) = Y20
      WRITE (5,190) X,Y(1),Y(2)
  190 FORMAT ( 3F12.6)
C-----THE MAIN LOOP
      DO 1 I=1,NCALL
      CALL EULER (NEQ,X,Y,K,DX,F)
      WRITE (5,190) X,Y(1),Y(2)
    1 CONTINUE
      WRITE (5,200)
  200 FORMAT ( ' IF YOU WANT TO CHANGE THE STEPSIZE, TYPE 1')
      READ (5,170) IFF
      IF (IFF.EQ.1) GO TO 10
      STOP
      END

      FUNCTION F(I,X,Y)
      DIMENSION Y(10)
      PI = 3.14159
      IF (I.EQ.2) GO TO 10
      F = -PI * Y(2)
      RETURN
   10 F = PI * Y(1)
      RETURN
      END
```

```
      SUBROUTINE EULER (N,X,Y,K,H,F)
C-----THIS SUBROUTINE ADVANCES THE SYSTEM OF N ODE'S
C-----DY(I)/DX = F(I,X,Y(1),Y(2),....Y(N)) BY K STEPS OF SIZE H
C-----USING THE EXPLICIT EULER METHOD.  THE OTHER PARAMETERS ARE:
C-----X = INITIAL VALUE OF THE INDEPENDENT VARIABLE; IT IS INCREASED
C-----      BY THE SUBROUTINE.
C-----Y = INITIAL VALUE OF THE DEPENDENT VARIABLE(S); THE ROUTINE
C-----     OVERWRITES THIS WITH THE NEW VALUE(S).
      EXTERNAL F
      DIMENSION Y(10),T(10)
C-----THE MAIN LOOP
      DO 1 I=1,K
C-----A TEMPORARY ARRAY IS NEEDED IN ORDER TO AVOID USING THE
C-----UPDATED VALUES OF Y(I).
      DO 2 J=1,N
      T(J) = F(J,X,Y)
    2 CONTINUE
      DO 3 J=1,N
      Y(J) = Y(J) + H * T(J)
    3 CONTINUE
      X = X + H
    1 CONTINUE
      RETURN
      END
```

Example 3.4

As a final example, we take an equation with nonconstant coefficients

$$y' = -2xy$$

with $y(0)=1$. The exact solution is e^{-x^2}.

Table 3.2. Solution of $y' = -2xy$ by Euler's Method with $h=0.4$.

x	y	Exact Solution
0	1.	1.
0.4	1.	0.8521
0.8	0.6800	0.5273
1.2	0.2448	0.2369
1.6	0.0098	0.0773
2.0	−0.0027	0.0183
2.4	0.0016	0.0032
2.8	−0.0015	3.9×10^{-4}
3.2	0.0019	3.6×10^{-5}
3.6	−0.0029	2.4×10^{-6}
4.0	0.0055	1.1×10^{-7}
4.4	−0.0121	3.9×10^{-9}
4.8	0.0305	9.9×10^{-11}

Computing the solution with $h=0.4$ gives the results shown in Table 3.2. Due to the large step size the solution is not very accurate but, at least up to $x=1.6$, it bears some resemblance to the correct solution. Beyond $x=1.6$, the point at which hy'/y first exceeds unity, the numerical solution oscillates, while the exact solution is monotonic. Beyond $x=3.2$ (where hy'/y first exceeds 2), the solution not only oscillates but it grows to amplitude as well. This accords with the result stated in the discussion and is valid for equations nonlinear in y as well (see the problems at the end of this chapter).

4. BACKWARD OR IMPLICIT EULER METHOD

At this point it is worth investigating another method that is similar in a number of respects to the Euler method. The differences, however, turn out to be very illuminating. This method is based on the backward difference formula shown in Eq. 3.1.3 and is known either as the *backward Euler* or *implicit Euler method*. The difference equation for this method is

$$y_{n+1}-y_n=hf(x_{n+1},y_{n+1}) \tag{3.4.1}$$

Like the Euler method, this method is first order accurate and we will find the error formula shortly. Before doing so, we note an important difference between this method and the preceding one. We cannot solve Eq. 3.4.1 for y_{n+1} in a simple manner unless the function f is extremely simple. In general, a nonlinear algebraic equation must be solved and this can be quite a task. Any method that requires the solution of an equation for the new value of the function is called *implicit*; the backward Euler method is the simplest member of this class.

One might be tempted to reject implicit methods outright because they require more computation per point than explicit methods, but they have the important redeeming virtue of stability. We do, therefore, spend some time studying the backward Euler method as a representative of the class. The error analysis is a little more complicated than that for the Euler method of Section 2, so we will go directly to the application of the implicit method to Eq. 3.3.1. The difference equation then is

$$y_{n+1}-y_n=\alpha hy_{n+1} \tag{3.4.2}$$

which is solved easily, and we have

$$y_{n+1}=(1-\alpha h)^{-1}y_n \tag{3.4.3}$$

so that

$$y_n=(1-\alpha h)^{-n}y_0 \tag{3.4.4}$$

Thus the use of this method is equivalent to replacing $e^{\alpha h}$ by $(1-\alpha h)^{-1}$. Then, comparing the Taylor series expansion of $(1-\alpha h)^{-1}$

$$(1-\alpha h)^{-1}=1+\alpha h+(\alpha h)^2+\cdots \tag{3.4.5}$$

with the Taylor series expansion 3.3.6 for $e^{\alpha h}$, we see that this method also reproduces only the first two items of the Taylor series and is, therefore, first order accurate. It is interesting to note, however, that the error to lowest order is,

$$y(x_n)-y_n=-\frac{(\alpha h)^2}{2} \tag{3.4.6}$$

which is equal to the error of the Euler method but of opposite sign.

Next, let us consider the stability of this method. It is not necessary to look very far to see that if the real part of αh is negative, then the real part of $(1-\alpha h)$ is always greater than unity and no matter what the imaginary part of αh may be the magnitude of $(1-\alpha h)$ is always greater than 1. Consequently, the magnitude of $(1-\alpha h)^{-1}$ is less than 1 for all αh in the left-half plane and, by the definition introduced above, the backward Euler method is *unconditionally stable*. We emphasize that this does not imply anything about the accuracy of the method or its overall usefulness.

This result turns out to be fairly general. The only methods that are unconditionally stable are implicit methods; implicit methods, therefore, become the methods of choice when stability becomes the overriding consideration. Not all implicit methods are unconditionally stable, but they are almost always (if not always) more stable than explicit methods of similar accuracy and type. This brings us to the major tradeoff that occupies us for much of the remainder of this chapter. Explicit methods offer more speed per step, but in some problems stability may require such a small step that the advantage is lost. Implicit methods place much weaker restrictions on the size of the step allowed by stability, but the computation cost per step will probably be higher.

Example 3.5

We repeat the problem of Example 3.1 using the implicit Euler method. The resulting solution is

$$y_n=(1+h)^{-n}$$

and is plotted in Fig. 3.7. It is clear that the error is in the opposite direction to that produced by the explicit Euler method. This is expected from the truncation error formula given in the discussion. More important, we see that for large step sizes the error continues to increase but the result is still quite smooth. There is no tendency for the solution to oscillate or amplify.

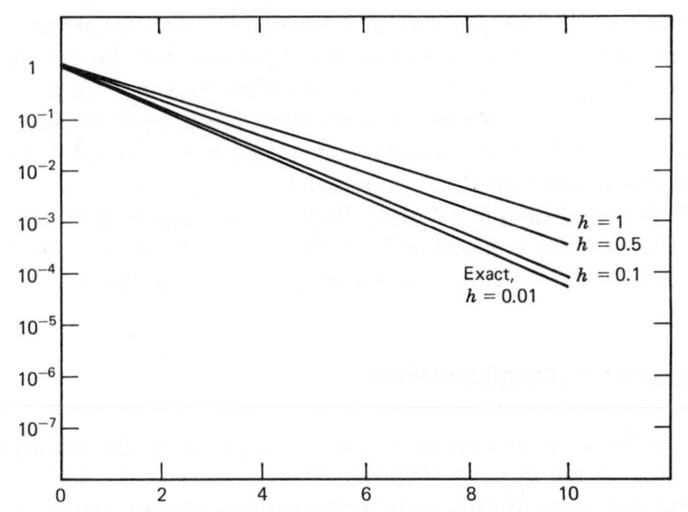

Fig. 3.7. Solution of $y' = -y$ by the implicit Euler method using various step sizes.

A word of warning should be issued here. The inaccurate solution obtained by using the backward Euler method with a large step size may, in fact, be a curse rather than a blessing. Because the solution seems "reasonable," there is sometimes a temptation to accept it as correct even though it contains considerable error. The danger is obvious and one can never be careful enough about a numerically obtained solution. The easiest way to assure that the solution obtained is accurate enough is to repeat the procedure using half the step size. This (for a first order method) should halve the error and the accuracy is then known to a sufficient approximation.

Example 3.6

Solving the problem of Example 3.2 with backward Euler gives the solution

$$y_n = (1 - i\pi h)^{-n}$$

Now we can write

$$(1 - i\pi h)^{-1} = Ae^{i\theta}$$

with $A = (1 + \pi^2 h^2)^{-1/2}$ and $\theta = \tan^{-1} \pi h$. The phase is exactly the same in the explicit Euler method, while the amplitude is always less than unity. Thus the method is again stable as expected. For small step size the implicit method produces an amplitude error of approximately the same size as the explicit method. The "stable" solution produced is thus no better than the "unstable" one that the explicit method gave. In fact the choice of preferred method can be a matter of judgment by the user.

Note that to solve the problem of Example 3.3 with the implicit method requires the solution of a system of two equations for the two dependent variables at the new step. For this linear problem we can solve the system of equations by hand or by means of a standard linear equation solver. In the general case we would need something quite a bit more complicated such as a multidimensional Newton–Raphson method.

The problem of Example 3.4 also offers no challenge to the implicit Euler method. The problem is linear and a well behaved solution is produced. At large x, however, y'/y becomes large and accuracy is sacrificed.

5. ACCURACY IMPROVEMENT

In a great many cases (probably a majority of them) the desired accuracy of the solution fixes the step size. Typically the user wants the solution for some range of the independent variable to within some predetermined accuracy. It is difficult to predict (and, therefore, to control) the global accuracy of a method, but we can frequently make an educated guess based on the single step error. More sophisticated approaches to error control will be discussed later. For the first order methods we have considered so far, the single step error is approximately $h^2y''/2$. As the step size is decreased, the number of steps required increases in proportion to $1/h$. If we use the sum of the single step errors as an estimate of the global error (rather than the more accurate but harder to apply Eq. 3.2.4), we see that the global error is proportional to h. Thus the step size must be reduced linearly with the error requirement. This can become very expensive. On the other hand, if we had a more accurate method, for example, one in which the error scaled like h^p, the problem would be much less severe because the required step size then would be proportional to $\varepsilon^{1/p}$ if ε is the desired error. When ε is small, it is desirable that p be fairly large. Thus there is considerable incentive to seek more accurate methods and we spend much of the rest of this chapter looking at some possibilities. Naturally, increased accuracy comes at a price.

There are a number of ways in which one can obtain increased accuracy. In this section we discuss some of the simpler ones and only mention the more complicated ones, which are investigated later.

a. Richardson Extrapolation

The idea of using Richardson extrapolation is simple. Suppose we use Euler's method to compute y_{n+1}:

$$y_{n+1}^{(1)} = y_n + hy_n' \qquad (3.5.1)$$

The superscript (1) indicates that this value will not be accepted as final. Now we repeat the calculation with step size $h/2$, using two steps to go the same distance:

$$y_{n+\frac{1}{2}}^{(2)} = y_n + \frac{h}{2} y_n' \tag{3.5.2a}$$

$$y_{n+1}^{(2)} = y_{n+\frac{1}{2}} + \frac{h}{2} y_{n+\frac{1}{2}}' \tag{3.5.2b}$$

Then we combine these results by the usual rules of Richardson extrapolation (cf. Section 2 of Chapter 2) to get

$$y_{n+1} = 2y_{n+1}^{(2)} - y_{n+1}^{(1)} \tag{3.5.3}$$

Applied to the test differential equation $y' = \alpha y$ this gives

$$y_{n+1} = 2\left(1 + \frac{\alpha h}{2}\right)^2 y_n - (1+\alpha h) y_n = \left(1 + \alpha h + \frac{(\alpha h)^2}{2}\right) y_n \tag{3.5.4}$$

which is a second order accurate approximation to the exact result. Assuming that evaluation of y' is the most expensive item, the cost of the total computation is only a little more than the cost of Eq. 3.5.2, since the value of y' needed in Eq. 3.5.2a is the same one needed by Eq. 3.5.1.

This method has not found widespread popularity. Apparently one reason is that Richardson extrapolation was not discovered until after some of the more popular methods were discovered. The particular version we have just given turns out to be a rearrangement of the second order Runge–Kutta method, which is discussed in Section 6.

b. Higher Order Methods

Another approach to increased accuracy is to use a higher order method, but, as we have seen, accurate estimation of derivatives requires using more data points. Methods based on the use of data from multiple points are among the best methods, but they require a somewhat more involved analysis, so we defer discussion of them until Section 7.

If we are restricted to using data only at points x_n and x_{n+1}, the simplest means of obtaining a second order formula is to use the trapezoid rule, which is based on the second order accurate central difference formula

$$y_{n+1} - y_n = \frac{h}{2}\left[f(x_n, y_n) + f(x_{n+1}, y_{n+1})\right] \tag{3.5.5}$$

This method is implicit and suffers from the difficulties of the backward Euler method. To understand it better, let us apply the method to the test differential equation 3.3.1. With little difficulty we find

$$y_n = \left(\frac{1+\alpha h/2}{1-\alpha h/2} \right)^n y_0 \tag{3.5.6}$$

So the trapezoid rule approximates $e^{\alpha h}$ by $(1+\alpha h/2)/(1-\alpha h/2)$. The Taylor series for this function is

$$\frac{1+\alpha h/2}{1-\alpha h/2} = 1+\alpha h+\tfrac{1}{2}(\alpha h)^2+\tfrac{1}{4}(\alpha h)^3+\cdots \tag{3.5.7}$$

which is certainly more accurate than the corresponding result in Eq. 3.4.5 for the backward Euler method. Because the trapezoid rule provides more accuracy than backward Euler at almost no increase in cost, it is almost always the preferable method. There is one property that makes use of backward Euler attractive in a few cases. When αh is large and negative, the trapezoid rule approximation to $e^{\alpha h}$ goes to -1, while that for the backward Euler method goes to zero as $(\alpha h)^{-1}$. Although $(\alpha h)^{-1}$ is very different from the exponential function it is supposed to approximate, it is better than the -1 of the trapezoid rule. This property sometimes mediates in favor of the backward Euler method for stiff problems (see Section 9).

Another approach to finding higher order methods makes direct use of the Taylor series:

$$y_{n+1} = y_n + hy_n' + \frac{h^2}{2} y_n'' + \frac{h^3}{6} y_n''' + \cdots \tag{3.5.8}$$

In fact all high order methods can be derived from this equation by using appropriate approximations for derivatives higher than the first. Alternatively, the derivatives can be obtained directly by differentiating the differential equation 3.1.15. One need only be careful to take the total (as opposed to the partial) derivative of f with respect to x, that is,

$$y'' = \frac{df}{dx} = \frac{\partial f}{\partial x} + \frac{\partial f}{\partial y} y' = \frac{\partial f}{\partial x} + f\frac{\partial f}{\partial y} \tag{3.5.9}$$

In some cases the differentiation is easy to perform and the evaluation of y'' is no more difficult than the evaluation of y'. For these cases this is a useful approach. More commonly, the evaluation of the higher derivatives becomes successively more difficult and this method is not practical. One must then use a method that approximates the higher derivatives by a finite difference. Some of these, not necessarily derived in this way, are studied in the sections that follow.

As we saw in Section 1, the estimation of higher order derivatives (or the higher order estimation of the first derivatives) requires more than 2 data points. Thus all explicit methods of an order higher than one (and implicit methods of an order higher than two) must reference points other than x_n and x_{n+1}. There are two approaches to doing this and they lead to two major types of high order methods. In *multistep methods* the value of y_{n+1} is computed from previously computed data, that is, $y_n, y_{n-1}, y_{n-2}, \ldots$ and so on. These methods have the disadvantage that special methods are needed at the first few points, but they are usually simple to program and fast. In *Runge–Kutta methods* temporary use is made of the values of the function at points between x_n and x_{n+1}. Thus we might first compute the value of y at $x_{n+1/2}=x_n+h/2$ and then use this value to compute y_{n+1}. Runge–Kutta methods pose no problem in getting started and can be very accurate, but they tend to be more expensive in computation time. They are also closely related to predictor-corrector methods, which are discussed in Section 6.

Example 3.7

The trapezoid rule is much more accurate than the first order methods discussed earlier. Results for the test problem are shown in Fig. 3.8. No results are shown for $h<1$ because they are close to the exact solution. For $h=2$ the trapezoid rule produces a solution that decays to zero on the first step and stays there. For $h>2$ the result is a damped oscillation. The decay rate decreases as h increases. Hence, even though the method is stable, it should not be used with a large step size.

Comparing this method with first order methods, we note that for small step size the error in either of the Euler methods with stepsize h_e is $h_e y''/2=h_e y/2$ for $y'=-y$. The trapezoid rule error with step size h_t is $h_t^2 y'''/12=-h_t^2 y/12$

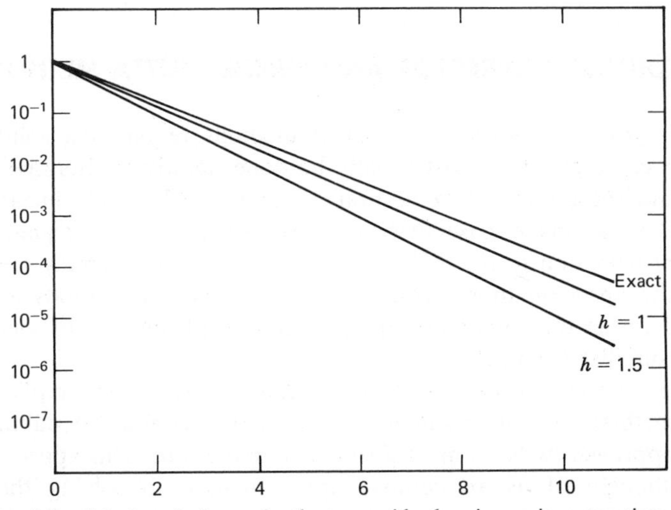

Fig. 3.8. Solution of $y'=-y$ by the trapezoid rule using various step sizes.

for $y'=-y$. So to equal the accuracy of the trapezoid rule, the Euler methods would have to use a step size of $h_e=h_t^2/6$. The higher the accuracy required, the higher the relative cost of the Euler methods.

Example 3.8

Applying the trapezoid method to the solution of $y'=i\pi y$ produces the solution

$$y_n\left(\frac{1+i\pi h/2}{1-i\pi h/2}\right)^n$$

It is easy to see that

$$\frac{1+i\pi h/2}{1-i\pi h/2}=e^{i\theta}$$

with $\theta=2\tan^{-1}\pi h/2$. The amplitude is exactly unity, so there is in fact no amplitude error. The phase error is considerably smaller than was the case for the Euler methods. The lack of amplitude error makes the trapezoid rule an excellent choice for problems with oscillatory solutions. This is the reason that the trapezoid rule is the basis for some of the best methods of solving partial differential equations. For the application of these methods see Section 2 of Chapter 4.

Applying the trapezoid rule to systems of equations or nonlinear equations poses difficulties similar to those described for the implicit Euler method. Since these methods are only occasionally used in solving ordinary differential equations, details of these applications are not given here.

6. PREDICTOR-CORRECTOR AND RUNGE–KUTTA METHODS

When $y'=f(x, y)$ is nonlinear, implicit methods require the solution of a nonlinear equation. This must usually be done iteratively. Iterative equation solvers find successively more accurate solutions and usually linearization is used. It does not make sense, however, to solve the difference equation, which is an approximation to the differential equation, to an accuracy greater than that of the approximation itself. So one uses implicit methods in the way suggested earlier only when the equations are simple enough to solve directly or when stability is crucial.

We are thus led to consider a compromise between the simplicity of the explicit method (with its attendant lack of stability) and the stability of the implicit approach (with its more difficult computation). This approach, which can be thought of as an iterative approximation to solving the implicit equation is the *predictor-corrector* approach. As the name suggests, the method consists of the application of an explicit method to estimate (predict) the new value of the dependent variable, and the subsequent application (or applications) of an implicit method to improve (correct) it. As is the case with explicit and implicit methods, there are a great many predictor-corrector methods. It is not our purpose to survey exhaustively this class of methods; we examine a few to discover their general properties and give a few of the better methods of the class.

The simplest predictor-corrector scheme consists of using the Euler method as a predictor:

$$y_{n+1}^{*}=y_n+hf(x_n, y_n) \tag{3.6.1a}$$

followed by the trapezoid rule as a corrector:

$$y_{n+1}=y_n+\frac{h}{2}\left[f(x_n, y_n)+f(x_{n+1}, y_{n+1}^{*})\right] \tag{3.6.1b}$$

This method has been so commonly used that it is sometimes called *the* predictor-corrector method. It is also known as *Heun's method*.

Applying this method to the differential equation in Eq. 3.3.1 we find that

$$y_{n+1}^{*}=(1+\alpha h)y_n \tag{3.6.2a}$$

$$y_{n+1}=\left(1+\frac{\alpha h}{2}\right)y_n+\frac{\alpha h}{2}y_{n+1}^{*}=\left(1+\alpha h+\frac{(\alpha h)^2}{2}\right)y_n \tag{3.6.2b}$$

Thus this method inherits the second order accuracy of the trapezoid rule corrector. This result is fairly general—predictor-corrector methods, when designed properly, have the accuracy of the final corrector step. Also note that when applied to the test equation $y'=\alpha y$, it is identical to the Richardson extrapolation method of the preceding section. For nonlinear equations or equations with nonconstant coefficients these two methods differ.

The stability limit of this method is not found easily, since the boundary curve is a quartic. For almost all higher order methods the stability limit is best found by direct calculation. The problem is to find the complex values of αh for which the characteristic function (in the present case, the polynomial multiplying y_n in the last part of Eq. 3.6.2b) has unit magnitude. We are interested only in values of αh in the left-half plane. Also, since the coefficients appearing in the characteristic function are real (they are in most cases of interest), the boundary curve is symmetric about the imaginary axis. Thus it is sufficient to consider only the second quadrant in the complex αh plane. The calculation may be done by choosing a phase of αh between 90° and 180° and increasing the magnitude of αh until the magnitude of the characteristic function is unity. There are other methods that produce substantially the same result but since the calculation is straightforward, it is not important to give

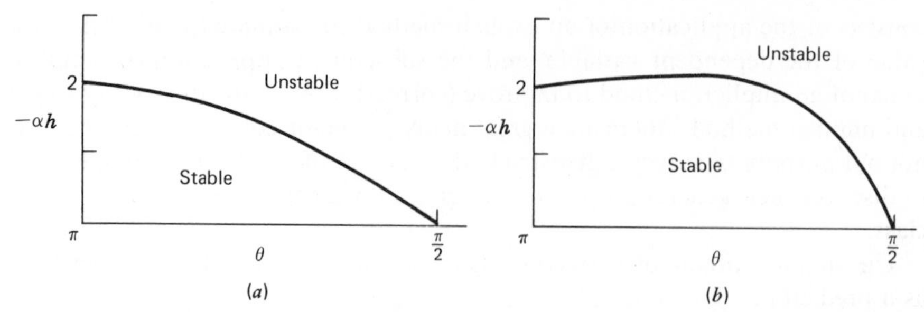

Fig. 3.9. Stability limit for Eulers and Heuns (predictor-corrector) methods (*a*) Euler's method; (*b*) Heun's method.

them here. In this way we find the stability of the method shown in Eq. 3.6.1 as given in Fig. 3.9. This result is presented in a slightly different format from that in Fig. 3.3 to familiarize the reader with different methods of representing the results. For real values of αh, the method in Eq. 3.6.1 has the same stability limit as Euler's method but the stability is improved for complex values of αh.

We can improve the method by treating the result in Eq. 3.6.1b as a new predicted value and using the same formula again as a further corrector. Applying this method to Eq. 3.3.1 yields

$$y_{n+1}=\left(1+\alpha h+\frac{(\alpha h)^2}{2}+\frac{(\alpha h)^3}{4}\right)y_n \tag{3.6.3}$$

in place of Eq. 3.6.2b. This formula is still only second order accurate but it is more accurate than Eq. 3.6.2b. The coefficient of y_n in Eq. 3.6.3 is the sum of the first four terms of the Taylor series expansion of $(1+\alpha h/2)/(1-\alpha h/2)$, the characteristic function of the trapezoid rule. One might think that this method of iteration would produce the trapezoid rule if continued, but the Taylor series for $(1+\alpha h/2)/(1+\alpha h/2)$ converges only for $|\alpha h|<2$. There is no point in continuing to iterate, as we will achieve neither increased accuracy nor improved stability. Furthermore, each iteration requires another evaluation of the function $f(x, y)$ and increases the expense.

An alternative approach to obtaining increased accuracy introduces points between x_n and x_{n+1}. Methods that do this are related to predictor-corrector methods but they are usually regarded as a separate class—Runge–Kutta methods. There are a great many of these, only a few of which we review. The simplest method of this class uses the Euler method to predict the value at $x_{n+1/2}=x_n+h/2$:

$$y_{n+1/2}=y_n+hf(x_n, y_n) \tag{3.6.4a}$$

This is then corrected using the midpoint rule:

$$y_{n+1}=y_n+\frac{h}{2}f(x_{n+1/2}, y_{n+1/2}) \tag{3.6.4b}$$

Application of this method to Eq. 3.3.1 shows that it has exactly the same characteristic function as the method shown in Eq. 3.6.1. In fact this method is identical to the Richardson extrapolation method of Section 5a. It is not at all uncommon that two apparently different methods are actually different ways of writing the same method.

The most commonly used Runge–Kutta methods are fourth order accurate and there are a number of these. The best known such method (sometimes called *the* fourth order Runge–Kutta method) is

$$y^*_{n+1/2}=y_n+\frac{h}{2}f(x_n, y_n) \tag{3.6.5a}$$

(Euler predictor—half step)

$$y^{**}_{n+1/2}=y_n+\frac{h}{2}f(x_{n+1/2}, y^*_{n+1/2}) \tag{3.6.5b}$$

(Backward Euler corrector—half step)

$$y^{***}_{n+1}=y_n+hf(x_{n+1/2}, y^{**}_{n+1/2}) \tag{3.6.5c}$$

(Midpoint rule predictor—full step)

$$y_{n+1}=y_n+\frac{h}{6}\left[f(x_n, y_n)+2f(x_{n+1/2}, y^*_{n+1/2}) +2f(x_{n+1/2}, y^{**}_{n+1/2})+f(x_{n+1}, y^{***}_{n+1})\right] \tag{3.6.5d}$$

(Simpson's rule corrector—full step)

Looking at this method one can see that derivation of such methods is not an easy task. An analysis of it for the general case is also difficult. It is not too difficult to analyze, however, when applied to Eq. 3.3.1. We find that

$$y_{n+1}=\left(1+\alpha h+\frac{(\alpha h)^2}{2}+\frac{(\alpha h)^3}{6}+\frac{(\alpha h)^4}{24}\right)y_n \tag{3.6.6}$$

so that the method is indeed fourth order accurate and the error is of order $(\alpha h)^5/120$. It is interesting to note that the steps that comprise this method are of order one, one, two, and four, respectively, and the method has inherited the accuracy of the final corrector.

The stability limit for this method is shown in Fig. 3.10. We see that it is both considerably more accurate and more stable than the methods already discussed. When combined with the fact that it is easy to program and use, we see why it has been a very popular method until recently. Its only disadvantage is that it requires the evaluation of the derivative function four times. This makes it expensive and is the reason why it has been partially supplanted.

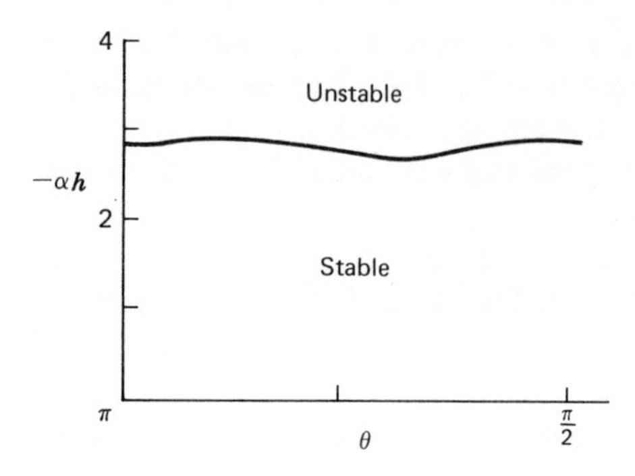

Fig. 3.10. Stability limit for Fourth order Runge–Kutta method.

One problem with Runge–Kutta methods of the past was the difficulty in estimating the error in the results during the calculation, making automatic error control difficult. A number of methods have been developed that overcome this deficiency by including a means of calculating an estimate of the error in the results, among which are the Merson, Scraton and Fehlberg variations.

The Runge–Kutta–Merson method is

$$y^*_{n+1/3} = y_n + \frac{\Delta x}{3} f(x_n, y_n) \tag{3.6.7a}$$

(Euler predictor—one-third step)

$$y^{**}_{n+1/3} = y_n + \frac{\Delta x}{6}\left[f(x_n, y_n) + f(x_{n+1/3}, y^*_{n+1/3})\right] \tag{3.6.7b}$$

(Trapezoid corrector—one-third step)

$$y^*_{n+1/2} = y_n + \frac{\Delta x}{8}\left[f(x_n, y_n) + 3f(x_{n+1/3}, y^{**}_{n+1/3})\right] \tag{3.6.7c}$$

(Adams–Bashforth type predictor—half step)

$$y^*_{n+1} = y_n + \frac{\Delta x}{2}\left[f(x_n, y_n) - 3f(x_{n+1/3}, y^{**}_{n+1/3}) + 4f(x_{n+1/3}, y^*_{n+1/2})\right] \tag{3.6.7d}$$

(Adams–Bashforth type predictor—full step)

$$y^{**}_{n+1} = y_n + \frac{\Delta x}{6}\left[f(x_n, y_n) + 4f(x_{n+1/2}, y^*_{n+1/2}) + f(x_{n+1}, y^*_{n+1})\right] \tag{3.6.7e}$$

(Simpson's rule corrector—full step)

One can show (the easiest way is by application to Eq. 3.3.1) that both y^*_{n+1} and y^{**}_{n+1} are fourth order accurate. In fact one can show that if $y_n = y(x_n)$ is the exact value of y at x_n then

$$y_{n+1}^{*} \simeq y(x_{n+1}) - \frac{(\Delta x)^5}{120} y^{(v)} + 0(\Delta x^6) \tag{3.6.8a}$$

$$y_{n+1}^{**} \simeq y(x_{n+1}) - \frac{(\Delta x)^5}{720} y^{(v)} + 0(\Delta x^6) \tag{3.6.8b}$$

Thus y_{n+1}^{**} contains approximately one-sixth the error of y_{n+1}^{*}. The error in the latter may be estimated as one-fifth the difference between y_{n+1}^{*} and y_{n+1}^{**}. This knowledge allows the routine to adjust the step size automatically to maintain the desired accuracy. The cost of this improvement is an extra evaluation of the derivative.

Example 3.9

The predictor-corrector method in Eq. 3.6.1, when applied to the equation $y' = -y$, yields results that are similar to those obtained earlier. The error is in fact in a direction opposite to that of the trapezoid rule and is approximately twice as large. For step sizes larger than approximately 1.23 the solution begins to oscillate and, finally, for steps larger than 2 the kind of oscillatory growth observed earlier for the Euler method occurs.

Since the results for predictor-corrector methods are similar to those for earlier methods, we shall compare the two. Although there are several ways of doing this, we use one that is somewhat uncommon. All methods, when applied to $y' = -y$, produce exponential solutions of the form $y_n = \beta^n$ where β is an approximation to e^{-h} and can be written $\beta = e^{\gamma h}$ and $\gamma - 1$ is a measure of the accuracy of a method. We call $\gamma = h^{-1} \ln \beta(h)$ the *effective exponent*, which is plotted for the methods considered so far in Fig. 3.11.

For small h, $(\gamma - 1)$ is proportional to h^n for an nth order method and all of the curves are straight lines whose slopes are the order of the method to which they pertain. At larger h the curves deviate from straight line behavior. For the Euler, trapezoid, and predictor-corrector methods, all of which give $\beta = 0$ at some value of h, the inaccuracy is very obvious on this plot.

The fourth order Runge–Kutta method goes unstable by producing exponential growth rather than decay. This contrasts with the oscillatory behavior of the other methods studied and is the reason for the behavior of the effective exponential of the Runge–Kutta method.

More important, Fig. 3.11 can be used to determine the most effective method for a given accuracy. Thus for $(\gamma - 1)$ to be 0.01 or less, the allowable step size for the Runge–Kutta method is approximately 2.5 times as large as that for the trapezoid rule, 4 times as large as the predictor-corrector method, and 30 times as large as the Euler methods. Although Runge–Kutta requires twice as much computation per point as predictor-corrector and four times as much as Euler, it still produces the required results at the lowest cost. The

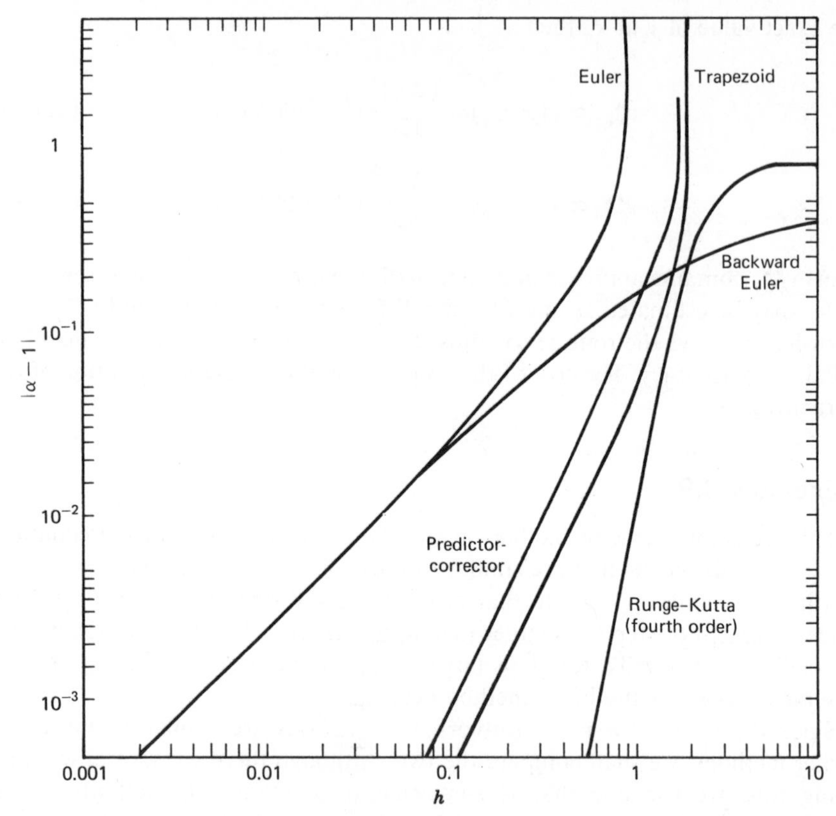

Fig. 3.11. Effective exponents of five methods of solving ordinary differential equations.

advantage is still greater if higher accuracy is required. This combination of accuracy, relatively low cost, and ease of programming (see Example 3.10) have made this one of the most popular methods.

Example 3.10

It is also interesting to compute the solution to $y' = i\pi y$ by the various methods. Since the amplitude error, when there is any, is exponential, we present the results for the amplitude in a manner similar to that used in the preceding example. The numerical solution has the form $y_{n+1} = \beta y_n$ where β is complex. In Fig. 3.12 we have plotted $h^{-1}|\ln|\beta(h)||$ as a function of h.

There are several points worthy of note in Fig. 3.12. The trapezoid rule gives no amplitude error and does not appear in the plot. The two Euler methods have errors in opposite directions but of equal magnitude. The Runge–Kutta method has a negative (dissipative) amplitude error and is stable until $h = 2\sqrt{2}/\pi = 0.90$; beyond that it is unstable. The change in sign of $\ln|\beta(h)|$ means that the result could only be plotted for a value of $h < 0.9$. It is also interesting

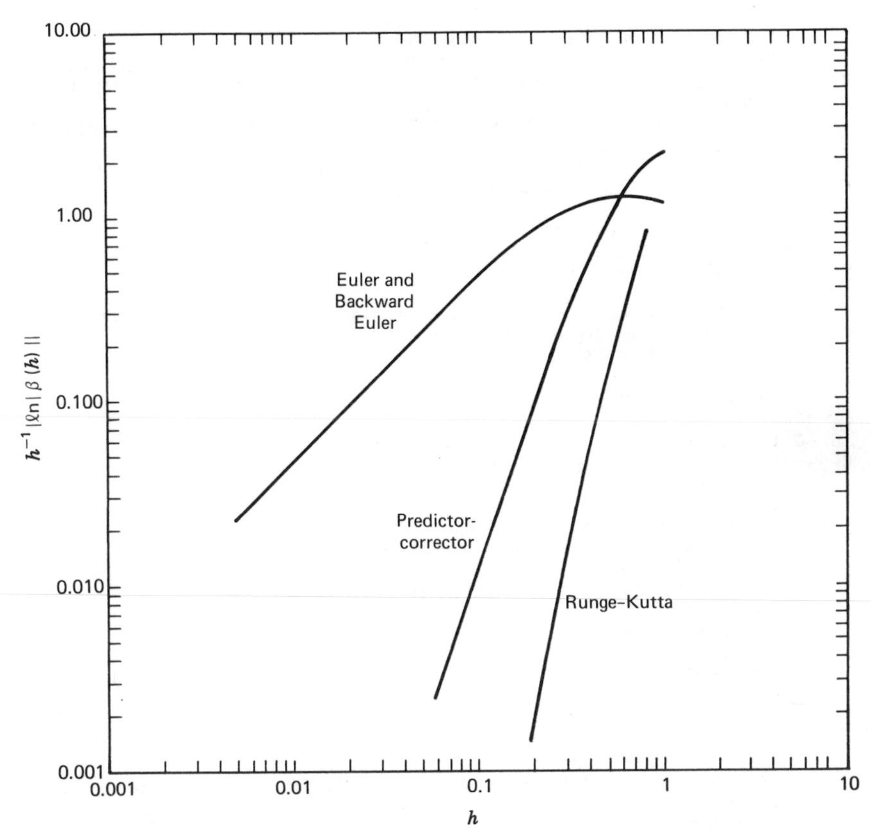

Fig. 3.12. Effective exponent of amplitude in the solution of $y''+\pi^2 y=0$ using various methods.

to note that the slopes of the curves in Fig. 3.12 are smaller than those of the corresponding curves in Fig. 3.11—in other words, the amplitude error of these methods is actually of higher order than the order of the method.

Phase errors are not plotted, since both the predictor-corrector and Runge–Kutta phase errors change sign as h is increased. The Euler method gives a negative phase error indicating that the solution lags the exact solution and the phase error is second order for small step size. The trapezoid rule produces a lagging phase error that, for small step sizes, is about one-quarter of the phase error of the Euler methods. The predictor-corrector method yields a lagging error whose magnitude is half that of the Euler methods at small step sizes and which changes to a leading error at large step sizes. Finally, the fourth order Runge–Kutta method gives very small lagging phase errors at small step sizes but leading errors for $h \gtrsim 0.5$.

Runge–Kutta again shows itself to be a superior method. It is capable of handling solutions of both the exponential and oscillatory types with great ease. For this reason we give a Runge–Kutta subroutine below. It can be used in the same way as subroutine EULER given in Section 3 by changing only the name in the calling statement.

```
      SUBROUTINE RUNGE4(N,X,Y,K,H,F)
C-----THIS SUBROUTINE ADVANCES THE SOLUTION TO THE SYSTEM OF N ODE'S
C-----DY(I)/DX = F(I,X,Y(1),Y(2),,.....Y(N)) USING THE FOURTH ORDER
C-----RUNGE-KUTTA METHOD.  THE OTHER INPUT VARIABLES ARE
C-----X = INITIAL VALUE OF THE INDEPENDENT VARIABLE; IT IS INCREASED
C-----    BY THE SUBROUTINE.
C-----Y = INITIAL VALUE OF THE DEPENDENT VARIABLE(S); THE ROUTINE
C-----    OVERWRITES THIS WITH THE NEW VALUE(S).
      EXTERNAL F
      DIMENSION Y(10),YS(10),YSS(10),YSSS(10),T1(10),T2(10),
     1          T3(10),T4(10)
C-----THE MAIN LOOP
      DO 1 I=1,K
C-----TEMPORARY ARRAYS ARE NEEDED FOR THE FUNCTIONS TO SAVE THEM
C-----FOR THE FINAL CORRECTOR STEP.
C-----FIRST (HALF STEP) PREDICTOR
      DO 2 J=1,N
      T1(J) = F(J,X,Y)
      YS(J) = Y(J) + .5 * H * T1(J)
    2 CONTINUE
C-----SECOND STEP (HALF STEP CORRECTOR)
      X = X + .5 * H
      DO 3 J=1,N
      T2(J) = F(J,X,YS)
      YSS(J) = Y(J) + .5 * H * T2(J)
    3 CONTINUE
C-----THIRD STEP (FULL STEP MIDPOINT PREDICTOR)
      DO 4 J=1,N
      T3(J) = F(J,X,YSS)
      YSSS(J) = Y(J) + H * T3(J)
    4 CONTINUE
C-----FINAL STEP (SIMPSON'S RULE CORRECTOR)
      X = X + .5 * H
      DO 5 J=1,N
      T4(J) = F(J,X,YSSS)
      Y(J) = Y(J) + (H/6.)*(T1(J) + 2.*(T2(J)+T3(J)) + T4(J))
    5 CONTINUE
    1 CONTINUE
      RETURN
      END
```

7. MULTISTEP METHODS

Runge–Kutta methods were the standard methods for the numerical solution of ordinary differential equations for many years. The reasons for the increased use of other methods are probably the availability of large computers and the extreme accuracy required by a number of advanced technologies. It is possible to develop Runge–Kutta methods of any desired accuracy, but from a practical point of view it is very difficult. Their other major flaw was mentioned in the preceding section. They require a large number of function evaluations per step. We now seek methods that provide higher accuracy without the large number of function evaluations. Naturally the advantages come at a price, but in many instances the benefit is worth the price.

It is not difficult to construct methods of the type we seek and a large number of them have been constructed. This is in fact an area of current research. One simple way of finding some of these methods is to begin with Eq. 3.5.8 and approximate the second and higher derivatives by difference formulas of the kind discussed in Section 1 of this chapter. We can use information at any of the mesh points $1,2,\ldots n-1, n$ as well as the data at the new point $n+1$. Both the function and the derivative at these points can be used in the formulas, so there are many possibilities. The resulting methods are based on difference formulas that include data from points $n-1$, $n-2$, and so on, as well as n and $n+1$, and are called *multipoint* or *multistep methods*. Both explicit and implicit multipoint methods are possible.

We treat one of the simplest multistep methods in some detail in order to discover some of the properties of this class of methods. Then some of the more important members of this class are given without derivation and, finally, means of using them in practical computations are presented.

As the method to examine in detail we select the leapfrog method for which the basic formula is

$$y_{n+1}=y_{n-1}+2hf(x_n, y_n) \tag{3.7.1}$$

This method is a bit different from those described above, but can be made to fit the format with a little juggling. It can be regarded as the midpoint rule applied to the interval (x_{n-1}, x_{n+1}). It is second order accurate and is a two-step formula because data from the last two data points are required to produce the value of y_{n+1}. The major difficulty shared by all multipoint schemes is immediately obvious; one cannot start a calculation with this formula. If we are given only the initial value y_0, Eq. 3.7.1 does not provide a means of computing y_1. Once we have both y_0 and y_1, there is no problem getting y_2. Another method (one of the methods described earlier could be used) is needed to provide y_1 and thus allow us to begin using the leapfrog technique. The choice of starting method can have an important bearing on the results and is discussed later. For now let us assume that values of y_0 and y_1 are available and see what the leapfrog method does. We again use Eq. 3.3.1 as a simple test case but the difference equation is not as readily solved as the ones for the methods described earlier. We can use our knowledge of the differential equation to guess that the difference equation

$$y_{n+1}=y_{n-1}+2\alpha h y_n \tag{3.7.2a}$$

might have exponential solutions of the form

$$y_n=\rho^n \tag{3.7.2b}$$

(There are other, equally good methods of solving Eq. 3.7.2a. We chose this one for its close resemblance to methods used for differential equations.) Plugging this into the difference equation and dividing by ρ^{n-1}, we find that ρ satisfies a quadratic equation

$$\rho^2-2\alpha h\rho-1=0 \tag{3.7.3}$$

that has two solutions

$$\rho_{1,2} = \alpha h \pm \sqrt{\alpha^2 h^2 + 1} \tag{3.7.4}$$

This is in contrast to the earlier methods, which had only a single exponential solution. To see what these roots represent, let us expand the square root in a series in αh. We find

$$\rho_1 = 1 + \alpha h + \frac{(\alpha h)^2}{2} - \frac{(\alpha h)^4}{8} + \cdots \tag{3.7.5a}$$

$$\rho_2 = -1 + \alpha h + \cdots \tag{3.7.5b}$$

Clearly the first root is a second order accurate approximation to $e^{\alpha h}$, while the second bears no relation at all to the solution of the differential equation. This root is called the *computational* or *parasitic* root and does nothing but cause trouble. In fact if αh is real and negative, ρ_2 has magnitude greater than 1 and the method may be unstable. Let us look further into the solution of the difference equation.

Since the difference equation in Eq. 3.7.2a is linear, the general solution is a linear combination of the two solutions given by Eq. 3.7.4.

$$y_n = a_1 \rho_1^n + a_2 \rho_2^n \tag{3.7.6}$$

The two coefficients a_1 and a_2 can be found by requiring the solution to match the starting values y_0 and y_1. We easily find

$$a_1 = \frac{y_1 - \rho_2 y_0}{\rho_1 - \rho_2} \qquad a_2 = \frac{y_1 - \rho_1 y_0}{\rho_1 - \rho_2} \tag{3.7.7}$$

Note that if $y_1 = \rho_1 y_0$, then $a_2 = 0$ and the computational root will play no role. In other words, we can choose the starting values for the simple test equation so as to suppress the parasitic root. Also, if y_1 is a second order accurate approximation to $y(x_1)$, then it is also a second order accurate approximation to $\rho_1 y_0$. In this case a_2 will be of order h^3 and for small h it may not cause much of a problem (see below). Thus the method used to start the computation can have an important effect on the results.

There is another difficulty with the leapfrog method that also occurs in other multistep methods. As the step size is increased, either of the roots may increase in magnitude. In fact as we saw, if the real part of αh is negative, the magnitude of ρ_2 is greater than 1. Even if we start the method in a manner that totally suppresses the parasitic root, roundoff error can be interpreted as introducing some of this component. This component then will be amplified and may eventually overwhelm the accurate part of the solution. In a calculation using the leapfrog method, the signal that this problem is arising is that the solution will begin to oscillate about the correct solution with a period of two steps. This is a consequence of a parasitic root being negative. The problem can be corrected in a number of ways. One simple way is to average the results at successive steps every once in a while to remove the parasitic component. From a strict point of view the leapfrog method should be regarded as unconditionally unstable, but since the difficulty is due to the parasitic root and may be suppressed, the method is called weakly unstable and has seen considerable service. Another reason for the use of the leapfrog method is found in the numerical examples that follow.

Example 3.11

Let us compute the solution of $y'=-y$ using the leapfrog method. With $h=1$, the two roots in Eq. 3.7.4 are

$$\rho_1=0.414 \qquad \rho_2=-2.414$$

The first is an approximation to the correct solution. (The exact result is $e^{-1}=0.368$ and ρ_1 is reasonably accurate when the large step size is taken into account.) The second root is the troublemaker. With the starting values given as the exact values, that is, $y_0=1$, $y_1=e^{-1}$, we obtain the results shown in Table 3.3. It is obvious that despite the most accurate starting values possible, the results are garbage. Once the instability has taken over, $y_{n+1}\approx -2.414\, y_n$ and the "solution" is a rapidly growing oscillation.

This case is extreme, of course, because of the large step size. For smaller step sizes the instability, although present, is not as severe. It is possible to use the leapfrog method if the range of integration is not large, but caution is needed. "Reasonable"-looking results are frequently obtained and one is sometimes tempted to accept as valid results that are significantly in error. The need for caution in accepting numerical results is emphasized.

Table 3.3 Leapfrog Solution of $y'=-y$ with Exact Starting Values

x	y
1	0.368
2	0.264
3	−0.161
4	0.585
5	−1.33
6	3.24
7	−7.28
8	18.91
9	−45.64
10	110.2

We solved the same problem with the starting value $y_1 = 0.414$, the exact "physical" root of the leapfrog method for this problem. This calculation was done in double precision (to reduce roundoff error) with a step size of 0.1. With this method of starting, leapfrog behaves like a stable second order method and the error at $x = 10$ is only approximately 8.10^{-7} or 2%. In most calculations, however, the coefficients are not constant and it is not possible to make leapfrog behave this well. Roundoff error will eventually creep in and destroy the calculation. Using a proper starter delays the onset of disaster; it does not prevent it.

This example displays the most important features of multistep methods: the importance of the method of starting the calculation and the trouble caused by the parasitic root. The next example shows that a method that has difficulty in problems with exponential-like solutions may do very well when the solutions are oscillatory.

Example 3.12

When the leapfrog method is applied to the solution of $y' = i\pi y$, the results are excellent but the existence of a parasitic solution makes it impossible to express the results in terms of a simple amplitude and phase error. We will discuss them in a qualitative manner. The reason that excellent results are obtained is that when α is pure imaginary, both roots in Eq. 3.7.4 have unit magnitude and there is no instability. Furthermore, as we have seen, with a proper starting method the coefficient of the parasitic component of the solution is quite small. Consequently, the method behaves very much like the second order methods discussed earlier.

For problems with oscillatory solutions the leapfrog method is nearly *neutrally stable*. By neutral stability we mean that a method produces neither unwanted growth of the solution (instability) nor artificial dissipation. This desirable property, which is possessed by few methods, and the fact that it is explicit make the leapfrog method the basis for some important methods for solving partial differential equations.

The leapfrog method is not exactly neutrally stable because the parasitic term, although small, has a different phase from the physical term and the interference between the two produces small amplitude errors.

All multistep methods when applied to $y' = \alpha y$ have more than one solution. In fact the number of solutions is equal to the number of steps that the method uses. One of these roots represents the accurate solution; the others are parasitic and a computational inconvenience. All explicit multistep methods are at most conditionally stable, but unlike the leapfrog method, many of them have regions in which they are truly stable. Implicit multistep methods may, of course, be unconditionally stable.

The most general k step method is

$$\sum_{m=0}^{k} \alpha_m y_{n+1-m} + \sum_{m=0}^{k} \beta_m f(x_{n+1-m}, y_{n+1-m}) = 0 \tag{3.7.8}$$

and is defined by the number of steps k and the parameters α_m and β_m. We can let $\alpha_0 = 1$ with no loss of generality. If $\beta_0 = 0$, the method is explicit; otherwise it is implicit. From this it is clear that there are a great many methods of this type, only a few classes of which are treated here.

Methods in which $\alpha_0 = 1$ and $\alpha_1 = -1$ are called *Adams methods* and are probably the most widely used members of this class. The explicit methods are called *Adams–Bashforth methods* and the implicit ones are called *Adams–Moulton methods*. They can be derived in a number of ways, one of which is from difference formulas. For the Adams–Bashforth case backward formulas are needed and for the Adams–Moulton case formulas with one forward point are required. They can also be derived by direct use of interpolation formulas. Fitting a Lagrange polynomial to the *derivative* y' at the points $x_{n+1-k}, x_{n-k}, \ldots x_n$ equation and integrating from x_n to x_{n+1} yields the Adams–Bashforth formulas. Doing the same calculation with y'_{n+1} included in the interpolation yields the Adams–Moulton methods. The various derivations are, of course, completely equivalent.

The coefficients for the Adams–Bashforth methods are given in Table 3.4. As might be expected, the first order method is simply the Euler method, but the higher order methods are new.

Similarly, the coefficients for the Adams–Moulton methods are given in Table 3.5. To keep the order of the method equal to the index k, we have terminated the series at $k-1$. The first order Adams–Moulton method is backward Euler and the second order method is the trapezoid rule, but the higher order methods are new.

The stability of these methods can be analyzed in the manner described earlier and is obviously a chore for all but the lowest order methods. The

Table 3.4. Coefficients for the Adams–Bashforth Methods

$$\left(y_{n+1} = y_n + h \sum_{m=1}^{k} \beta_{km} f_{n+1-m}\right)$$

	$m=1$	2	3	4	5	6
β_{1m}	1					
$2\beta_{2m}$	3	-1				
$12\beta_{3m}$	23	-16	5			
$24\beta_{4m}$	55	-59	37	-9		
$720\beta_{5m}$	1901	-2774	2616	-1274	251	
$1440\beta_{6m}$	4277	-7923	9982	-7298	2877	-475

Table 3.5. Coefficients for the Adams–Moulton Methods

$$\left(y_{n+1}=y_n+h\sum_{m=0}^{k-1}\beta^*_{km}f_{n+1-m}\right)$$

	$m=0$	1	2	3	4	5
β^*_{1m}	1					
$2\beta^*_{2m}$	1	1				
$12\beta^*_{3m}$	5	8	-1			
$24\beta^*_{4m}$	9	19	-5	1		
$720\beta^*_{5m}$	251	646	-264	106	-19	
$1440\beta^*_{6m}$	475	1427	-798	482	-173	27

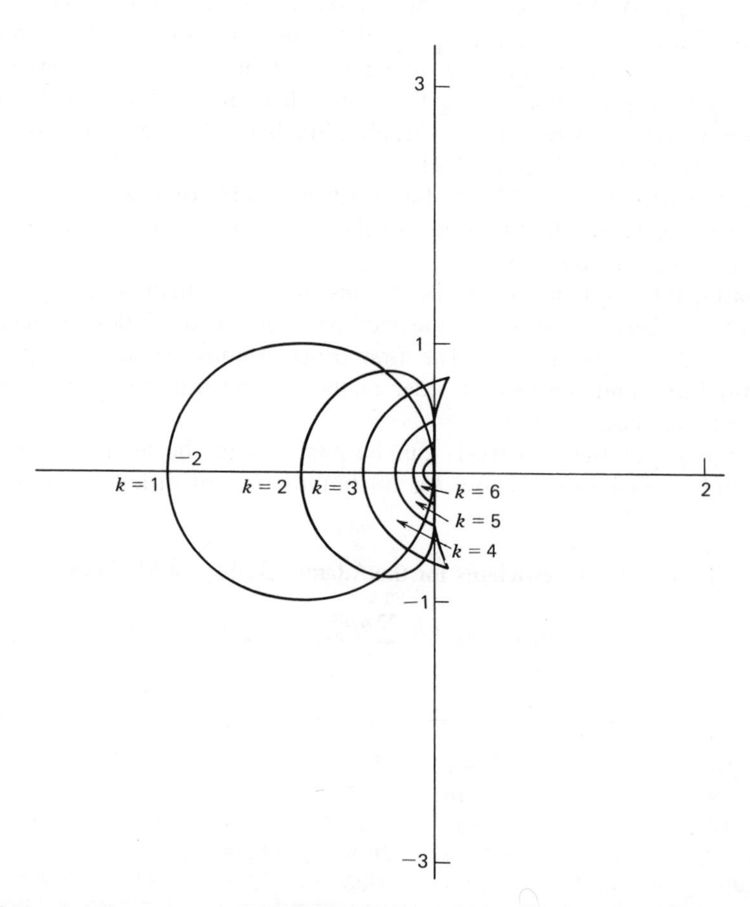

Fig. 3.13. Stability regions for Adams–Bashforth methods. The method of order k is stable inside the region indicated.

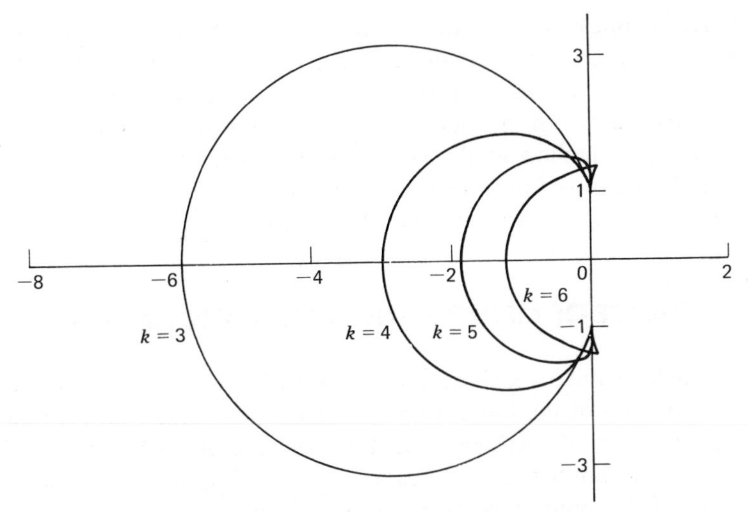

Fig. 3.14. Stability regions for Adams–Moulton methods. The method of order k is stable inside region indicated.

results are shown in Figs. 3.13 and 3.14, respectively. In Table 3.6 we give the stability limits for real values of $hy'/y = \alpha h$, which show clearly that the implicit methods are always considerably more stable than the corresponding explicit methods as expected.

A great many variations on these basic methods are possible and many have been developed by various authors. Since it is not our purpose here to give an exhaustive survey of methods and since the Adams methods are among the best ones, additional methods are not treated.

Adams methods have the same properties that were found for the simpler methods. The explicit ones are easy to use but are not as stable as the more difficult-to-use implicit methods. An obvious improvement is a predictor-corrector method based on the Adams family. Methods of this kind are among the most commonly used differential equation solvers when high accuracy is required. Adams methods allow easy change of the order of the method and it is relatively simple to construct computer routines that provide for automatic error control. Generally, these methods consist of an Adams–Bashforth predic-

Table 3.6. Stability Limits for the Adams Methods

Order	Adams–Bashforth	Adams–Moulton
1	2	∞
2	1	∞
3	0.5	6
4	0.3	3
5	0.2	1.9

tor followed by one or more corrector steps of the Adams–Moulton type. In a well-designed method the overall accuracy is the same as that of the last corrector step. One of the simplest and most popular of these uses the nth order Adams–Bashforth method as the predictor and the $(n+1)$st order Adams–Moulton method as the corrector. The overall accuracy is $(n+1)$st order and the stability limit is somewhere between that of the predictor and that of the corrector. We elaborate on this in Section 8.

8. THE CHOICE OF METHOD AND AUTOMATIC ERROR CONTROL

It is obvious from the preceding sections that there exist many methods for the numerical solution of the initial value problem for ordinary differential equations. The number of methods is in fact so large as to become confusing. Usually any of a number of methods will do the job well at a reasonable cost. There are a few points, however, that should be considered in choosing a method in order to avoid a really poor choice.

Except for very crude calculations, first order methods are never used; they simply do not provide the accuracy needed. Second order methods find considerable service as both ordinary and partial differential equation solvers. (As we will see, in the latter application they have the important advantage of requiring less memory than higher order methods.) For much engineering work, in which relative accuracies of 10^{-2} to 10^{-4} are required, fourth order methods usually prove to be about as efficient (if not slightly more efficient) as second order methods and are generally recommended. Their flexibility in being able to provide higher accuracy with only modest cost increases has probably made them so popular. For very accurate work in which errors as small as 10^{-10} may be required methods of very high order (sixteenth order methods have been used in space applications) are necessary.

Most modern computer subroutines for differential equation solving have built-in devices for the control of order and step size. The user need only state the error requirement and, of course, the information necessary to define the differential equation (or set of equations) to be solved. The program will then select the optimum order and step size to achieve the user's desires at lowest cost. A number of such programs are available and they use several different algorithms. Rather than looking very deeply into these algorithms, a short overview of what is contained in some of them is presented.

One popular scheme developed by Bulirsch and Stoer uses a second order method (usually leapfrog) and extrapolation (which may be Richardson but more commonly is polynomial or rational extrapolation) to achieve high order accuracy. In order to provide the requested accuracy, the method contains a built-in error estimator. If the error is not within the allowable limit after the maximum amount of extrapolation allowed has been done, step size is reduced and the calculation repeated.

Another common approach uses Adams methods. Since high order Adams methods do not allow easy adjustment of the step size, this parameter is not changed very often. Rather, the order of the method is selected to provide the desired error. Only when the order has been increased to the maximum limit allowed by the program is the step size adjusted. Naturally, routines based on this approach have special starting algorithms built in.

Subroutines based on these methods generally contain many logic and control statements and their structure cannot be described meaningfully in a short space. A good description of all of the details of one of the better subroutines (ODE, which is used in the examples below) can be found in the work of Shampine and Gordon (1975).

Example 3.13

The results of applying two of the better subroutines currently available to the solution of $y'=-y$ will be given; we will integrate from $x=0$ to $x=10$. (The solution to this problem is obviously very smooth and does not show these methods to the best advantage.) Both of these routines are designed to produce a solution with the desired accuracy using the minimum number of function evaluations. Naturally, the control strategies used must be general purpose so they cannot be optimal for every particular problem.

ODE is a program written by Shampine, Gordon, and colleagues and is based on the Adams methods. The strategy is basically that described above, in which the order of the method is the primary adjustment needed to achieve the desired accuracy. Methods up to twelfth order are built in. The step size is changed only when the error criterion cannot be met by the highest order.

The RKF45 routine is based on the Fehlberg modification of the Runge–Kutta method. It contains fourth and fifth order accurate algorithms and the error estimation is similar to that described for the Merson version of the Runge–Kutta method. Since the order of the method is fixed, the main error control device is step size management.

Results are shown in Fig. 3.15. Recall that the test problem chosen does not show these methods at their best. Thus for this problem RKF45, a sophisticated Runge–Kutta routine is only slightly better than the simple fourth order Runge–Kutta method with a fixed step size. The curve also shows clearly what we had anticipated—when relative low accuracy is required, RKF45, which is of a much lower order than ODE, is the more effective method but the reverse is true when high accuracy is demanded. In the range of accuracy most frequently required there is in fact little difference between the two methods.

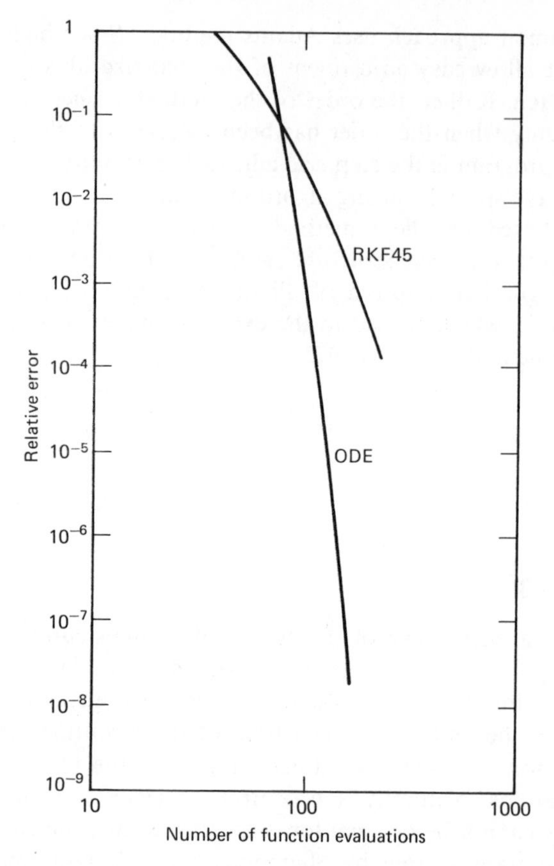

Fig. 3.15. Error versus number of function evaluations for ODE and RKF45; solution of $y'=-y$.

Example 3.14

An example that shows the variable order and step methods to better advantage is the differential equation used in Example 3.4:

$$y'=-2xy$$

We will integrate up to $x=5$ for which $y \simeq 1.39 \times 10^{-11}$. The fixed step Runge–Kutta method requires a step size of 0.2 or less to insure that the method is stable for all x, so a minimum of 100 steps is required.

ODE and RKF45 request both absolute and relative error tolerances. The less stringent limit is satisfied. Best results (in the sense of maximum accuracy for a given number of function evaluations) are obtained when the two tolerances are of approximately equal stringency. For the equation we wish to solve this means that the absolute error must be 10^{-11} times the relative error. In Fig. 3.16 we give the relative error actually obtained (i.e., compared to the exact solution) versus the number of function evaluations for this relationship between the parameters. The advantage of the step size and order controls is

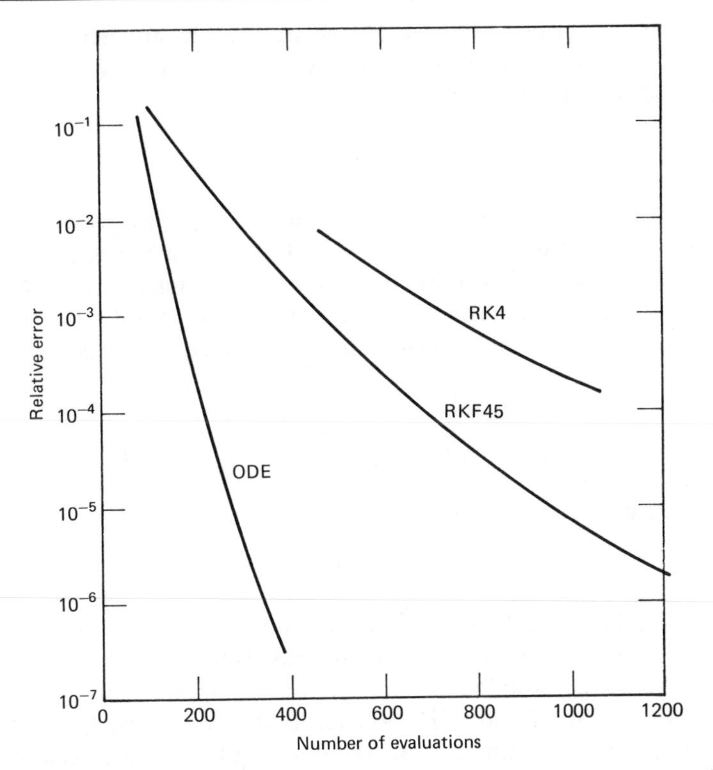

Fig. 3.16. Relative error versus number of function evaluations for solution of $y'=-y$ by various methods.

obvious and ODE appears to be the better method. Note that although ODE has a 2 : 1 advantage in function evaluations, it is a more complicated code and the ratio of running times is something less than this.

9. SYSTEMS OF EQUATIONS—STIFFNESS

We now turn to the problem of dealing with systems of ordinary differential equations. As high order equations can almost always be reduced to systems of first order equations, these can also be included here. Systems of equations are most easily treated by reducing them to a form to which we can apply the results obtained for single first order equations. Almost all the preceding results can be taken over directly. The only new issue that arises is the problem of stiffness, to which we devote most of this section.

The basic problem is to solve the system of ordinary differential equations

$$y_i' = f_i(x, y_1, y_2, \ldots y_m) \qquad i=1,2,\ldots m \tag{3.9.1}$$

subject to the initial conditions

$$y_i(0) = c_i \qquad i=1,2,\ldots m \tag{3.9.2}$$

The methods described earlier in this chapter can be applied to this system of equations and the subroutines given have this capability. For example, the Euler method would give

$$y_{i,n+1}=y_{i,n}+hf_i(x_n,\mathbf{y}_n) \tag{3.9.3}$$

where we introduce the notation $\mathbf{y}_n=(y_{1,n},y_{2m},\ldots y_{m,n})$ for convenience. Any of the other methods discussed earlier, including the implicit ones, can also be applied to this problem. To study the difficulties that might arise, we do what we did for the case of a single equation—that is, we consider the simple but important case of a system of linear equations with constant coefficients:

$$y_i'=\sum_{j=1}^{m}a_{ij}y_j \qquad i=1,2,\ldots m \tag{3.9.4}$$

which can be written in matrix notation

$$\mathbf{y}'=A\mathbf{y} \tag{3.9.5}$$

If we linearize Eq. 3.9.1 locally, we will obtain a system of equations of the form in Eq. 3.9.4 with an inhomogeneous term. This last term is important in producing the correct solution but is relatively unimportant in the analysis of accuracy and stability. Hence it is sufficient to deal with Eqs. 3.9.4 or 3.9.5.

We assume that the numbers of equations and unknowns are equal so that A is an $m\times m$ matrix. From linear algebra it is known that a matrix possesses a set of eigenvalues defined by

$$A\boldsymbol{\phi}_k=\lambda_k\boldsymbol{\phi}_k \tag{3.9.6}$$

The number of eigenvectors may be less than or equal to m, but in essentially every case that arises from a physical problem the number of eigenvectors is precisely equal to m, and we shall assume this to be the case. When this is true, one can construct a matrix S the columns of which are the eigenvectors of A, which has the property that the matrix product

$$S^{-1}AS=\Lambda \tag{3.9.7}$$

produces a diagonal matrix Λ the elements of which are the eigenvalues of A. Then premultiplying Eq. 3.9.5 by S^{-1} and defining $\mathbf{z}=S^{-1}\mathbf{y}$, we have

$$S^{-1}\mathbf{y}'=\mathbf{z}'=S^{-1}A\mathbf{y}=(S^{-1}AS)S^{-1}\mathbf{y}=\Lambda\mathbf{z} \tag{3.9.8}$$

which, in component form, is the uncoupled set of linear differential equations

$$z_k'=\lambda_k z_k \qquad k=1,2,\ldots n \tag{3.9.9}$$

Each member of this set of equations has precisely the form of Eq. 3.3.1, the equation we used for testing methods for single first order differential equations. Thus *everything that we found for methods for a single first order differential equation applies with equal validity to systems of equations or higher order equations.*

There is a difficulty, however. Except for systems with constant coefficients, the method used above is impractical. In general the system must be solved as written; that is, we must solve the system in the form of Eq. 3.9.4 rather than of Eq. 3.9.9. Solving Eq. 3.9.4 by any method produces, however, the same result as solving each of Eqs. 3.9.9 by the same method (and the same step size) and then finding $\mathbf{y}=S\mathbf{z}$. For conditionally stable methods there will be a largest step size h that yields a stable solution for each of Eq. 3.9.9. In particular, for the kth equation we must have $h<c|\lambda_k|^{-1}$ where c is a constant dependent on the method chosen. The most restrictive of these conditions is one based on the largest eigenvalue (we assume that they all have negative real part). The step size chosen (which must be the same for all equations) must satisfy this criterion. In many cases this causes no problem. In others we want the solution over a large span of the independent variable, but one of the eigenvalues is so large that an undesirably small step size (and large number of steps) is required. This happens when the magnitudes of the eigenvalues cover a wide range. A problem of this type is called *stiff*. From a mathematical point of view the problem of stiffness arises whenever the ratio of the largest to the smallest eigenvalue is very large; in fact this ratio is called the *stiffness ratio*. Physically, stiffness occurs when the problem contains widely disparate time or length scales. Some common problems in which stiffness might be a factor are:

1. Any problem that displays boundary layer-type behavior. A system will have boundary layer behavior when there is a small length scale that is important in a small region and a longer scale that is important in the remainder of the system. Such problems are common in fluid mechanics, heat transfer, and the kinetic theory of gases.
2. Problems in which there are a great variety of rate constants. A very common source of problems of this type is any system in which chemical kinetics plays an important role. Chemical rate constants may easily vary over orders of magnitude and almost all problems of this type turn out to be stiff.
3. Many problems arising from parabolic partial differential equations. When finite differenced, these turn out to be equivalent to solving a stiff system of ordinary differential equations. We examine this in Chapter 4.

Stiffness is the one of the most difficult problems to deal with in the solution of differential equations. It is very closely connected with the problem of ill-conditioning that arises in the numerical solution of linear algebraic equations. (An ill-conditioned system is one in which the solution is very sensitive to small changes in either the coefficient matrix or the known vector. The

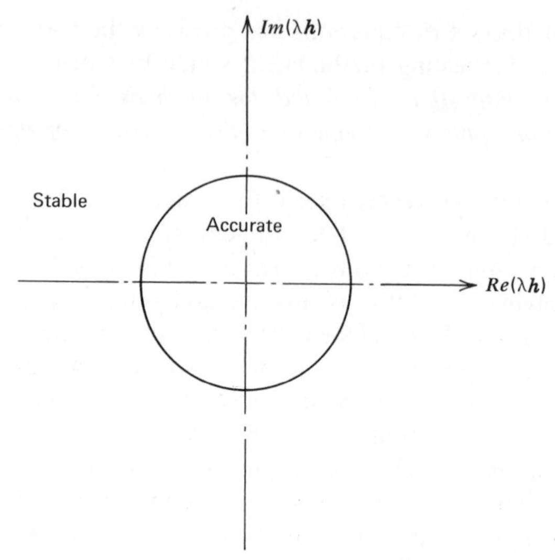

Fig. 3.17. Desired accuracy and stability regions of a stiff method.

problem arises when the matrix has eigenvalues whose magnitudes span a large range.) There are a number of methods of dealing with stiffness but none is completely satisfactory and the development of better methods for stiff systems is a topic of current research interest. A few of the better methods in current use are covered.

In many stiff problems we are primarily interested in the relatively smooth behavior of the solution at large values of the independent variable. This part of the solution is controlled by the small eigenvalues; the large eigenvalues are merely a nuisance. Unconditionally stable implicit methods with a large step size treat the part of the solution due to the large eigenvalue very inaccurately, but prevent the solution from blowing up and, therefore, are good candidates. On the other hand, explicit methods run the risk of blowing up. Specifically, we want an implicit method that has the accuracy and stability regions indicated in Fig. 3.17—in other words one that is accurate for small λh (so that it follows the slowly varying part of the solution well) and stable for large λh (so that the large eigenvalues cause no trouble). A class of methods of this type was found by Gear (1971). For a single equation they are given by:

$$y_n = \sum_{i=1}^{k} \alpha_i y_{n-i} + h\beta_0 f(x_n, y_n) \tag{3.9.10}$$

The coefficients for this method are given in Table 3.7. For details of how they are chosen, see Gear (1971).

The regions of stability of these methods are shown in Fig. 3.18. The method is stable outside the region shown and k indicates the order of accuracy of the method.

Table 3.7. Coefficients for Gear's Method

k	2	3	4	5	6
β_0	$\frac{2}{3}$	$\frac{6}{11}$	$\frac{12}{25}$	$\frac{60}{137}$	$\frac{60}{147}$
α_1	$\frac{4}{3}$	$\frac{18}{11}$	$\frac{48}{25}$	$\frac{300}{137}$	$\frac{360}{147}$
α_2	$-\frac{1}{3}$	$-\frac{9}{11}$	$-\frac{36}{25}$	$-\frac{300}{137}$	$-\frac{450}{147}$
α_3		$\frac{2}{11}$	$\frac{16}{25}$	$\frac{200}{137}$	$\frac{400}{147}$
α_4			$-\frac{3}{25}$	$-\frac{75}{137}$	$-\frac{225}{147}$
α_5				$\frac{12}{137}$	$\frac{72}{147}$
α_6					$-\frac{10}{147}$

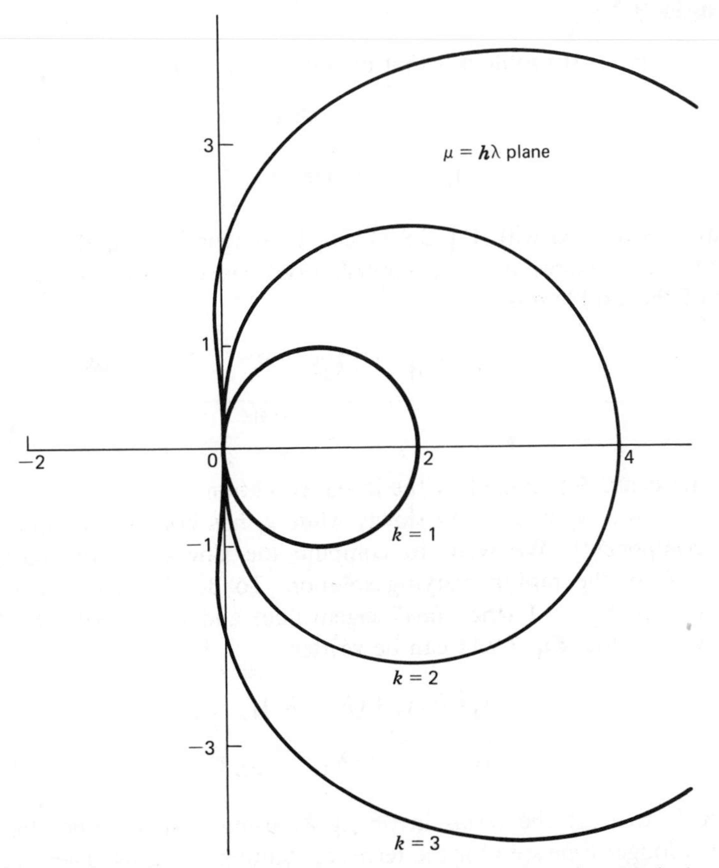

Fig. 3.18. Regions of stability for Gear's method of orders 1 through 3. Methods are stable *outside* of closed contours.

A number of other approaches to the solution of stiff problems have been suggested by various authors. Many of these use explicit methods that sacrifice some of the potential accuracy of the method in order to achieve improved stability, but none of them seems to be fully satisfactory. Gear's method is probably the most popular one in use today and variations on it have been developed by various authors. As is pointed out below, it is far from ideal.

In recent years methods have been developed for use in solving partial differential equations that contain difficulties analogous to stiffness, but which work only in certain cases. An introduction to some of these ideas is given because they are simpler when applied to ordinary differential equations. They have the advantage that they can be used to produce either the entire solution accurately or just the long-term part of the solution. The idea is to treat the terms in the equations that cause the problem separately from the well-behaved terms. These are known as *splitting methods*. They cannot, however, be applied in every case. The method is best illustrated by an example.

Example 3.15

A very simple stiff problem is that of solving the pair of equations

$$\begin{aligned} y_1' &= -y_1 + 0.999 y_2 \\ y_2' &= \qquad -0.001 y_2 \end{aligned} \tag{3.9.11}$$

The matrix associated with Eq. 3.9.11 can be diagonalized and it is easy to separate the eigenvalues and their contributions and to verify that the general solution of this problem is

$$\begin{aligned} y_1 &= c_1 e^{-x} + c_2 e^{-0.001x} \\ y_2 &= \qquad\quad c_2 e^{-0.001x} \end{aligned} \tag{3.9.12}$$

The constants are determined by the initial conditions.

In this problem y_2 varies very slowly while y_1 has both slowly and rapidly varying components. We want to compute the slowly varying part of y_1 separately from the rapidly varying solution. To do this we note that the eigenvalues are $\lambda_1 = -1$ (the "fast" eigenvalue) and $\lambda_2 = 0.001$ (the "slow" eigenvalue) and that Eq. 3.9.11 can be written

$$\begin{aligned} y_1' &= \lambda_1 y_1 + (\lambda_2 - \lambda_1) y_2 \\ y_2' &= \qquad\quad \lambda_2 \quad y_2 \end{aligned} \tag{3.9.13}$$

The trick is to treat the terms involving λ_1 using a small time step while reserving a longer time step for the terms containing λ_2. Thus given the initial values $y_{1,0}$ and $y_{2,0}$, we compute

$$\tilde{y}_{1,n+1} = \tilde{y}_{1,n} + \lambda_1 h(\tilde{y}_{1,n} - y_{2,0}) \qquad (3.9.14a)$$

that is, we advance the solution using only those terms that depend on λ_1. We have used the Euler method as an illustration; the advantages of better methods are obvious. Note that the initial value of y_2 is used at every step and since y_2 is not being updated, it is the only value we can use. The step size h must be small enough that $|\lambda_1 h| < 2$. We continue using Eq. 3.9.14a until we have computed y_N. Generally, N should be of the order of the stiffness ratio (1000 in the problem above). Then we compute

$$y_{1,N} = \tilde{y}_{1,N} + \lambda_2 N h y_{2,0} \qquad (3.9.14b)$$

$$y_{2,N} = y_{2,0} + \lambda_2 N h y_{2,0} \qquad (3.9.14c)$$

in which we have applied the Euler method to the parts of the system of equations that involve the slow eigenvalue, using a step of Nh. The cycle is then repeated using the values $y_{1,N}$ and $y_{2,N}$ as the initial values.

This method can, in principle, be applied to any stiff problem. The method used above is designed to produce a solution that is accurate on both the short and long scales, in other words, an accurate solution to the entire problem. Should accuracy be necessary on only the long-term scale, one can use an implicit method with a large step size for the part of the solution related to the large eigenvalue. The method is useful when the equations be solved using the small time step are simpler than the full set of equations. In a number of important cases deriving from partial differential equations, this turns out to be the case and this approach is increasing in popularity. We should note in passing that this method is similar to analytical methods for deriving asymptotically valid solutions.

The method of splitting the system of equations can be difficult to find. In most cases there is more than one good way to do it. One method that is frequently successful is actually to carry out the diagonalization described earlier in this section (assuming there are not too many equations) and to set the small eigenvalues equal to zero. When the diagonalization is inverted, the equations that result are the ones that should be solved on the fast scale; the equation for the slowly varying solution can then be found by difference. It is also possible to construct methods using more than two scales.

Example 3.16

To see the problem that stiffness can cause, let us solve the system of equations 3.9.11. If $c_1 = 0$ and $c_2 = 1$, the initial conditions are $y_1 = y_2 = 1$ and the exact solution (Eq. 3.9.12) becomes $y_1 = y_2 = \exp(-0.001x)$, a slowly varying and well-behaved function. It should be computed well by any method as long as the step size is less than 100.

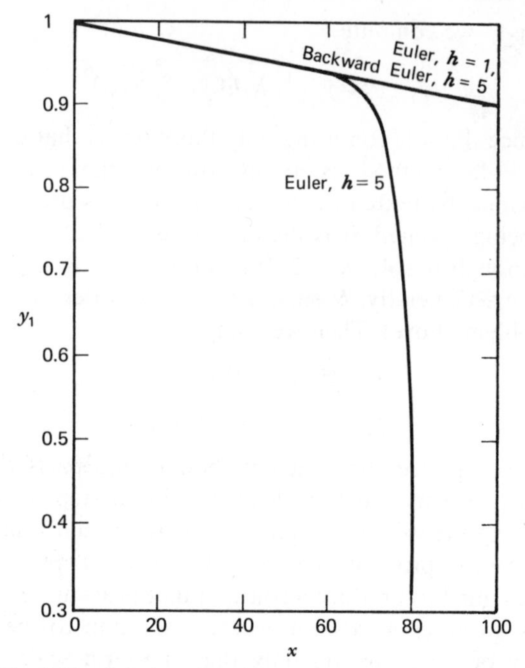

Fig. 3.19. Solution of a stiff problem using Euler methods.

Since the method used for demonstration is not important, we use the Euler method. Figure 3.19 shows the solution obtained with $h=1$. For this case the product of the step and the large eigenvalue is 1. The Euler method is not unstable and the solution obtained is almost exactly what we would have gotten if we were solving $y'=0.001y$ with the same method.

The same problem with $h=5$ is also shown in Fig. 3.19. The solution proceeds quite well until about $x=50$. Until this point the error increases linearly but is not obvious on the figure. In fact we probably would accept the solution up to $x=70$. After $x=70$, however, something is obviously very wrong and the difficulty cannot be overlooked.

What has happened is that since there was no initial error, it took a while for roundoff error to enter and be amplified. Had there been significant error in the initial condition, the instability would have occurred much sooner. Once the error becomes significant, it is multiplied by approximately -4 at each step and the results deteriorate rapidly. The greatest danger is that unless either the instability is so bad that it cannot be missed or the stiffness problem is recognized in advance, a bad solution may be accepted as accurate. It is also important to note that these results depend on the arithmetic accuracy of the machine used and on whether single or double precision is used. A more accurate machine (or double precision arithmetic) would delay the onset of trouble.

The solution of the same problem using the backward Euler method with $h=5$ is reasonably accurate and is about what we would have gotten if we had solved $y_1'=0.001y_1$ with this method. The relative error increases linearly and there is no indication of any problem arising from the stiffness. We thus see that implicit methods can compute effectively the slowly varying portion of the solution of a stiff problem.

The problem is a little different if the initial conditions are such that the rapidly varying component plays a role. Thus if in Eq. 3.9.12 $c_1=c_2=1$, the corresponding initial conditions are $y_1(0)=2$, $y_2(0)=1$ and

$$y_1 = e^{-x} + e^{-0.001x}$$

$$y_2 = \qquad e^{-0.001x}$$

Now the problem is that, in order to obtain the correct long-term solution, we are forced to compute accurately the decay of the fast part of the solution. To do this with a fixed step length method would require a small step size to produce an accurate solution. For the Euler method a step size of order 0.01 would be needed to yield at 1% error. Higher order methods would do the job with a much larger step but would still require many steps to compute to $x=10$.

It is for problems of this kind that variable step size and variable order methods really show their worth. They are able to take small steps in the initial stages where the rapidly decaying component demands it, but can take larger steps in the later stages where the slowly varying part of the solution dominates. The results of running this problem using ODE are shown in Fig. 3.20. The number of function evaluations for a given error is considerably greater than it was for the nonstiff problem, but it is also very much smaller than the number required by fixed step length methods.

The results obtained using RKF45 are also shown in Fig. 3.20. Note that the use of variable step size allows the program to reduce the error much faster as a function of the number of function evaluations than would a fourth order method with fixed step size. In fact for this problem the Runge–Kutta program RKF45 performs nearly as well as the higher order Adams program, ODE. Note that the behavior of the actual error is not always smooth as a function of the requested error and the number of function evaluations.

Example 3.17

To see what a program based on Gear's method for stiff systems can do, we have run both of the problems in the preceding example using a code named GEAR written by Hindmarsh of Lawrence Livermore Laboratory. The results are shown in Fig. 3.20 and are disappointing. The control strategy used in

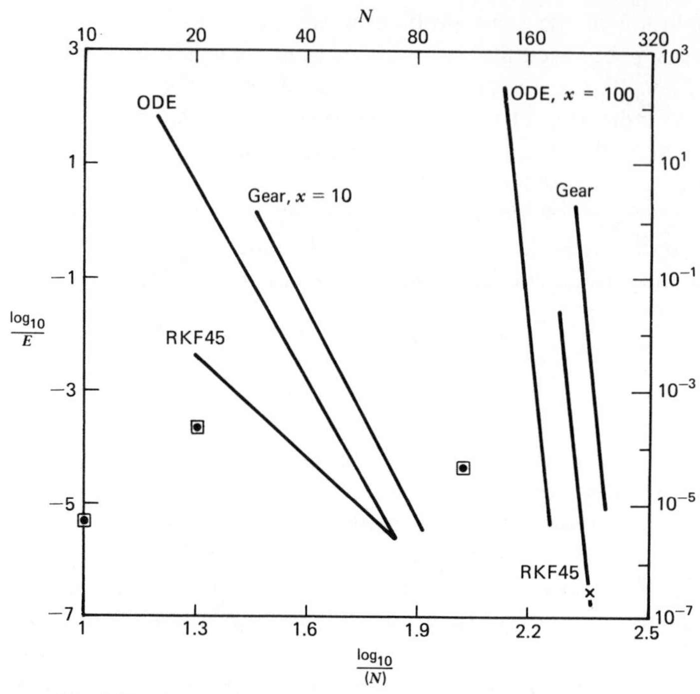

Fig. 3.20. Integration of a stiff problem using various methods.

GEAR is apparently not as good as that in the other programs; a wider range of tests is needed to draw definitive conclusions. As we have seen earlier, it is possible for a program to do an excellent job on one problem and do a poor job on another.

10. SYSTEMS OF EQUATIONS—INHERENT INSTABILITY

A computational difficulty that is sometimes overlooked is that higher order equations in general have many linearly independent solutions. It often happens that the desired solution is well behaved but the equation has other solutions that increase without bound in the direction of integration. Consider the equation that might arise in a heat transfer problem:

$$y'' - k^2 y = 0 \tag{3.10.1}$$

with boundary conditions $y(0) = y_0$ and $y'(0) = -ky_0$. The solution is, of course, $y = y_0 e^{-kx}$. The problem is that the differential equation has a second solution e^{+kx}, which is suppressed by the initial conditions. If this problem is solved numerically, part of the truncation and roundoff errors may be interpreted by the computer as being the undesired solution. Once this happens, the error will be exponentially amplified and the calculation is quickly reduced to rubble. Such a problem is called *inherently unstable*; none of the standard numerical methods discussed so far will produce the correct solution of it. In complicated equations this problem is not easy to recognize and sometimes is seen only after exhaustive searching for a method fails to produce a reasonable solution. In this case there are a couple of tricks that sometimes work:

1. Compute backward. If it is feasible to reverse the direction of integration, the growing solution becomes a decaying one and vice versa. This will work if it is possible to find reasonable starting conditions for the backward integration.
2. If the equation set is linear, start the computation with any arbitrary initial conditions; the computation will produce a growing solution. Supposing we call this solution $y_1(x)$, we can then assume that the desired solution has the form $y(x)=y_1(x)g(x)$ and find a lower order system of equations for $g(x)$. The resulting solution may be bounded; if it is not, this is an indication that the original system had more than one growing solution and the technique must be repeated. This method is called *filtering*. For nonlinear equations, the equations can be linearized locally and the method applied in a small interval.

11. BOUNDARY VALUE PROBLEMS: I. SHOOTING

Many, perhaps even a majority of the problems that arise in applications are not initial value problems but boundary value problems. In the latter, some data are given at two different values of the independent variable. Also, many problems arising from partial differential equations are similar to boundary value problems for ordinary differential equations. The results of the next few sections play an important role in Chapter 4.

There are two commonly used approaches to the numerical solution of boundary value problems. Each has its advantages and disadvantages and the choice is often more a matter of taste than of persuasive argument. The first approach is based on the use of methods for the initial value problem. Since the original application of this method was adjustment of artillery settings to hit a target, the method is known as *shooting*. The second approach, taken up in Section 12 and succeeding sections, deals with the entire problem directly by writing a finite difference approximation to the differential equations and boundary conditions. The resulting problem is then solved by means of standard algebraic equation solving routines.

A boundary value problem cannot arise from a first order equation. To illustrate the shooting method, therefore, we choose the simplest possible boundary value problem

$$y''=f(x, y, y') \qquad \begin{array}{l} y(0)=A \\ y(1)=B \end{array} \tag{3.11.1}$$

These are the simplest boundary conditions, but the technique applies equally

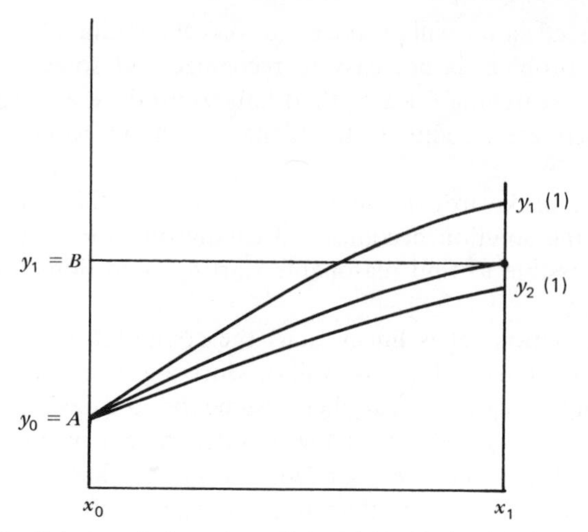

Fig. 3.21. Solution of a boundary value problem by the shooting method.

well to other cases. The idea is that if the first derivative at the initial point $y'(0)$ corresponding to the desired solution were somehow known, the problem could be reduced to an initial value problem. However, $y'(0)$ can be found only by solving the problem, suggesting an iterative procedure. We guess a value of $y'(0)$ and compute the solution of the resulting initial value problem to $x=1$. The value of $y(1)$ will not in general agree with the value B that the boundary condition demands. Consequently, we need to adjust the initial slope and try again. The process is repeated until the computed value at the final point agrees with the boundary condition. The process is illustrated in Fig. 3.21. The initial slope $y_1'(0)$ is too large and the solution overshoots the mark. Reducing the slope to $y_2'(0)$ undershoots the target, and on the third try an accurate result is achieved.

Shooting can be regarded as a procedure that defines a functional relationship between the endpoint value of the function [i.e., $y(1)$] and the initial slope [i.e., $y'(0)$] as shown in Fig. 3.22. Looked at in this way the problem becomes one of finding the root of an equation, and the differential equation solver is nothing more than a complicated algorithm for evaluating the functional relationship.

For a linear differential equation, shooting is especially simple. In this case we have the superposition principle. Any linear combination of solutions to the differential equation is also a solution. The functional relationship between $y'(0)$ and $y(1)$ becomes linear and the solution is straightforward. Suppose we compute two solutions to the differential equation $y_1(x)$ and $y_2(x)$. Both solutions are given the same initial value, in other words, $y_1(0)=y_2(0)=A$, but different initial slopes $y_1'(0)$ and $y_2'(0)$, respectively. Then, provided that

$$c_1 + c_2 = 1 \tag{3.11.2}$$

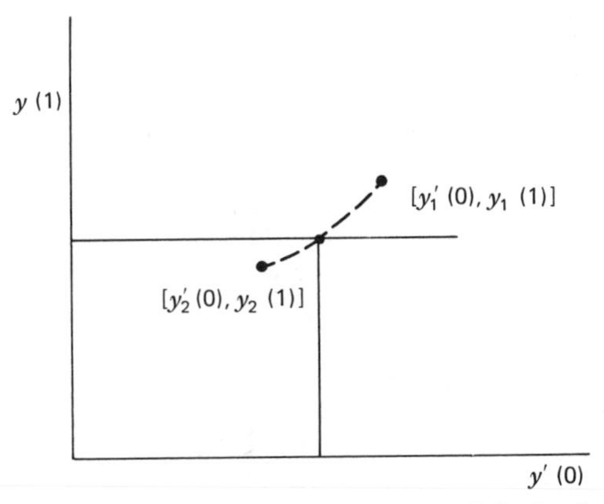

Fig. 3.22. Functional relationship between initial slope and final value used in the shooting method.

the linear combination

$$y(x)=c_1 y_1(x)+c_2 y_2(x) \tag{3.11.3}$$

is also a solution of the differential equation satisfying the initial condition. Its final value is

$$y(1)=c_1 y_1(1)+c_2 y_2(1) \tag{3.11.4}$$

and setting this equal to the desired final condition, that is, $y(1)=B$, and using Eq. 3.11.2, we have two linear algebraic equations to solve for the constants c_1 and c_2. Solving them we find

$$c_1=\frac{B-y_2(1)}{y_1(1)-y_2(1)} \qquad c_2=1-c_1 \tag{3.11.5}$$

With these values of the constants, Eq. 3.11.3 provides the solution to the problem.

This approach also works for higher order linear systems. If we had a higher order problem in which two conditions were given at the final point, it would be necessary to produce three solutions and take linear combinations to achieve the desired result. In general, if n conditions are specified at the final point, $n+1$ solutions must be generated. For this reason if the number of conditions at the two endpoints are unequal, it is best to start at the side on which the larger number of constants is given.

In the nonlinear case we must use an iterative technique. Furthermore, since the relationship described by Fig. 3.22 is implicit, the Newton–Raphson method cannot be used but use of the secant method is feasible. How this can be done is indicated in Fig. 3.23. The first guess produces a value of $y(1)$ larger

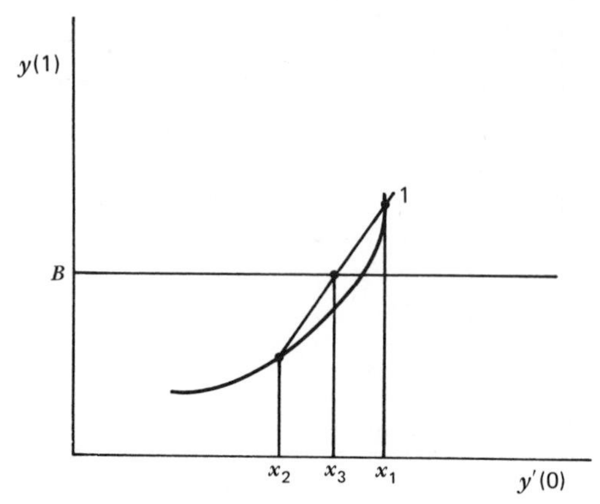

Fig. 3.23. Application of the secant method to the shooting method.

than B, while the second produces a value that is too low. Then approximating the relationship between $y(0)$ and $y(1)$ as the straight line joining the sets of values generated in the first two attempts, and noting the value of $y'(0)$ for which this straight line gives $y(1)=B$, we have the estimate of $y'(0)$ to use in the next attempt. Since the secant method is usually rapidly convergent, an accurate result normally is obtained in a reasonable number of iterations.

This method is easily extended to higher order systems in which several data elements are given at each of the endpoints. We regard all of the endpoint values required to fit the given data as functions of the missing initial data. Then the solution can be obtained iteratively. The difficulty is that the secant method involves quite a bit more computation for several equations than for a single equation.

Although the method is quite straightforward, there are many physical problems in which shooting is either unstable or slowly convergent. This happens when the final value of the function is sensitive to the initial slope. Generally, problems in which the differential equation is stiff, that is, ones in which there is the possibility of boundary layer behavior, give this type of trouble. In these cases a method that sometimes works is to shoot from both sides; in other words, one guesses information at both boundaries, solves both initial value problems, and tries to make them match at some intermediate point. For example, in a second order problem one would try to match the function and first derivative at midpoint.

Example 3.18

In many problems in fluid mechanics the mathematical difficulty is caused by the coefficient of the highest order derivative (usually the viscosity) being very small. In the limit as the coefficient goes to zero, the solution behaves irregularly. To illustrate the point, consider the problem

$$\varepsilon y'' + y' = 0$$

with the boundary conditions

$$y(0) = 0 \qquad y(x) \to 1 \quad \text{as } x \to \infty$$

This problem is one of the simplest that displays boundary layer-type behavior. The exact solution is

$$y(x) = 1 - e^{-x/\varepsilon}$$

This solution rises from $y=0$ at $x=0$ to $y=1$ in a distance of order ε. As $\varepsilon \to 0$, the solution approaches a step function. Consider what happens when we try to solve the problem with $\varepsilon = 0$. The differential equations reads $y' = 0$ and the solution $y = \text{constant}$ is capable of matching the boundary condition at infinity, but not the one at $x=0$.

This results from the differential equation being reduced from second to first order when ε is set to zero. As a result, the equation has only one solution and the ability to match both boundary conditions is lost. Problems with this property are called *singular perturbation problems* and have been extensively investigated. When one deals with them numerically, the equations are found to be stiff and the boundary value problems will be difficult to deal with. We now solve a singular perturbation problem by shooting. (This problem is used in Section 12 as well.)

Since this problem is so simple, we can solve it without the computer. If the problem is solved as an initial value problem with $y(0)=0$ and $y'(0)=A$, the

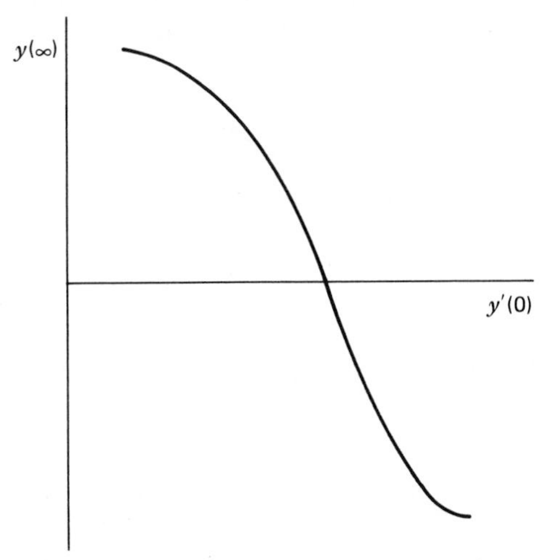

Fig. 3.24. Illustration of the dependence of $y(\infty)$ on $y'(0)$ that might arise in singular perturbation problems.

solution is

$$y(x)=A\varepsilon(1-e^{-x/\varepsilon})$$

Thus with a given value of A, the asymptotic value of y is $A\varepsilon$. The correct value of A is $1/\varepsilon$ and is very large. Because of this, it is easy to miss it by a wide margin. In this linear case nothing very serious will result, but in a nonlinear case the problem could behave in an unstable way. Figure 3.24 shows how the difficulty might arise. The sensitivity of $y(\infty)$ to $y'(0)$ produces a functional behavior that makes convergence slow and in some cases causes divergence.

12. BOUNDARY VALUE PROBLEMS: II. DIRECT METHODS

The problems with shooting make it advisable to have another method available. The most successful alternative is a "sledgehammer" approach. In this approach we make a finite difference or equivalent approximation to the differential equation at a number of mesh points, use the boundary data where appropriate, and solve the resulting set of algebraic equations. This may seem a difficult task, but the algebraic equations that result from approximating differential equations in this way always have simple structures and the method is quite competitive with the shooting method.

The method is illustrated by taking Eq. 3.11.1 as the example. Since there is no reason to do otherwise, we approximate this equation on a uniform grid of mesh points. The solution interval $(0,1)$ is broken into N equally spaced subintervals each of size h, so that $Nh=1$. The mesh points are thus $x_i=ih$, $i=0,1,\ldots N$. Again because there is no reason to do otherwise, we use central difference approximations. Thus y' at the ith mesh point is approximated by

$$y_i'=\frac{y_{i+1}-y_{i-1}}{2h} \tag{3.12.1}$$

and the second derivative is approximated by

$$y_i''=\frac{y_{i+1}-2y_i+y_{i-1}}{h^2} \tag{3.12.2}$$

Thus the differential equation in Eq. 3.11.1 at the ith mesh point can be approximated by

$$y_{i+1}-2y_i+y_{i-1}=h^2 f\left(x_i, y_i, \frac{(y_{i+1}-y_{i-1})}{2h}\right) \tag{3.12.3}$$

This approximation will be enforced at all of the interior mesh points x_i, $i=1,2,\ldots N-1$. At $i=1$, the first interior mesh point, y_0, wherever it occurs, can be replaced by the given boundary value A. In the same manner, we can substitute B for y_N where it occurs in the equation at mesh point $N-1$. We thus wind up with $N-1$ equations in $N-1$ unknowns. It is also important to note that finite difference equations inherit the character of the original equation. If the differential equation is linear, so are the finite difference equations; if the differential equation is homogeneous, the difference equations are homogeneous.

The set of equations 3.12.3 must be solved. We deal with the linear case first; not only are the equations in Eq. 3.12.3 linear, they are in fact tridiagonal. That is, the ith equation contains only the variables y_{i-1}, y_i, and y_{i+1}. A system of linear equations of this form is extremely easy to solve numerically and efficient algorithms and subroutines exist for this purpose (see Appendix B). In fact this approach is usually preferred over shooting because it is computationally more efficient and because the approximations used tend to distribute the error more evenly. The connection between tridiagonal systems of equations and differential equations is the major reason that efficient solvers for tridiagonal systems were developed.

In the nonlinear case we are faced with the problem of solving a coupled system of nonlinear algebraic equations. We are fortunate here, too. All of the standard methods for solving nonlinear systems use some kind of linearization and the linearized equations will always be tridiagonal and thus easy to solve.

It is necessary to deal with boundary conditions other than the simple ones in the problem above. Let us assume that a boundary condition involves the first derivative as well as the function. The best way of handling this is to write the difference equation 3.12.3 at the boundary point as well as at the interior points. This requires the introduction of an artificial point outside the boundary. For the left boundary we denote this point as $x_{-1}=-h$. For a boundary condition of the form

$$ay'(0)+by(0)=c \tag{3.12.4}$$

we can write a finite difference approximation

$$\frac{a(y_1-y_{-1})}{2h}+by_0=c \tag{3.12.5}$$

Equation 3.12.5 can be solved for y_{-1} and substituted into the finite difference equation at $x=x_0$. Carrying out a similar procedure at x_N, we obtain a set of $N+1$ equations in $N+1$ unknowns. Alternatively, we can regard y_{-1} and y_{N+1} as additional variables, include Eq. 3.12.5 and its counterpart at the right boundary of the equation set, and solve a set of $N+3$ equations in $N+3$ unknowns. The equations, however, are not quite tridiagonal.

There are many other methods of forming difference approximations and any of the formulas introduced in Section 1 can be used to do this. One popular method is to reduce the equation to a system of first order equations and finite difference those.

Example 3.19

For the first example we take a problem with few difficulties. In the problem of the neutron distribution in a material, the following equation arises:

$$\frac{d^2\phi}{dx^2}-\kappa^2\phi=\frac{S(x)}{D} \tag{3.12.6}$$

where ϕ is the neutron flux, κ is a nuclear property (inverse diffusion length), D the diffusion coefficient, and $S(x)$ represents a neutron source. For simplicity we take S to be constant. The boundary conditions are

$$-D\frac{d\phi}{dx}=\pm\lambda\phi \qquad x=\pm\frac{t}{2}$$

where t is the thickness of the slab and λ is a constant.

The problem can be simplified by introducing dimensionless variables

$$\alpha^2=k^2t^2 \qquad \beta=\frac{\lambda t}{D} \qquad \theta=\frac{\phi}{\phi_0} \qquad \xi=\frac{x}{t} \qquad Q=\frac{St^2}{D\phi_0}$$

where ϕ_0 is an arbitrary flux level, to give

$$\frac{d^2\theta}{d\xi^2}-\alpha^2\theta=Q \tag{3.12.7a}$$

$$\frac{d\theta}{d\xi}=\pm\beta\theta \text{ at } \xi=\pm\tfrac{1}{2} \tag{3.12.7b}$$

This problem can be solved analytically, so we can determine the error in the numerical solution directly.

Applying the second order central difference approximation gives

$$\theta_{j-1}-(2+\alpha^2\,\Delta\xi^2)\theta_j+\theta_{j+1}=-\Delta\xi^2Q \qquad j=0,1,\ldots N$$

where $x_j=jh-\frac{1}{2}$. At the boundaries $j=0$ and $j=N$ the boundary conditions give

$$\theta_1-\theta_{-1}=2\beta\Delta\xi\theta_0$$

$$\theta_{N+1}-\theta_{N-1}=-2\beta\Delta\xi\theta_N$$

These can be substituted into the difference equations for $j=0$ and $j=N$, respectively, to yield the tridiagonal system that is to be solved.

The program below illustrates the solution of this problem by use of the subroutine for the solution of tridiagonal systems given in Appendix B.

```
      PROGRAM BDYVAL
C-----THIS PROGRAM SOLVES THE PROBLEM OF NEUTRON TRANSPORT IN A
C-----FUEL ELEMENT.  SEE TEXT FOR DETAILS.
      DIMENSION THETA(51),A(51),B(51),C(51),R(51),X(51),EXACT(51)
C-----REQUEST INPUT
      WRITE (5,100)
  100 FORMAT ( ' TYPE ALPHA')
      READ (5,110) ALPHA
  110 FORMAT (F)
      WRITE (5,120)
  120 FORMAT ( ' TYPE BETA')
      READ (5,110) BETA
   10 WRITE (5,130)
  130 FORMAT ( ' TYPE THE NUMBER OF INTERVALS')
      READ (5,140) N
  140 FORMAT (I)
      NP1 = N + 1
      DX = 1./FLOAT(N)
C-----SET UP THE MATRIX FOR THE INTERIOR POINTS
      DO 1 J=1,NP1
      A(J) = 1.
      B(J) = -2. - ALPHA*ALPHA*DX*DX
      C(J) = 1.
      R(J) = -DX*DX
      X(J) = (J-1) * DX - .5
      EXACT(J) = -((BETA * COSH(ALPHA*X(J))) / (ALPHA * SINH(.5*ALPHA)
     1 + BETA * COSH(.5*ALPHA)) - 1.) / (ALPHA * ALPHA)
    1 CONTINUE
C-----NOW SET UP THE BOUNDARY CONDITIONS
      C(1) = A(1) + C(1)
      B(1) = B(1) - 2.* BETA * DX * A(1)
      A(NP1) = A(NP1) + C(NP1)
      B(NP1) = B(NP1) - 2.* BETA * DX * C(NP1)
C-----SOLVE THE PROBLEM USING THE TRIDIAGONAL SYSTEM SOLVER; SEE
C-----APPENDIX B FOR DETAILS.
      CALL TRDIAG (NP1,A,B,C,THETA,R)
C-----COMPUTE THE ERROR AND WRITE THE RESULTS
      WRITE (5,150)
  150 FORMAT ( 5H X(J),9X,5HEXACT,9X,5HTHETA,9X,5HERROR)
      DO 2 J=1,NP1
      ERROR = EXACT(J) - THETA(J)
      WRITE (5,160) X(J),EXACT(J),THETA(J),ERROR
  160 FORMAT ( 1F6.2,3X,3(E12.6,3X))
    2 CONTINUE
      WRITE (5,170)
  170 FORMAT ( ' ANOTHER CASE?, IF SO, TYPE 1')
      READ (5,140) IFF
      IF (IFF.EQ.1) GO TO 10
      STOP
      END
```

Table 3.8. Solution of a Boundary Value Problem by a Direct Method

N	$\Delta\xi$	Error	Error$/\Delta\xi^2$
2	0.5	2.76×10^{-3}	0.0110
4	0.25	7.03×10^{-4}	0.0112
6	0.167	3.14×10^{-4}	0.0113
8	0.125	1.77×10^{-4}	0.0113
10	0.100	1.13×10^{-4}	0.0113
16	0.0625	4.41×10^{-5}	0.0113
20	0.0500	2.82×10^{-5}	0.0113

The results are sufficiently close to the exact solution so that plotting them will not show very much. Instead we give the error at $\xi=0$ for various grid sizes with $\alpha=\beta=1$ in Table 3.8. The exact solution at $\xi=0$ is $\theta=0.393$, so using just two intervals produces an accuracy of better than 1%. As stated earlier, this is an easy problem. The last column in Table 3.8 clearly shows the second order nature of the method.

Example 3.20

A more difficult problem is the one used in the preceding section:

$$\varepsilon y''+y'=0 \qquad y(0)=0 \qquad y\rightarrow 1 \text{ as } x\rightarrow\infty$$

One difficulty is that one of the boundary conditions is imposed at infinity. There are several ways of getting around this, one of the easiest of which is to set $y=1$ at some large but finite value of x such as $x=b$. As long as $b\gg 1/\varepsilon$, there will be no problem. In more general cases, in which the behavior of the solution is not known a priori, one can apply the condition at some x, compute the solution, repeat the calculation with a larger final value of x, and see whether the solution changes by more than an acceptable amount.

The method used is essentially the one used for the preceding problem. Specifically, the difference equations are

$$(2\varepsilon-\Delta x)y_{i+1}-4\varepsilon y_i+(2\varepsilon+\Delta x)y_{i-1}=0 \qquad i=1,2,\ldots N-1$$

with $y_0=0$ and $y_N=1$.

Since the rapid rise of the solution occurs within a distance of order ε, the key parameter in determining the accuracy of the solution is $\Delta x/\varepsilon$; small values of it are needed for accuracy.

An example of how things go wrong is shown in Table 3.9. The boundary condition $y=1$ was applied at $x=10$ and we used $\Delta x=1$ and $\varepsilon=0.001$. It is obvious that the solution is complete nonsense. An explanation is that when $\Delta x/\varepsilon$ is large, the diagonal elements in the matrix A are very much smaller than the off-diagonal elements. Consequently, the odd-index variables, that is, y_1, y_3, and the like, are closely coupled as are the even-index variables. The two sets, however, are nearly uncoupled. When an odd number of points is used, the even-index variables are the only ones that are tightly coupled to the boundary conditions and they represent an approximation to the solution that is smooth but not accurate. The odd-index variables are not tied to anything definite and simply reflect the fact that the matrix is almost singular; in fact for odd N, the problem becomes singular as $\varepsilon\rightarrow 0$.

When the number of points is even, the matrix no longer becomes singular since $\varepsilon\rightarrow 0$, but the results are still not good. The even numbered points are

Table 3.9. Direct Solution of $\varepsilon y''+y'=0$ ($\Delta x=1$, $\varepsilon=0.001$)

x	y
1	50.9
2	0.20
3	50.7
4	0.40
5	50.5
6	0.60
7	50.3
8	0.80
9	50.1
10	1.00

coupled to the boundary condition at $x=0$ and since $\varepsilon \to 0$, $y_{2j} \to y_0$. The odd points are coupled to the condition at x_N and $y_{2j-1} \to y_N$.

The solution can be obtained more economically with variable mesh spacing. What is needed is close spacing of the points near $x=0$ where the solution is growing rapidly and wider spacing where the solution is changing more slowly. This is discussed in Section 14.

13. BOUNDARY VALUE PROBLEMS: III. HIGHER ORDER METHODS

For the initial value problem we found that high accuracy is more efficiently obtained by using high order methods than by using small steps. The same should be true for boundary value problems. It should be recalled, however, that most higher order methods for initial value problems are difficult to start. For the boundary value problem the analogous difficulty is that boundary conditions are difficult to deal with.

An example illustrates the problem. Suppose that we try to solve Eq. 3.11.1 with the fourth order central difference scheme given in Table 3.1. The resulting finite difference equations would then be

$$-y_{j-2}+16y_{j-1}+-30y_j+16y_{j+1}-y_{j+2}=12h^2f_j \qquad (3.13.1)$$

(Uniform spacing is used to avoid introducing too many complications at once and the issue of how f_j is evaluated is ignored for now.) Consider what happens near the left boundary. The value $y(0)=y_0=A$ is known and can be used whenever it occurs in the equation set. This quantity occurs in Eq. 3.13.1 with

$j=2$ and no problem results from substituting it into the equation. For $j=1$, however, there is a problem. The equation contains y_0, which causes no difficulty, but it also contains y_{-1}, which represents a value at a point not in the solution domain. There is no way to estimate this value without further approximation.

There are a number of ways of dealing with this problem. Most require that different difference formulas be used near the boundary than in the center of the region. A fourth order approximation to the second derivative at x_1 would include y_0, y_1, y_2, y_3, y_4, and y_5 in the formula, while the approximation used in Eq. 3.13.1 requires only y_0, y_1, y_2, y_3, and y_4 to estimate $y''(x_2)$. (The difference approximation used in Eq. 3.13.1 is fourth order because it is a central difference on a uniform grid. The symmetry causes the third order error terms to cancel out.) Using a difference approximation at x_1 that contains points beyond those included in the approximation at x_2, causes the results to be unreliable. It also violates our sense that the solution at x_1 should be less affected by what happens at x_5 than the solution at x_2. Hence this is not a good approach.

The best method, and the recommended one, is to use a difference approximation at x_1 that uses the values y_0, y_1, y_2, y_3, and y_4 and simply to take the highest order we can get. This leads to using the third order formula

$$y''(x_1) \simeq \frac{1}{12h^2}(11y_0 - 20y_1 + 6y_2 + 4y_3 - y_4) \tag{3.13.2a}$$

The analogous formula at the other end is

$$y''(x_{N-1}) \simeq \frac{1}{12h^2}(-y_{N-4} + 4y_{N-3} + 6y_{N-2} - 20y_{N-1} + 11y_N) \tag{3.13.2b}$$

The truncation error in these formulas is $-h^3 f^v/12$. Strictly speaking, using these formulas makes the approximation only third order accurate everywhere, although as we demonstrate in the example below, the method behaves more like a fourth order method than a third order one.

The system of equations 3.13.1, if we ignore the boundaries for the moment, is pentadiagonal—that is, the matrix of the problem has nonzero elements only on the main diagonal and two diagonals on either side of it. Solution of a pentadiagonal system can be achieved with a generalization of the method used for a tridiagonal system; however, the computational work required is almost three times as much. The boundary conditions of Eq. 3.13.2 introduce two extra elements in the first and last rows of the matrix and a modified pentadiagonal solver is required.

Example 3.21

The problem of Example 3.19 is solved using the method recommended above. In matrix form the problem is

$$\begin{bmatrix} (-20-12k^2h^2) & 6 & 4 & -1 & 0\ldots \\ 16 & (-30-12k^2h^2) & 16 & -1 & 0\ldots \\ -1 & 16 & (-30-12k^2h^2) & 16 & -1\ldots \\ \vdots & \vdots & \vdots & \vdots & \vdots \\ \vdots & \vdots & \vdots & \vdots & \vdots \end{bmatrix}$$

$$\times \begin{bmatrix} y_1 \\ y_2 \\ y_3 \\ \vdots \\ \vdots \end{bmatrix} = 12h^2 \begin{bmatrix} Q \\ Q \\ Q \\ \vdots \\ \vdots \end{bmatrix}$$

with extra elements in the first and last rows. The matrix is almost pentadiagonal and the last row is identical to the first row with the order of the elements reversed. The method is sufficiently accurate that double precision is required in order to compute the difference between this solution and the exact one. The error is shown in Fig. 3.25 in which this is designated method 1.

For comparison, another method in which the second derivative at x_1 is approximated by the second order central difference approximation

$$y''(x_1) \simeq \frac{1}{2h^2}(y_2 - 2y_1 + y_0) \tag{3.13.3}$$

is also tried. This produces a standard pentadiagonal problem. This is called method 2 in Fig. 3.25.

The solution and the error produced by each of these methods with 40 points is shown in Fig. 3.25. Method 1 produces an error that is larger near the boundary than in the center of the region. The error near the boundaries is positive, that is, tending to make the solution smaller than the true value, and is due to the approximation, Eq. 3.13.2, used at the boundary. The error in the center is negative and is due to the finite difference approximation, Eq. 3.13.1. As the result of the "conflict" between the errors produced near the boundary and near the center, the error in this method does not scale with a power of the mesh size.

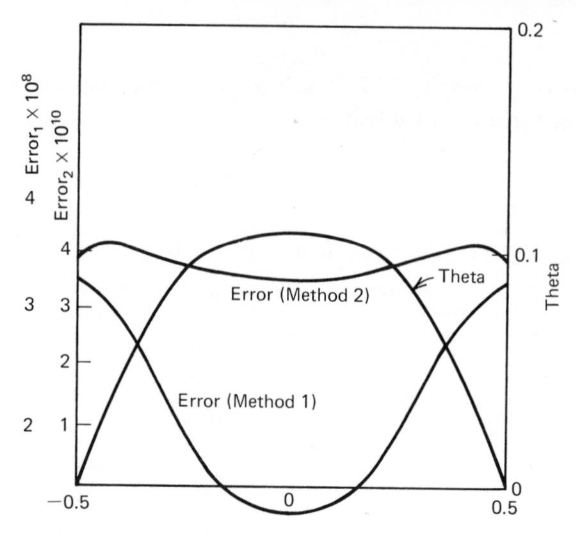

Fig. 3.25. Fourth order solutions of a boundary value problem.

The use of the approximation in Eq. 3.11.3 (Method 2) produces an error that for this problem is nearly constant across the region. As Table 3.10 shows, the error behaves in a fourth order manner but is larger than the error of Method 1. The last column in Table 3.10 relates to a method described below.

This example shows that multipoint difference formulas are an effective but somewhat difficult way to achieve accuracy in boundary value problems. The difficulty increases as still higher orders are demanded. An alternative approach is to seek methods that use only three points but which are more accurate than the standard formulas.

Table 3.10. Errors in Fourth Order Solutions of a Boundary Value Problem $(\varepsilon(0)/\Delta x^4)$

No. of Intervals	Method 1	Method 2	Method 3
4	0.00164	0.0773	−0.000426
6	0.00234	0.0797	−0.000426
8	0.00213	0.0806	−0.000427
10	0.00180	0.0813	−0.000427
20	0.00066	0.0828	−0.000427
30	0.00013	0.0834	−0.000427
40	−0.00016	0.0837	−0.000427
50	−0.00034	0.0839	−0.000427

One possibility comes from looking at the standard central difference approximation

$$\frac{y_{j+1}-2y_j+y_{j-1}}{h^2}=y''(x_j)+\frac{h^2}{12}y^{iv}(x_j) \tag{3.13.4}$$

On the other hand, using Taylor series we can show

$$\tfrac{1}{12}(y''_{j+1}+10y''_j+y''_{j-1})=y''(x_j)+\frac{h^2}{12}y^{iv}(x_j) \tag{3.13.5}$$

The right-hand sides of these equations are identical. This means that obtaining y''_j by solving the system of equations

$$\tfrac{1}{12}(y''_{j+1}+10y''_j+y''_{j-1})=\frac{y_{j+1}-2y_j+y_{j-1}}{h^2} \tag{3.13.6}$$

produces an approximation of fourth order accuracy.

In applying this method to the solution of differential equations, it is simplest to write Eq. 3.13.6 in matrix form

$$A\mathbf{y}''=B\mathbf{y} \tag{3.13.7}$$

where $\mathbf{y}''$ and $\mathbf{y}$ are vectors the components of which are the second derivatives and functions at the mesh points and A and B are tridiagonal matrices obtained by inspection of Eq. 3.13.6. Eq. 3.13.7 can be written

$$\mathbf{y}''=A^{-1}B\mathbf{y} \tag{3.13.8}$$

and the differential equation in Eq. 3.11.1 becomes

$$A^{-1}B\mathbf{y}=\mathbf{f} \tag{3.13.9}$$

Multiplying by A and writing the result in component form we have

$$y_{j+1}-2y_j+y_{j-1}=\frac{h^2}{12}(f_{j+1}+10f_j+f_{j-1}) \tag{3.13.10}$$

If f does not contain the first derivative (or if y' appears linearly), there is no problem. This approach is almost as inexpensive as the second order method and boundary conditions are no problem. This is known as the *compact fourth order method*.

Example 3.22

We now solve the problem of the preceding example using the compact fourth order scheme. The job is as easy as using a second order method; the only difference is in the matrix elements and the right-hand side. For this particular problem the right-hand side is in fact unchanged and the equations are

$$\left(\frac{1-\alpha^2h^2}{12}\right)\theta_{j-1}-\left(\frac{2+5\alpha^2h^2}{6}\right)\theta_j+\left(\frac{1-\alpha^2h^2}{12}\right)\theta_{j+1}=-Qh^2$$

Thus the problem can be solved by method used for the second order method but requires a double precision tridiagonal solver. As Table 3.10 shows, this method behaves in a very stable fourth order fashion and is considerably more accurate than either of the two earlier methods. Furthermore, for this problem the error is distributed much like the solution itself—that is, the relative error is approximately constant, a very desirable property.

This example shows the efficiency of the method but it is difficult to extend it to nonuniform grids. This is the subject of current research.

Another approach to the achievement of higher order methods is based on recognizing that Hermite interpolation is a high order interpolation based on the use of data from a small set of points. Using Hermite interpolation as the basis of a numerical method means that not only the dependent variable and its first derivative must be treated as unknowns. This doubles the number of equations, but the accuracy may be worth the extra cost required. These methods are also undergoing development at the present time.

14. BOUNDARY VALUE PROBLEMS: IV. NONUNIFORM GRIDS

As we demonstrated in the preceding section, higher order methods are difficult to apply to boundary value problems and the problem grows worse as the order is increased. An alternative is to use a nonuniform grid or, what is nearly the same thing, a transformation of the independent variable.

The concept is simple. In any finite difference approximation the local error is proportional to the product of a power of the step size and a derivative of the function. An optimum computation is one in which the error is more or less evenly distributed over the solution domain. Since we have no control over the derivatives and prefer not to use different approximations in different regions, the only way of equalizing the error is to vary the step size. We want the grid spacing to be small where the function is highly curved and large where it is smooth. In this way we should be able to obtain maximum accuracy for a given amount of computation, assuming that the cost of the computation is proportional to the number of mesh points. This concept is similar to the one used in developing adaptive quadrature methods but in the present case the grid has to be selected prior to the computation.

A point about which there has been some misunderstanding is that of choosing the difference approximation. Although our interest is in boundary

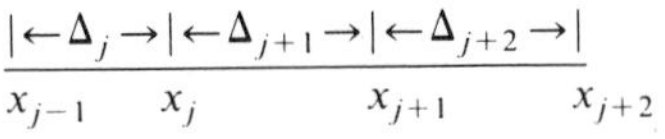

$$|\leftarrow\Delta_j\rightarrow|\leftarrow\Delta_{j+1}\rightarrow|\leftarrow\Delta_{j+2}\rightarrow|$$

x_{j-1} x_j x_{j+1} x_{j+2}

Fig. 3.26. A section of a nonuniform grid.

value problems for which the differential equations are always of order higher than one, we can learn something by looking at first derivatives. Specifically, let us consider some possibilities.

First, suppose that we use a nonuniform mesh, a section of which is shown in Fig. 3.26. The derivative at x_j can be approximated by the simple central difference formula

$$y_j' \simeq \frac{y_{j+1}-y_{j-1}}{\Delta_j+\Delta_{j+1}} \tag{3.14.1}$$

A formal Taylor series analysis of the right-hand side shows that

$$\frac{y_{j+1}-y_{j-1}}{\Delta_j+\Delta_{j+1}} \simeq y_j' + \left(\frac{\Delta_{j+1}-\Delta_j}{2}\right) y_j'' \tag{3.14.2}$$

and the method is formally only first order accurate. But if the mesh spacing changes slowly, $\Delta_{j+1}-\Delta_j$ is quite small and the error is more like that of a second order method than a first order one. A formally second order accurate approximation is:

$$\frac{(\Delta_j/\Delta_{j+1})(y_{j+1}-y_j)-(\Delta_{j+1}/\Delta_j)(y_{j-1}-y_j)}{\Delta_j+\Delta_{j+1}} = y_j' + \frac{\Delta_j\Delta_{j+1}}{6} y_j''' \tag{3.14.3}$$

The use of a nonuniform grid is equivalent to using a coordinate transformation. Suppose that the independent variable were transformed from x to

$$\xi = g(x) \tag{3.14.4}$$

Then

$$\frac{d}{dx} = \frac{d\xi}{dx}\frac{d}{d\xi} = g'\frac{d}{d\xi} \tag{3.14.5}$$

and a uniform grid in ξ corresponds to a nonuniform grid in x and vice versa. The situation is shown in Fig. 3.27. Then dy/dx at x_j can be approximated by

$$y'(x_j) = g'(x_j)\left[\frac{y(x_{j+1})-y(x_{j-1})}{2\,\Delta\xi}\right] \tag{3.14.6}$$

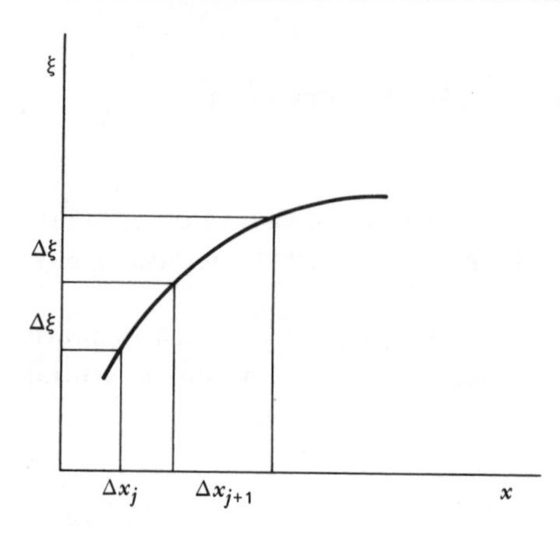

Fig. 3.27. Coordinate transformation used in a boundary value problem.

The term in brackets is an approximation to $dy/d\xi$, which has an error proportional to $\Delta\xi^2$, and the approximation is thus second order accurate in terms of ξ. Since the x and ξ coordinates system are equally valid, we can regard (3.14.6) as a second order approximation.

With the proper choice of transformation, it is possible to make the solution a very smooth function of the transformed coordinate. To go to extremes, the equation $dy/dx=f$ becomes $dy/d\xi=1$ if $g'=f$. Of course this transformation is possible only if the solution is known, but it shows that a good choice of transformation can make the problem much simpler.

In practice, the choice of a coordinate transformation is based on previous experience or, when available, on approximate solutions of the problem.

Example 3.23

In this example a variable grid is used to solve the problems of the earlier examples. In particular, we use a grid system that is often chosen—the so-called *compound interest grid* in which each interval is a constant multiple of the preceding one:

$$x_{j+1}-x_j=c(x_j-x_{j-1}) \tag{3.14.7}$$

With $c>1$, this gives more resolution at small x, while $c<1$ places more mesh points at larger values of x. For the second derivative the finite difference approximation 3.1.8 was used, while for the first derivative we used Eq. 3.14.1.

For the problem of Example 3.19, which has a smooth solution, the variable grid helps but only relatively little; we use the grid generated by Eq. 3.14.7 for $0<x<0.5$, while for $-0.5<x<0$ the reflection is used. We then vary the constant in Eq. 3.14.7 to minimize the mean square difference between the exact solution and the finite difference approximation. The best results are

Table 3.11. Variable Grid Solution of Equation 3.12.7

N	Grid Constant (c)	Root Mean Square Error	Uniform Grid Error
10	0.825	1.14×10^{-5}	5.5×10^{-5}
20	0.91	2.62×10^{-6}	1.57×10^{-5}
30	0.95	1.56×10^{-6}	6.96×10^{-6}

obtained with c slightly less than 1. The best value of c increases as the number of mesh points increases. As Table 3.11 shows, the variable grid reduces the error by a factor of 5 or 6. The compound interest grid is not optimal for this problem but it does yield improved results at little increase in cost.

A much more dramatic effect is found when one does the problem of Example 3.20. Here more resolution is clearly needed for small x so that $c>1$. Some results with $\varepsilon=0.01$ are shown in Fig. 3.28. For $c=1$ (the uniform grid) the solution displays the kind of spurious oscillations that we saw in Example 3.20, but they are less exaggerated here because this case is not quite as severe as the earlier one. It is severe enough, however, that $c=2$ is needed to give the best results, producing a highly distorted grid and the results shown in Fig. 3.28. The errors are too small to be shown clearly in the figure. In fact the maximum error is approximately 0.02 and occurs at $x=0.03$ in this case. The error reduction achieved by the nonuniform grid has turned a totally unacceptable solution into one that is probably sufficiently accurate for engineering purposes.

With more mesh points, a smaller value of c is needed to produce the optimum results, although we find that the improvement is about two orders of

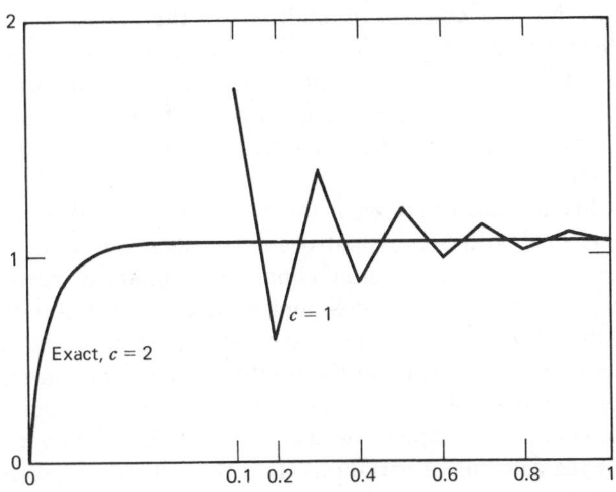

Fig. 3.28. Solution of $0.01y''+y'=0$ using a variable grid. Note the break in scale at $x=0.1$.

magnitude in each case. If a uniform grid is used, it requires 80 intervals to reduce the root mean square (r.m.s.) error to what was achieved with a nonuniform mesh of 10 intervals, and 500 intervals to reduce the maximum error to what the nonuniform 20-interval calculations produced.

The compound interest grid is well suited to this problem and the results are spectacular. The optimum grid is problem dependent and a little bit of foreknowledge about the solution can be an excellent aid in producing accurate results with relatively little effort. In the case of partial differential equations (for which the costs are much higher) the use of a well-fitted nonuniform grid may make the difference between a practical tool and a much-too-costly curiosity.

15. BOUNDARY VALUE PROBLEMS: V. FINITE ELEMENT METHODS

In Section 3.1c we found that it is possible to find approximations to differential equations by integrating them rather than by approximating the derivatives. When this technique is applied to partial differential equations, it is called the *finite volume method*. Many problems encountered in engineering are conservation laws expressed in the form of differential equations. The importance of the finite volume method lies in the fact that approximating the equations by integration makes it easier to assure that the integrated (or global) form of conservation laws is accurately reproduced.

Finite volume methods may in turn be regarded as the simplest of a more general set of methods called *finite element methods*. The nomenclature comes from the mechanics of solids for which the method was first developed. One version of the method consists of multiplying the equation by a "weight" function (which may be a known function or the solution itself) and integrating the result over a finite region of the solution domain. Both the solution and the weight function are then approximated by polynomials in order to make quadrature possible and one obtains a system of equations for the coefficients of the polynomials.

Although this approach may seem rather arbitrary, it takes on true significance when we deal with nondissipative linear systems. For such systems one can show that solving the differential equation (ordinary or partial) is equivalent to finding the minimum or maximum value of the integral of a quadratic function of the solution. In a typical case the integral might represent the potential energy of the system and the solution is the function that minimizes this quantity. An example of this method is the Lagrangian approach to classical mechanics. A complete development of this theory requires some knowledge of the calculus of variations. It is the author's experience that this requires nearly the equivalent of a separate course on the subject and entire books have been devoted to it. For this reason only a short introduction by means of an example is presented here.

Solving the differential equation

$$\frac{d^2y}{dx^2} - k^2y = f(x) \tag{3.15.1}$$

together with the boundary conditions

$$y'(0) = y'(1) = 0 \tag{3.15.2}$$

is equivalent to minimizing the quantity

$$F = \int_0^1 (y'^2 + k^2y^2 + 2yf)\, dx \tag{3.15.3}$$

To see this, substitute $y = \tilde{y} + \delta y$, where $\tilde{y}$ is the function for which F is minimal and δy is a small "variation," into Eq. 3.15.3. After neglecting terms quadratic in δy and integrating by parts to eliminate the derivative of δy, one finds that the functional F is minimized if $\tilde{y}$ is a solution to Eqs. 3.15.1 and 3.15.2. The reader unfamiliar with the calculus of variations is advised to try this.

Now we can attempt to solve the problem by finding the piecewise linear function, that is, the function of the type

$$y(x) = \frac{x - x_j}{x_{j+1} - x_j} y_{j+1} + \frac{x_{j+1} - x}{x_{j+1} - x_j} y_j \qquad x_j < x < x_{j+1} \tag{3.15.4}$$

That produces the smallest value of F. This is presumed to be the piecewise linear function, which gives the best approximation to the exact solution of the original problem. We also approximate f by a piecewise linear function. Computing the integrals we find

$$F \simeq \frac{1}{\Delta} \sum_{j=0}^{N} (y_{j+1} - y_j)^2 + \frac{k^2\Delta}{6} \sum_{j=0}^{N} \left(y_{j+1}^2 + y_j y_{j+1} + y_j^2 \right)$$

$$+ \frac{\Delta}{6} \sum_{j=0}^{N} (2y_{j+1} f_{j+1} + y_j f_{j+1} + y_{j+1} f_j + 2y_j f_j) \tag{3.15.5}$$

We have thus evaluated F for any piecewise linear function. Just as the exact solution of the differential equation is the function that minimizes the exact F (Eq. 3.15.3), it is reasonable to assume that the best piecewise linear approximation is the one that minimizes the approximate F (Eq. 3.15.5). This minimization is easily accomplished by differentiating with respect to each of the y_j. We then obtain the system of linear equations

$$(y_{j+1}-2y_j+y_{j-1})-\frac{k^2\Delta^2}{6}(y_{j+1}+4y_j+y_{j-1})$$
$$=\frac{\Delta^2}{6}(f_{j+1}+4f_j+f_{j-1})\qquad j=1,2,\ldots N-1 \tag{3.15.6}$$

For $j=0$ the terms involving y_{-1} and f_{-1} are missing and for $j=N$ the terms involving y_{N+1} and f_{N+1} are also missing. These equations are different from the ones obtained by the standard finite difference method but are similar to the equations obtained by methods described in Section 13. We can regard Eq. 3.15.6 as a finite difference approximation and it is formally second order accurate. In fact any approximation derived by finite element methods can be regarded as a finite difference method. The finite element approach, however, has produced new and valuable approximations that had not been considered earlier.

There are a great variety of finite element methods. Their principal advantages are that (1) the spacings of the grid points can be made arbitrary with no increase in complication, and (2) the accuracy of the approximation can be increased by using higher order polynomials without much difficulty other than an increase in the amount of calculation to be done. In dealing with partial differential equations in two and three dimensions, the first advantage becomes the very powerful one that the computational points need not be arranged on a rectangular grid (triangles are used quite commonly), allowing the method to deal with irregular boundaries very effectively, and the second advantage is retained as well. The major disadvantages of finite element methods are that they require a bit more work to derive and the systems of equations are not quite as regular as those produced by finite differences.

These advantages have made the finite element method extremely popular over the last decade. The power of the method shows itself best, however, in systems for which the solution is equivalent to maximizing (or minimizing) a functional, and it has become the preferred method for such problems. The method may be used for other systems (for example, dissipative systems) and it is often quite powerful in these applications. Whether its advantages are worth the difficulties is a matter of debate at the present time.

16. BOUNDARY VALUE PROBLEMS: VI. EIGENVALUE PROBLEMS

Eigenvalue problems are simply the linear homogeneous versions of the problems treated in the preceding sections and in principle can be treated by same methods. We discuss them separately only because we wish to derive some specific results. Experience shows that the solutions of eigenvalue problems are very sensitive to small changes when the parameter is close to the exact eigenvalue. For this reason the shooting method is not well suited to eigenvalue computation and the direct method is almost always used. We illustrate the main ideas by a particular case.

Again we take the simplest problem of the class as an example. The classic eigenvalue problem in differential equations is the one associated with free vibrations:

$$y''+k^2y=0 \tag{3.16.1}$$

with the boundary conditions

$$y(0)=y(1)=0 \tag{3.16.2}$$

The solution to this problem is well known. There are no nontrivial solutions unless k is one of the eigenvalues:

$$k_n=n\pi \qquad n=1,2,\ldots \tag{3.16.3}$$

and the corresponding eigenfunctions are

$$y_n(x)=\sin n\pi x \tag{3.16.4}$$

To obtain a finite difference approximation to this problem, we use the second order central difference approximation to the second derivative. At the intermediate points, the finite difference equations take the form (after a little rearrangement)

$$y_{j-1}+(k^2h^2-2)y_j+y_{j+1}=0 \qquad i=1,2,3\cdots N \tag{3.16.5}$$

At the points x_1 and x_N, the equations are identical to these except that the terms involving y_0 and y_{N+1}, respectively, are missing. This set of equations can be written in matrix form as

$$A\mathbf{y}-\lambda\mathbf{y}=(A-\lambda I)\mathbf{y}=\mathbf{0} \tag{3.16.6}$$

where A is the tridiagonal matrix

$$A=\begin{pmatrix} -2 & 1 & 0 & & \cdots & 0 \\ 1 & -2 & 1 & & \cdots & 0 \\ & & & & & \\ & & & 1 & -2 & 1 \\ & & & & 1 & -2 \end{pmatrix}=\mathrm{Tr}(1,-2,1) \tag{3.16.7}$$

and $\lambda=-k^2h^2$.

Equation 3.16.6 is the standard eigenvalue problem of linear algebra. Thus the finite difference version of the eigenvalue problem for a differential equation becomes the eigenvalue problem of linear algebra. This is not surprising, but it is the major reason why the eigenvalue problem is always treated by direct finite differencing and not by shooting. Other methods of approximating the differential equation, for example, finite elements, will also lead to algebraic eigenvalue problems but the matrix will be different.

Eigenvalue problems arising from second order equations always lead to tridiagonal matrix eigenvalue problems when second order differences are used. In this form the problem is ideally set up for the standard method of solving matrix eigenvalue problems—the tridiagonal symmetric QR algorithm. This routine is quite fast, is easy to use, and can produce eigenvectors as well as eigenvalues. Subroutines embodying the method are available at most computer centers. One that is in common use is SYMEIG.

In the particular example we have chosen it is actually possible to solve the matrix eigenvalue problem in closed form! This is an unusual situation, but this solution proves valuable later. The clue is provided by the solution of the differential equation. It suggests that we try

$$y_j = \sin j\alpha \tag{3.16.8}$$

as a solution to the difference equation. The cosine could also be used, but it will not satisfy the boundary condition as is pointed out shortly. Plugging Eq. 3.16.8 into Eq. 3.16.5 for $j=2,3,\ldots N-1$ and using the trigonometric identity

$$\sin(j\pm 1)\alpha = \sin j\alpha \cos\alpha \pm \cos j\alpha \sin\alpha \tag{3.16.9}$$

we get

$$(2\cos\alpha - 2 - \lambda)\sin j\alpha = 0 \tag{3.16.10}$$

so that 3.16.8 is the solution, provided the term in parentheses is zero. Since $\lambda = -k^2h^2$, this is equivalent to

$$k^2 = \frac{2(1-\cos\alpha)}{h^2} \tag{3.16.11}$$

and we have the solution if we can find the allowed values of α. Now we note that the equations in Eq. 3.16.5 for $j=1$ and $j=N$ are the same as the other equations if we agree to set $y_0 = y_{N+1} = 0$. The condition $y_0 = 0$ is satisfied by Eq. 3.16.8 and is the reason for choosing it.

To cause $y_{N+1} = 0$ we must have

$$\sin(N+1)\alpha = 0$$

The solutions to this equation are $(N+1)\alpha=\pi,2\pi,\ldots$ and thus

$$\alpha=\frac{\pi}{N+1},\frac{2\pi}{N+1},\ldots\frac{N\pi}{N+1},\ldots \tag{3.16.12}$$

It looks as if there are an infinite number of eigenvalues. This is impossible for a finite matrix. The difficulty is resolved by noting that $\alpha_k=k\pi/(N+1)$ and if we replace k by $2(N+1)-k$ we get:

$$y_j=\sin[2(N+1)-k]\pi j/[N+1]=-\sin k\pi j/(N+1) \tag{3.16.13}$$

which is just a negative of the solution belonging to α and is not, therefore, linearly independent. Thus only the first N of the values listed in Eq. 3.16.12 can be accepted and the eigenvalues to the finite difference equation of Eq. 3.16.5 are

$$k^2=\frac{2}{h^2}(1-\cos n\pi/N) \qquad n=1,2,\ldots n-1 \tag{3.16.14}$$

and it is interesting to compare them with the exact eigenvalues $k=n\pi$. This is done in Fig. 3.29, in which we have treated n as a continuous variable for display purposes. We see that for small n the exact and numerically derived eigenvalues agree quite well, but after $n\simeq N/2$ the two diverge considerably.

The ratio of the largest to the smallest eigenvalue of the finite difference equations is $\lambda_N/\lambda_1\simeq 4N^2/\pi^2$. This means that a system of differential equations in which the matrix A occurs is likely to be quite stiff. Unfortunately, this is precisely what happens when we deal with parabolic partial differential equations and is a major factor in choosing methods for solving such problems.

The situation can be improved somewhat by using higher order difference formulas. The methods and ideas of Section 3.13 apply here almost without change.

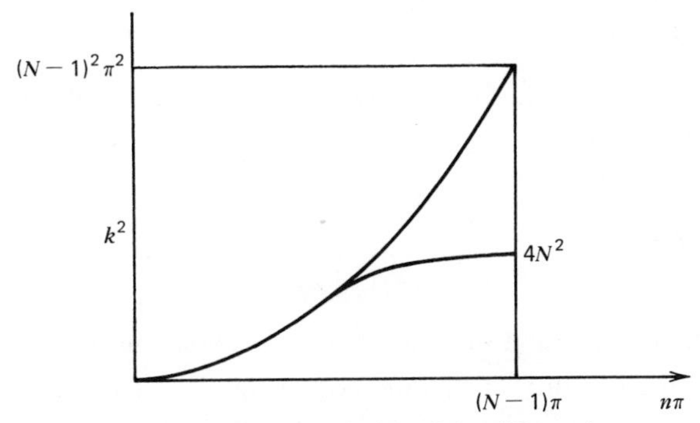

Fig. 3.29. Eigenvalues obtained by finite differencing.

Table 3.12. Computation of Eigenvalues of $y''+k^2y=0$

Index	Exact Eigenvalue	Computed Eigenvalues $N=5$	$N=7$	$N=10$
1	1	0.9836	0.9916	0.9959
2	2	1.8710	1.9335	1.9672
3	3	2.5718	2.7785	2.8902
4	4	3.0273	3.3218	3.7420
5	5		3.6090	4.5016
6	6		2.7951	5.1436
7	7			5.6723
8	8			6.0546
9	9			6.2878

Example 3.24

Typical numerical results are given in Table 3.12. The lowest eigenvalues are computed relatively accurately and the accuracy improves as the number of points is increased. For the approximation used here the errors are proportional to h^2 or $1/N^2$. As the number of points is increased, the number of eigenvalues is increased, but the "new" eigenvalues are poor approximations.

Problems

1. The numerical method

$$y_{n+1}=y_n+\Delta x[\beta f(x_n, y_n)+(1-\beta)f(x_{n+1}, y_{n+1})]$$

contains as special cases Euler ($\beta=1$), backward Euler ($\beta=0$), and trapezoid ($\beta=\frac{1}{2}$). Apply the method to the test equation $y'=\alpha y$.

a. For what values of β is this method unconditionally stable?

b. Find the truncation error as a function of β. For which β is the error smallest?

2. The following method was suggested as a means of solving ordinary differential equations:

$$y^*=y_n+\frac{h}{2}y_n'$$

$$y_{n+1}=y_n+hy^{*\prime}$$

a. Describe this method in as many ways as you can.

b. Discuss the stability of this method using $y'=\alpha y$ *for real* α *only*. *Hint*: Graph the function that determines the stability.

c. Would you recommend this method?

3. The method based on Simpson's rule:

$$y_{n+1}=y_{n-1}+\frac{\Delta x}{6}\left[f(x_{n-1}, y_{n-1})+4f(x_n, y_n)+f(x_{n+1}, y_{n+1})\right]$$

is used as the final corrector in some Runge-Kutta methods.

a. Classify this method in as many ways as you can, for example, accuracy (order), explicit/implicit, multipoint.
b. Apply this method to $y'=\alpha y$ and give an expression for

$$y_{n+1}/y_n$$

c. Investigate the stability of this method for
(i) α pure real
(ii) α pure imaginary
d. Would you recommend using this method and, if so, for what purposes?

4. Instead of using a differential equation solver (such as the Euler method) as the predictor in a predictor-corrector scheme, it has been suggested that we could find a predicted value by ordinary extrapolation (e.g., Lagrange) and then correct it using a standard implicit scheme. One of the simplest methods of this type would use the values of the solution at x_{n-1} and x_n to "predict" the value at x_{n+1} via linear extrapolation. We could then correct using the trapezoid rule.

a. Write the equations describing this method. Assume equal intervals $(x_{n+1}-x_n=x_n-x_{n-1})$ here and throughout this problem.
b. How accurate is this method (i.e., what order)?
c. How stable is it? (You may guess, or rely on the results in the text. If you work it out, do so only for the case of $y'=\alpha y$ and α real only).
d. What is your opinion of this method?

5. A dog chasing a rabbit runs so that he is always moving in the direction of the rabbit. If the rabbit runs at a speed of 15 m/sec and stays in a straight line (this problem is not very realistic) and the dog runs at 20 m/sec, how long does it take the dog to catch the rabbit if they are initially 150 m apart in the direction perpendicular to the rabbit's path?

Derive the differential equation for the dog's path and solve it numerically. The equations are singular at the instant of catch (so is what happens to the rabbit) and you will need to be careful.
Note: An analytical solution to this problem exists. Hotshots may want to look for it.

6. Two satellites are in circular Earth orbit. The manned one is at an altitude of 350 km and the one it wants to dock with is at 380 km. The astronaut wants to dock with the unmanned satellite by firing his engine only once; for purposes of this problem, assume that the rocket produces an

instantaneous change of speed without change of direction. If the target satellite is initially 2° of angle ahead of the astronaut's vehicle, by how much does the astronaut have to change his velocity to accomplish his mission? Try solving this problem by shooting. Assume that the Earth is spherical; the acceleration due to gravity is 9.8 m/sec^2 at the Earth's surface and the Earth's radius is 6370 km.

7. A simple predator-prey model is often used to simulate biological populations. One of the simplest such models is

$$\frac{dx}{dt}=\alpha x-\beta xy$$

$$\frac{dy}{dt}=-\gamma y+\eta xy$$

The behavior of the system depends on the set of constants chosen and the initial conditions. Assume $\alpha=1$, $\beta=0.01$, $\gamma=1$ and $\eta=0.001$. There are lots of interesting solutions to this system. Solve the system with the following initial conditions and discuss the behavior found.

a. $x=100$, $y=1$

b. $x=1100$, $y=120$

c. $x=20$, $y=150$

8. The differential equation describing ice build-up (on a lake for example) is given by:

$$R^2\frac{d^2R}{d\tau^2}+2\left(\frac{dR}{d\tau}\right)^2+4L^*\left[\frac{1+R}{2+R}\right]\frac{dR}{d\tau}-\frac{4L^*}{2+R}=0$$

where R is the thermal resistance of the ice slab (thickness divided by thermal conductivity), L^* is a nondimensionalized latent heat for which 0.5 is a reasonable value and τ is a nondimensional time. Appropriate initial conditions are $R(0)=0.01$, $R'(0)=\frac{1}{2}$. Solve the problem for all values of τ until $R=2$.

9. Consider the system of ordinary differential equations:

$$\frac{dx}{dt}=998x+1998y$$

$$\frac{dy}{dt}=-999x-1999y$$

a. Find the solution of this system of equations with the initial conditions $x(0)=y(0)=1$.

b. Would you anticipate any difficulty in finding the solution of this problem numerically?

c. Suggest a method (including a recommended value of Δt) which you would use to solve this problem numerically.

d. Advance the solution from $t=0$ to $t=2\Delta t$ using the method suggested in part (c) to show that your method works.

10. The Doppler broadened line shape occurs in problems in nuclear reactors, plasma physics, and astrophysics, among others. It can be represented by an integral but it is simpler to find it by solving the ordinary differential equation.

$$\frac{\partial^2\psi}{\partial x^2}+\xi^2 x\frac{\partial\psi}{\partial x}+\left[\frac{\xi^4}{4}(1+x^2)+\frac{\xi^2}{2}\right]\psi=\frac{-\xi^4}{4}$$

Appropriate initial conditions are:

$$\psi(\xi,0)=\sqrt{\pi}\left(\frac{\xi}{2}\right)\exp\left(\frac{\xi^2}{4}\right)\operatorname{erfc}\left(\frac{\xi}{2}\right)$$

$$\frac{\partial\psi}{\partial x}(\xi,0)=0$$

erfc is the complementary error function $(1-\operatorname{erf})$. Use this to generate $\psi(1,x)$.

11. The differential equation

$$\varepsilon y''+y'+y=0 \qquad y(0)=0,\ y'(0)=1$$

is stiff if ε is small. Solve it numerically until $x=2$ with 1% accuracy for $\varepsilon=0.001$.

12. Solve the problem

$$y''-k^2y=1$$

$$y(0)=0 \qquad y(1)=0$$

using three points (two boundaries and one at $x=\xi$). The idea is to see the effect of a non-uniform grid on the error. Use the two approximations:

$$y_1''=\frac{y_{j-1}-2y_j+y_{j+1}}{(\Delta_j+\Delta_{j+1})^2/4}$$

$$y_1''=\frac{2y_{j-1}}{\Delta_j(\Delta_j+\Delta_{j+1})}-\frac{2y_j}{\Delta_j\Delta_{j+1}}+\frac{2y_{j+1}}{\Delta_{j+1}(\Delta_j+\Delta_{j+1})}$$

And plot the *error* in $y(\xi)$ (the numerical result minus the exact result) as a function of ξ for $0\leqslant\xi\leqslant 0.5$. Which formula is better and why?

13. The temperature distribution in a nuclear reactor fuel element is described by the differential equation

$$k\left(\frac{d^2T}{dr^2}+\frac{1}{r}\frac{dT}{dr}\right)=-S(r)$$

where S is the energy source. Using $S(r)/k=1+0.5(r/R)^2$ and the boundary conditions $T'(0)=0$, $T(R)=T_0$, solve for $T(r)$. Work with a nondimensional temperature defined by $(T(r)-T_0)/T_0$.

14. The velocity model of a laminar hydrodynamic boundary layer is related to the solution of the ordinary differential equation

$$\frac{d^3f}{d\eta^3}+\tfrac{1}{2}f\frac{d^2f}{d\eta^2}=0$$

with the boundary conditions:

$$f(0)=1 \qquad f'(0)=0$$

$$f'(\eta)\to 1 \qquad \eta\to\infty$$

(f is the dimensionless stream function and the velocity u is proportional to $f'(\eta)$.) Solve this problem for $f(\eta)$ and plot the velocity profile.

4

Partial Differential Equations

Partial differential equations (PDEs) occur in almost all areas of science and engineering. In fact many ordinary differential equations (ODEs) actually arise from partial differential equations; only when approximations are introduced does the problem reduce to one involving ODEs.

As we have done in the preceding chapters, we begin by classifying the problems. Most of the classifications that were given in the Chapter 3 for ODEs apply to PDEs as well. Thus PDEs can be of first or higher order; there may be a single equation or a system of equations; they may be linear or nonlinear; and they may be homogeneous or inhomogeneous. These classifications are as important in the PDE case as they were in the ODE case. Much of what was said in the preceding chapter applies to PDEs as well as ODEs and need not be repeated here.

The distinction between initial and boundary value problems that is so important in the numerical approach to the solution of ODEs is more complicated for PDEs. The reason is that there are now two or more independent variables, so it is possible for a problem to be an initial value problem with respect to one variable and, simultaneously, a boundary value problem with respect to another variable. It is possible to have pure initial value problems, pure boundary value problems, and mixed initial-boundary value problems. The question of what appropriate initial and/or boundary conditions are for a given equation is intimately tied to the type of equation, so a short discussion of the classification of PDEs is given.

First order partial differential equations occur only occasionally in physics and engineering problems. Systems of first order equations are more common, but they can be discussed along with higher order equations. We need only note in passing that almost all first order equations have real characteristics (see below) and thus behave very much like hyperbolic equations of higher order. In fact the standard method for solving such equations (both analytically and numerically) is the method of characteristics (see Section 20).

Many PDEs occurring in applications are second order in at least one of the independent variables. Fourth order equations arise in the mechanics of solid bodies (and elsewhere). These are usually of a fairly simple type and frequently can be treated by methods similar to those that are applied to elliptic second order equations below. Thus we concentrate on second order equations and say only a bit about the others.

The classification scheme for second order PDEs depends on the nature of their characteristics, so we should review some of their properties.

In the case of two independent variables, characteristics are lines in the plane of the independent variables (in higher dimensions the characteristics become surfaces or hypersurfaces) along which "signals" can propagate. Thus they might represent the position of a wave front in space as a function of time. They are also the locations of possible discontinuities in the solution of the equation. These properties lead to the fact that along a characteristic, the PDE takes on a particularly simple form, and in two dimensions it reduces to an ordinary differential equation. The presence of real characteristics in a problem means that there are particular directions in which signals or information can flow and knowing this is often a valuable aid in finding the solution.

The most general second order PDE in two independent variables that is linear in the highest derivatives is

$$a\phi_{xx}+b\phi_{xy}+c\phi_{yy}=f \tag{4.1}$$

where a, b, c, and f may all be functions of x, y, ϕ, ϕ_x, and ϕ_y. The classification of this equation depends on the sign of b^2-4ac. If $b^2-4ac>0$, the equation is called *hyperbolic*; if $b^2-4ac=0$, the equation is *parabolic*; if $b^2-4ac<0$, the equation is *elliptic*. All three cases are common in physical problems. Equations with more than two independent variables may not fit as neatly into this classification scheme.

A hyperbolic equation possesses two families of real characteristics. Physical systems that are governed by hyperbolic equations are ones in which signals propagate at finite speed or over a finite region. The situation is illustrated in Fig. 4.1. The lines of α=constant and β=constant represent the two families of characteristics along which signals can propagate. An observer at point P can "feel" the effects of what has happened in the horizontally crosshatched region, but disturbances outside this region cannot be felt. This region is known, therefore, as the *domain of dependence* of the point P. Similarly, a disturbance created at point P can be felt only in the vertically crosshatched region known as the *domain of influence*. It is important that numerical methods for solving hyperbolic equations recognize the equally important fact that these domains are finite.

The most common hyperbolic equation is the wave equation

$$\phi_{tt}-c^2\phi_{xx}=0 \tag{4.2}$$

Many of its close relatives are also quite common. We thus use the terms "space" and "time" coordinates even though there are cases of interest in which all the coordinates are in fact spatial.

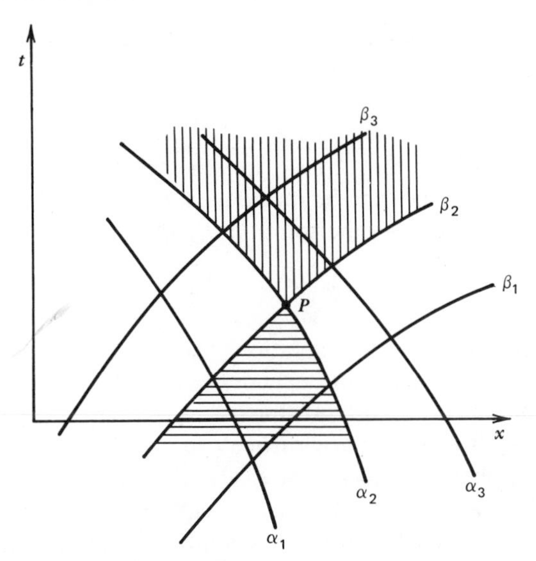

Fig. 4.1. The domains of dependence and influence for a hyperbolic equation. The lines are the characteristics and their slopes dx/dt are the speeds of propagation of signals.

Hyperbolic equations are always posed in domains that extend to infinity in the timelike coordinate and are thus open in this direction. The spatial coordinate may or may not be bounded. In either case one typically specifies *two initial* conditions at $t=0$. If the spatial region is bounded, there may also be boundary conditions (one at each boundary); otherwise, we have a pure initial value problem. For a more general treatment of the initial and boundary conditions that one is allowed to apply the reader is referred to such standard books on the subject as Garabedian (1964) and Courant and Hilbert (1962).

Parabolic equations can be regarded as the limit of hyperbolic equations in which the propagation speed of the signals goes to infinity. Physically, this might arise when the speed of light (or sound) is so large compared to any other parameter in the problem that it may be set to infinity (for example, the incompressible limit in gas dynamics). With respect to Fig. 4.1, this means that all characteristic lines become horizontal so that there is really only one independent family of characteristics. Consequently, the behavior of the solution at the point P is influenced by everything that has happened earlier and the numerical method should reflect this.

The most common parabolic equation is the heat equation

$$\phi_t = D\phi_{xx} \tag{4.3}$$

and it has many relatives. The initial and boundary conditions that are typically applied to parabolic equations are similar to those for hyperbolic problems with one important exception. Since Eq. 4.3 is only first order in time, only one initial condition at $t=0$ is necessary. This is also related to the fact that there is only one set of characteristics. The domain of solution is open in the time dimension and the spatial domain may be open or closed. In the case of the closed spatial domain, boundary conditions are essentially the same as those applied in the hyperbolic case.

Finally, we come to the elliptic case, in which there are no real characteristics. Complex characteristics can be introduced, leading to the well-known and important connection between the solutions of Laplace's equation

$$\phi_{xx} + \phi_{yy} = 0 \tag{4.4}$$

which is the best known elliptic equation, and complex variables. In elliptic problems there are no domains of dependence or influence and every point in the solution domain is affected by disturbances at every other point. This fact makes elliptic problems particularly difficult to solve.

It is possible to show (although not easily) that the proper domain for the solution of an elliptic equation is normally a closed region. Furthermore, it can be shown that necessary and sufficient boundary conditions are provided by giving one datum (the value of the dependent variable, its normal derivative, or some linear combination of the two) at each point on the surface. Laplace's equation describes the diffusion of heat in a conducting medium. It is obvious that the temperature distribution on a flat plate of any shape should depend only on the temperature distribution around the periphery or on the heat flux distribution, but not both. In fact the usual problem is that one of these is given and the other is to be calculated.

In this chapter parabolic equations are treated first because the methods used for solving these are closely related to the methods used for ODEs. Elliptic equations are usually solved by iterative procedures that are closely related to the methods used for parabolic equations, so they are treated next. Finally, hyperbolic equations, for which some of the methods (especially the method of characteristics) rely heavily on the properties of the PDE, are discussed.

1. PARABOLIC PDEs: I. EXPLICIT METHODS

Parabolic partial differential equations behave more like the initial value problem for ordinary differential equations than either elliptic or hyperbolic equations. With respect to the time variable, the diffusion or heat equation

$$\frac{\partial \phi(x,t)}{\partial t} = D\frac{\partial^2 \phi}{\partial x^2} \tag{4.1.1}$$

is first order, requires only a single initial condition, and is usually posed for all $t>0$. Equation 4.1.1 is used as an example of a parabolic PDE, but there are many others that arise in applications, including some important nonlinear ones. Fortunately, much of what we learn about the heat equation applies to these other cases, so it is sensible to begin here. The case in which there is more than one spatial independent variable is discussed later.

Let us suppose that the problem is to solve Eq. 4.1.1 with the initial condition

$$\phi(x,0)=f(x) \tag{4.1.2}$$

and the boundary conditions

$$\phi(0,t)=\phi(1,t)=0 \tag{4.1.3}$$

To get an idea of how things proceed, suppose we approximate the second spatial derivative with the second order central difference formula using a uniform grid. At grid point $x_j = jh$, the PDE is thus approximated by

$$\frac{d\phi_j(t)}{dt}=\frac{D}{h^2}(\phi_{j+1}-2\phi_j+\phi_{j-1}) \qquad j=1,2,\ldots N-1 \tag{4.1.4}$$

where $\phi_j(t) \equiv \phi(x_j, t)$. We see that the PDE is approximated by a system of ODEs. An approach such as this that uses finite differences in only one direction is called a *semi-discrete method*. The set of equations in Eq. 4.1.4 can be written in matrix-vector form as

$$\frac{d\boldsymbol{\phi}}{dt}=\frac{D}{h^2}A\boldsymbol{\phi} \tag{4.1.5}$$

where the matrix A is the tridiagonal matrix $Tr(1,-2,1)$ that was introduced in Section 3.15. We derived the eigenvalues and eigenvectors of this matrix, so we can diagonalize Eq. 4.1.5. It is interesting to do so only for pedagogical reasons because in practice diagonalization is a difficult procedure. Rather than using the standard linear algebra approach, we use a formal method more common in differential equation theory. Let the eigenvectors and eigenvalues of A be defined by

$$A\boldsymbol{\psi}^{(k)}=\lambda_k\boldsymbol{\psi}^{(k)} \qquad k=1,2,\ldots N-1 \tag{4.1.6}$$

Explicit expressions for these are given in Section 3.15. Since A is symmetric, its eigenvectors form a complete set and any vector of length $N-1$ can be represented as a linear combination of these eigenvectors. In particular,

$$\boldsymbol{\phi}=\sum_{k=1}^{N-1} c_k\boldsymbol{\psi}^{(k)} \tag{4.1.7}$$

Also, the eigenvectors are orthogonal; that is,

$$\sum_{j=1}^{N-1} \psi_j^{(k)} \psi_j^{(k')} = \delta_{kk'} \tag{4.1.8}$$

Now we substitute the expansion 4.1.7 (which is a discrete version of a Fourier series) into the finite difference form of the partial differential equation 4.1.5 and take the scalar product of the result with $\psi^{(l)}$, another eigenvector of A. Taking advantage of the orthogonality property, we find an ordinary differential equation for the Fourier coefficient c_l

$$\frac{dc_l}{dt} = \frac{D\lambda_l}{h^2} c_l \tag{4.1.9}$$

which has the solution $c_l = c_{lo} \exp(D\lambda_l t/h^2)$. The solution of Eq. 4.1.5 can thus be written

$$\phi(t) = \sum_k c_{ko} e^{D\lambda_k t/h^2} \psi^{(k)} \tag{4.1.10}$$

and we need only to find the constants c_{ko} to complete the solution. This can be done by applying the initial condition Eq. 4.1.2:

$$\phi_j(0) = \sum_k c_{ko} \psi_j^{(k)} = f(x_j) \tag{4.1.11}$$

The constants c_{ko} can be found by again using the orthogonality (Eq. 4.1.8) of the eigenvectors:

$$c_{ko} = \sum_j f(x_j) \psi_j^{(k)} \tag{4.1.12}$$

which completes the formal solution of the problem. This is only a formal solution because the cost of the computation required to carry out the indicated steps is prohibitive, except in a case as simple as the example we have chosen. This solution is useful for studying the behavior of numerical methods, however.

Now note that the eigenvalues λ_k are all negative and that they increase in magnitude as the index k increases. Thus although the initial conditions may produce constants c_{ko} that are appreciable, the terms in Eq. 4.1.10 with large values of k decay rapidly and, after some time, the solution will contain only the low k components. In most cases the long-term solution is all we want to compute; however, in some cases it is important to compute the precise short-term behavior of the system. The smallest eigenvalue λ_1, which is approximately $-\pi^2 h^2 \approx -\pi^2/N^2$, governs the long-term behavior of the solution. The largest eigenvalue λ_{N-1} is approximately -4. The ratio of these two eigenvalues is approximately $(2N/\pi)^2$, so that for large N the system of equations in Eqs. 4.1.4 or 4.1.9 is quite stiff. This situation is not at all unusual; *almost all parabolic PDEs are equivalent to stiff systems of ODEs*. This is the key point.

It should be obvious that explicit methods have difficulty with this problem. To make the point more strongly, suppose we try to solve the diffusion equation by the Euler method:

$$\phi_j^{n+1} = \phi_j^n + \frac{D\Delta t}{h^2}\left(\phi_{j+1}^n - 2\phi_j^n + \phi_{j-1}^n\right). \tag{4.1.13}$$

From the eigenvalues given above and what we know about the Euler method, we can guess that this method will be unstable unless $|\lambda_{max}|\, D\Delta t/h^2 < 2$ or since $|\lambda_{max}| = 4$,

$$\frac{D\Delta t}{h^2} < \frac{1}{2} \tag{4.1.14}$$

The fact that the stability depends on $D\Delta t/h^2$ could have been anticipated by dimensional analysis, and this will be true for any method. It also means that the time step allowed by stability is proportional to the square of the spatial mesh size used. Improvement of the spatial resolution will cost dearly because the number of steps required to reach a given time will increase enormously. This is a consequence of the stiffness problem discussed above.

It is clear that the Euler method, or any other explicit method, is not well suited to the solution of parabolic PDEs. One should use implicit methods, except possibly when the time accuracy necessary requires small time steps. We should also note that $|\lambda_{max}| = 4$ is a consequence of the spatial differencing method used. If we were able to compute $\partial^2\phi/\partial x^2$ accurately (by Fourier methods for example), λ_{max} would be π^2. The $\frac{1}{2}$ in Eq. 4.1.14 would be replaced by $2/\pi^2$ and the stability limitation on the time step would be two-and-a-half times tighter. Thus the stability criterion depends on both time and space differencing methods.

The stability condition is related to the largest eigenvalue of matrix A. The corresponding eigenvector is the one whose value changes sign between each pair of mesh points. It is this component that is most amplified when a calculation is unstable. Consequently, in an unstable calculation the solution oscillates spatially, and the presence of a rapid spatial oscillation is the best indication of instability.

An alternative approach to studying the stability numerical methods for PDEs was developed by von Neumann. This method is similar to the one used above, but it is a little simpler to apply in many cases. In the von Neumann method one ignores the boundary conditions, seeks a solution of the form $\phi(x,t) = e^{ikx}f(t)$, and substitutes this into the discretized partial differential equation. One then has an ODE for $f(t)$ containing k as a parameter. The stability criterion is determined by choosing the value of k that gives the strictest stability limit. For the heat equation the von Neumann approach applied to Eq. 4.1.4 leads to

$$\frac{df}{dt} = \frac{2\,D\Delta t}{h^2}(\cos kh - 1)f \tag{4.1.15}$$

The maximum absolute value of $\cos kh - 1$ is 2 and the stability result for Euler's method for ODEs gives Eq. 4.1.14.

We now turn to the equation of accuracy. From the results of Chapter 3 we have

$$\frac{\phi_{j+1}^n - 2\phi_j^n + \phi_{j-1}^n}{h^2} \simeq \frac{\partial^2 \phi_j^n}{\partial x^2} + \frac{h^2}{12}\frac{\partial^4 \phi_j^n}{\partial x^4} \tag{4.1.16}$$

and

$$\frac{\phi_j^{n+1} - \phi_j^n}{\Delta t} \simeq \frac{\partial \phi_j^n}{\partial t} + \frac{\Delta t}{2}\frac{\partial^2 \phi_j^n}{\partial t^2} \tag{4.1.17}$$

Note the sub- and superscripts carefully. These equations are based on Taylor series and when there are two or more independent variables, terms involving the derivatives in both directions occur. The error terms depend on the point about which one chooses to make the Taylor series [(x_j, t_n) in the above case]. All terms in the finite difference approximation must be expanded in Taylor series about the same point.

An interpretation of these results is that approximating the heat equation by using Euler's method in time and second order central differences in space is equivalent to replacing the original PDE by

$$\frac{\partial \phi}{\partial t} - D\frac{\partial^2 \phi}{\partial x^2} = \frac{Dh^2}{12}\frac{\partial^4 \phi}{\partial x^4} - \frac{\Delta t}{2}\frac{\partial^2 \phi}{\partial t^2} \tag{4.1.18}$$

This shows that the error is proportional to h^2 and Δt. Thus this method is first order in time and second order in space. It is typical that finite difference approximations to PDEs cannot be characterized by a single order parameter. Equation 4.1.18 is called the *modified equation*; it is essentially the PDE that the numerical method is solving and depends on both the original PDE and on the method used. Modified equations are interesting in their own right and they have been used to investigate stability.

The right-hand side of Eq. 4.1.18 represents the truncation error of the method. Note that $\partial^2\phi/\partial t^2$ can be evaluated by differentiating Eq. 4.1.1 with respect to time. The right-hand side of Eq. 4.1.18 can be rewritten

$$\left(\frac{Dh^2}{12} - \frac{D^2\,\Delta t}{2}\right)\frac{\partial^4 \phi}{\partial x^4} \tag{4.1.19}$$

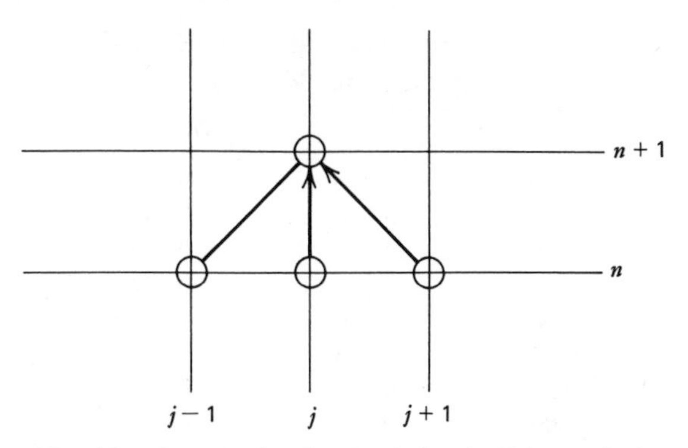

Fig. 4.2. Computational molecule for the Euler method.

which can be made zero by choosing $D\Delta t = h^2/6$. This is a fortuitous circumstance that is possible only for simple equations, but if this choice is made, the method becomes second order accurate in time and fourth order in space—these are the magnitudes of the next most significant error terms.

Generally, the time and space differencing errors cannot be made to cancel and, in fact, it is not always known in advance whether they reinforce or cancel each other. In such cases the best approach is to try to make the error arising from time differencing equal to the error arising from the space differencing. If one of the errors were much larger than the other, we could use a larger step size in the direction that gives the smaller error contribution without increasing the total error much and the computational cost can be reduced with little loss of accuracy. It is usually best to keep the errors in approximate balance.

Another way to describe numerical methods is by drawing a diagram (*computational molecule*) that displays the mesh points contributing to the difference formula, as shown in Fig. 4.2 for the Euler method applied to the heat equation. The value at time t_{n+1} at point x_j depends on the values of the function at earlier time t_n, but not on any of the values at the current time. This is the behavior we expect of hyperbolic equations, not parabolic ones. This can also be seen from Eq. 4.1.18. The presence of a second time derivative (albeit with a small coefficient) makes the modified equation hyperbolic.

Example 4.1

As our example, we take the heat equation (Eq. 4.1.1) with $D = 1$. For the initial condition we take

$$\phi(x,0) = 1 \qquad 0 \leqslant x \leqslant 1$$

and the boundary conditions are

$$\phi(0,t) = \phi(1,t) = 0$$

Physically, this problem represents the cooling of a heated slab of material

Table 4.1. Solution of the Heat Equation by the Euler Method

Δt	β	$\phi(0.5, 0.1)$
0.0005	0.05	0.4732
0.001	0.1	0.4721
0.002	0.2	0.4698
0.005	0.5	0.4744
0.00526	0.526	0.3766
0.00556	0.556	0.9779
0.00588	0.588	−0.2109
0.00667	0.667	−36.0
0.01	1.	1101

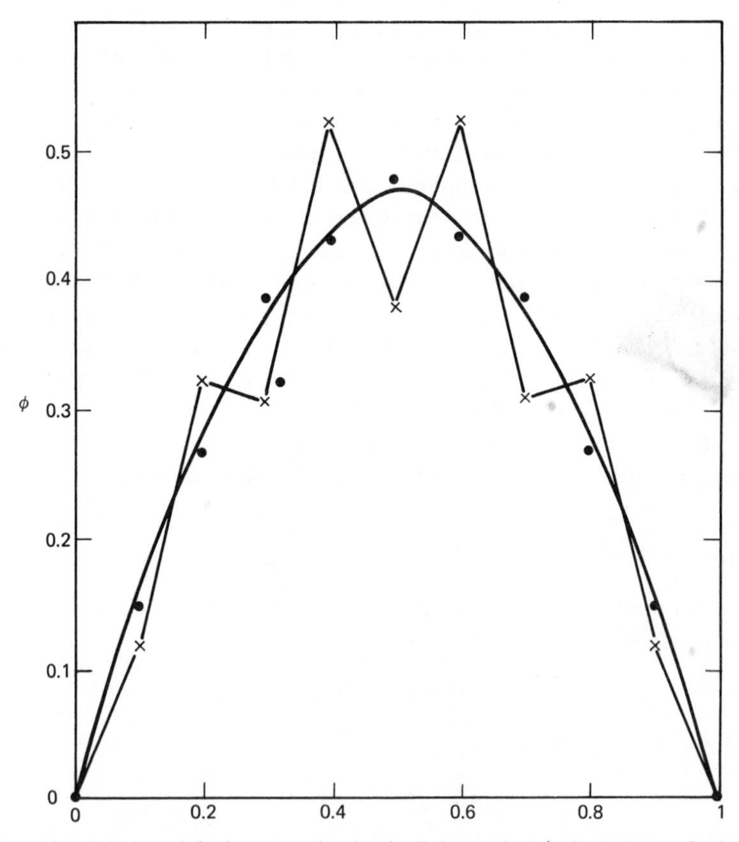

Fig. 4.3. Solution of the heat equation by the Euler method ($\bullet \beta=0.5000$; $\times \beta=0.5263$).

such as might occur in a metallurgical heat treatment process. The interesting result from an engineering point of view is the rate of cooling of the center of the slab. The problem of heating a cold slab is also of considerable interest (for example, in predicting the time to cook a roast).

We finite difference this problem using $\Delta x = 0.1$ and will vary Δt. The solution we obtain even for the smallest Δt will be slightly inaccurate due to the space differencing error.

The program is given below. As was the case in the preceding chapters the program is interactive and requests the data needed. Also, a major part of the calculation has been put into a subroutine so that the same program can be used in conjunction with subroutines given in the following sections, provided only that the subroutine name in the CALL statement is changed.

Table 4.1 gives the value of the centerline "temperature" as a function of $\beta = D\Delta t/h^2 = 100\,\Delta t$. For $\beta < 0.5$, the stability limit, there is not much change in the value computed. What happens is that since the exact solution at $t = 1.0$ is quite smooth, only the first few eigenvectors contribute much to it. For these, reasonable accuracy (better than 1%) is obtained for any time step within the stability limit. As soon as the stability limit is exceeded, the solution goes totally berserk.

An even better illustration of this is provided in Fig. 4.3. With $\beta = 0.5000$, the stability limit, a reasonably good result is obtained. Note, however, that at the stability limit, the component of the solution that changes sign at every mesh point is neutrally stable. As a result, the solution oscillates about the smooth curve (but only slightly). Increasing β to 0.5263, in other words, using 19 time steps instead of 20, produces the jagged curve shown in Fig. 4.3 and the instability is obvious. The amplification of the solution component that oscillates at every mesh point is obvious. Further increase in β produces results that are absolutely awful, as can be seen in Table 4.1.

Example 4.2

Meshes with variable grid size are frequently used when the solution is known to vary more rapidly in one region than another (see Section 3.15). In order to see the effect of a variable mesh on the behavior of the method, we take the simplest possible case: the problem of Example 4.1 with only one interior point that is not centered. Using the central difference formula of Eq. 3.1.8 and the boundary conditions $\phi_0 = \phi_2 = 0$, we find that the problem reduces to a single ODE

$$\frac{d\phi_1}{dt} = -\frac{2D}{h_1 h_2}\phi_1$$

where $h_1 = x_1 - x_0$ and $h_2 = x_2 - x_1$. From the results of Chapter 3, we know that the stability condition for Euler's method will be

$$\frac{D\Delta t}{h_1 h_2} < 1$$

```
      PROGRAM HEATEQ
C-----THIS PROGRAM SOLVES THE HEAT EQUATION IN ONE SPATIAL
C-----DIMENSION.  ZERO TEMPERATURE BOUNDARY CONDITIONS ARE
C-----USED.  THE ACTUAL TIME ADVANCEMENT OF THE SOLUTION
C-----IS DONE IN A SUBROUTINE SO THAT ANY OF A VARIETY OF
C-----METHODS CAN BE USED.  MOST OF THE PARAMETERS ARE
C-----REQUESTED BY THE PROGRAM.
      DIMENSION F(51)
C-----COLLECT THE DATA.
      WRITE (5,100)
  100 FORMAT ( ' GIVE THE DIFFUSION COEFFICIENT')
      READ (5,110) D
  110 FORMAT (F)
      WRITE (5,120)
  120 FORMAT ( ' GIVE THE WIDTH OF THE SLAB')
      READ (5,110) XL
      WRITE (5,130)
  130 FORMAT ( ' GIVE THE FINAL TIME')
      READ (5,110) TF
   10 WRITE (5,140)
  140 FORMAT ( ' GIVE THE NUMBER OF SPATIAL INTERVALS YOU WANT')
      READ (5,150) N
  150 FORMAT (I)
      WRITE (5,160)
  160 FORMAT ( ' GIVE THE TIME STEP')
      READ (5,110) DT
      WRITE (5,170)
  170 FORMAT ( ' RESULTS WILL BE PRINTED EVERY K TIME STEPS, GIVE K')
      READ (5,150) K
C-----CHOOSE YOUR METHOD.
      WRITE (5,175)
  175 FORMAT ( ' YOU HAVE A CHOICE OF METHODS OF SOLUTION.  TYPE A 1'
     1 ,' IF YOU WANT THE EXPLICIT EULER METHOD, A 2 IF YOU WANT THE'
     2 ,'DUFORT-FRANKEL METHOD, OR A 3 IF YOU WANT THE CRANK-NICOLSON'
     3 ,'METHOD.  OTHERS MAY BE ADDED LATER')
      READ (5,110) IMETH
C---- NOW COMPUTE THE IMPORTANT PARAMETERS.
      DX = XL/N
      TIME = 0.
      BETA = D * DT / (DX * DX)
      NM1 = N - 1
C-----SET THE INITIAL CONDITIONS.
      DO 1 I=1,NM1
      F(I) = 1.
    1 CONTINUE
      WRITE (5,180) D,DT,DX,BETA
  180 FORMAT ( 5H D = ,F8.4,2X,10HDELTA T = ,F8.4,2X,10HDELTA X = ,
     1 F8.4,2X,7HBETA = ,F8.4)
C-----WE NOW CALL THE SUBROUTINE TO ADVANCE THE SOLUTION K STEPS.
      IDUF = 1
   20 IF (IMETH.EQ.1) GO TO 30
      IF (IMETH.EQ.2) GO TO 40
      CALL CRANK (N,F,BETA,K)
      GO TO 50
   30 CALL PEULER (N,F,BETA,K)
      GO TO 50
   40 CALL DUFORT (N,F,BETA,K,IDUF)
C-----PRINT THE RESULTS.
   50 TIME = TIME + K * DT
      WRITE (5,190) TIME,(F(I), I=1,NM1)
  190 FORMAT ( F8.4,4X,51(E12.6,2X))
      IF (TIME.LT.TF) GO TO 20
C-----REQUEST ANOTHER CASE
      WRITE (5,200)
  200 FORMAT ( ' ANOTHER CASE?  IF SO, TYPE 1')
      READ (5,150) IFF
      IF (IFF.EQ.1) GO TO 10
      STOP
      END
```

```
      SUBROUTINE PEULER (N, F, BETA, NSTEP)
C-----THIS SUBROUTINE ADVANCES THE SOLUTION OF THE HEAT EQUATION
C-----BY NSTEP TIME STEPS USING THE EULER METHOD FOR PARABOLIC
C-----EQUATIONS.   THE VARIABLES ARE:
C-----N = THE NUMBER OF INTERVALS
C-----F = THE DEPENDENT VARIABLE (TEMPERATURE)
C-----BETA = D*DT/(DX*DX), COMPUTED IN THE CALLING ROUTINE
      DIMENSION F(51),T(51)
      NM1 = N - 1
      NM2 = N - 2
      DO 1 J = 1,NSTEP
C-----A TEMPORARY ARRAY IS NEEDED TO AVOID USING NEW VALUE OF F(I-1)
      DO 2 I = 2,NM2
      T(I) = F(I) + BETA * (F(I-1) - 2.*F(I) + F(I+1))
    2 CONTINUE
      T(1) = F(1) + BETA * (-2.*F(1) + F(2))
      T(NM1) = F(NM1) + BETA * (-2.*F(NM1) + F(NM2))
C---- WRITE THE NEW RESULTS INTO THE ORIGINAL ARRAY.
      DO 3 I=1,N
      F(I) = T(I)
    3 CONTINUE
    1 CONTINUE
      RETURN
      END
```

for this case. Since $h_1+h_2=1$, $h_1h_2=h_1(1-h_1)$, the stability parameter is smallest when $h_1=h_2$ and grows as h_1 becomes either smaller or larger than h_2. Thus the method becomes less stable as the mesh variation increases. Note that with $h_1=h_2$ this stability criterion is stricter than Eq. 4.1.14; the difference is due to the fact that Eq. 4.1.14 applies when the number of points N is large.

Problems with many nonuniformly spaced mesh points are obviously difficult to treat analytically. We can be sure, however, that the greater the variation in mesh size, the stiffer the problem and the more stringent the stability condition. The rule of thumb that one should always choose parameters so that $D\Delta t/h_{\min}^2 < c$ where c is a constant dependent on the method ($\frac{1}{2}$ for the Euler method) may be stricter than necessary, but it is not far from the actual limit and is on the safe side. This makes use of explicit methods in problems with variable meshes an even poorer choice than they are for fixed mesh problems.

2. PARABOLIC PDEs: II. THE CRANK–NICOLSON METHOD

An obvious cure for the stiffness problem of the preceding section is to use an implicit method. An additional factor that must be taken into consideration in choosing a method is that at each time step, the solution is not a single number as it was in the ODE case. Instead, it is the set of values of the solution at all the spatial mesh points. In PDEs with one spatial variable the problem is not very severe. In two or three dimensions, however, we need to keep large arrays representing the solution and the amount of data can easily exceed the memory capacity of even a large computer. Consequently, it is important that the method does not require too much data storage. From a practical point of view, this means that high order multistep methods must usually be ruled out.

A reasonable approach is to use one of the simpler implicit methods. A good choice, because it is second order accurate, is the trapezoid rule. Applying it to the diffusion equation Eq. 4.1.1 and again using second order central differences for the spatial derivative, we come to the formula

$$\left(\phi_j^{n+1}-\phi_j^n\right)=\left(\frac{D\Delta t}{2h^2}\right)\left[\left(\phi_{j+1}^n-2\phi_j^n+\phi_{j-1}^n\right)+\left(\phi_{j+1}^{n+1}-2\phi_j^{n+1}+\phi_{j-1}^{n+1}\right)\right] \tag{4.2.1}$$

which, with the definition $\beta=D\Delta t/h^2$, can be rewritten

$$-\beta\phi_{j+1}^{n+1}+2(1+\beta)\phi_j^{n+1}-\beta\phi_{j-1}^{n+1}=\beta\phi_{j+1}^n+2(1-\beta)\phi_j^n+\beta\phi_{j-1}^n \tag{4.2.2}$$

Thus finding the solution at time step $n+1$ requires the solution of a set of linear algebraic equations. Since the system is tridiagonal, the cost per time step for this method works out to be approximately double that of the Euler method. Moreover, this method, which is known as the *Crank–Nicolson method* when applied to PDEs, is unconditionally stable. The combination of accuracy and unconditional stability allows one to use a much larger time step with the Crank–Nicolson method than is possible with the Euler method (or higher order explicit methods, for that matter) and it almost invariably turns out that the implicit Crank–Nicolson method produces a solution at lower computational cost for the given accuracy. This statement remains true even for nonlinear equations. For the nonlinear case one has to solve a set of coupled nonlinear equations but the set is tridiagonal and a reasonable first guess of the solution is the one at the preceding time step, so iterative procedures converge rapidly. Usually only a few iterations are necessary and Crank–Nicolson remains the method of choice.

The computational molecule for this method is shown in Fig. 4.4. We see that not only does the Crank–Nicolson method have the advantages described

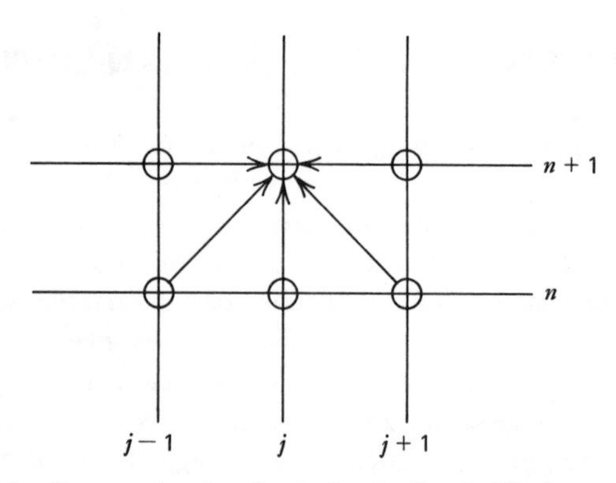

Fig. 4.4. Computational molecule for the Crank–Nicolson method.

above but it also retains the parabolic character of the original equation, in that the solution at point x_j at time t_{n+1} is influenced not only by all the earlier values but also by the solution at all other points at time t_{n+1} as well.

It is reasonably obvious from these results that the Crank–Nicolson method is second order with respect to both independent variables, but it is instructive to derive the modified equation. Note that the trapezoid rule is second order with respect to time $t+\Delta t/2$, so it is natural to try to expand in Taylor series about the point $(x_j, t_{n+1/2})$. The easiest way to do this is to make use of the results we have already obtained:

$$\frac{\phi_{j+1}^{n+1}-2\phi_j^{n+1}+\phi_{j-1}^{n+1}}{h^2} \simeq \frac{\partial^2\phi_j^{n+1}}{\partial x^2}+\frac{h^2}{12}\frac{\partial^4\phi_j^{n+1}}{\partial x^4} \tag{4.2.3a}$$

$$\frac{\phi_{j+1}^{n}-2\phi_j^{n}+\phi_{j-1}^{n}}{h^2} \simeq \frac{\partial^2\phi_j^{n}}{\partial x^2}+\frac{h^2}{12}\frac{\partial^4\phi_j^{n}}{\partial x^4} \tag{4.2.3b}$$

$$\frac{\phi_j^{n+1}-\phi_j^n}{\Delta t} \simeq \frac{\partial\phi_j^{n+1/2}}{\partial t}+\frac{\Delta t^2}{24}\frac{\partial^3\phi_j^{n+1/2}}{\partial t^3} \tag{4.2.3c}$$

Then noting that for any function ψ we have

$$\frac{\psi^{n+1}+\psi^n}{2} \simeq \psi^{n+1/2}+\frac{\Delta t^2}{8}\frac{\partial^2\psi^{n+1/2}}{\partial t^2} \tag{4.2.4}$$

and applying this to the terms $(\partial^2\phi^n/\partial x^2+\partial^2\phi^{n+1}/\partial x^2)$, we find that the modified equation for the Crank–Nicolson method is

$$\begin{aligned}\frac{\partial\phi}{\partial t}-D\frac{\partial^2\phi}{\partial x^2} &= -\frac{\Delta t^2}{24}\frac{\partial^3\phi}{\partial t^3}+D\left(\frac{\Delta t^2}{8}\frac{\partial^2}{\partial t^2}\frac{\partial^2}{\partial x^2}\phi+\frac{h^2}{12}\frac{\partial^4}{\partial x^4}\right)\phi \\ &= \frac{D}{12}\left(D^2\Delta t^2\frac{\partial^6\phi}{\partial x^6}+h^2\frac{\partial^4\phi}{\partial x^4}\right)\end{aligned} \tag{4.2.5}$$

And the second order accuracy is obvious. Note also that as h and Δt approach zero, the modified equation reduces to the original PDE. This property, which is shared by the methods discussed so far in this chapter, is called *convergence* and is clearly desirable. We point it out here because we encounter a method for which this property does not obtain shortly.

As suggested in the preceding section, one should attempt to balance the errors from time and space differencing. To do this properly requires that one estimate the relative magnitudes of $\partial^4\phi/\partial x^4$ and $\partial^6\phi/\partial x^6$. This cannot be done accurately a priori but it is possible to design a program to estimate the error and control it. A rule of thumb can be derived by noting that the ratio of these two derivatives is the square of a length scale. Unless the solution is sharply peaked, this length scale will be some fraction of the size of the computational region, in other words, of an order kNh where k will be approximately $\frac{1}{2}$. We then find that balancing the errors requires $(D\Delta t/h^2)\simeq N/2$. Although this is rather generous, it is clear that the Crank–Nicolson method can use time steps that would make an explicit method unstable without sacrificing accuracy.

The Crank–Nicolson method has been the overwhelming favorite for the solution of parabolic PDEs in one spatial dimension. It does run into trouble, however, in problems in more than one dimension. The difficulty is that finding the solution at the new time step requires solving an elliptic equation in two or three dimensions and this is not an easy task. For this reason we look at other methods for this in sections that follow.

It is also instructive to consider how the Crank–Nicolson method deals with a function that behaves like e^{ikx}. Thus let us assume that $\phi_j^n = \phi^n e^{ikx_j}$. Plugging this into Eq. 4.2.1 we have

$$\frac{\phi^{n+1}}{\phi^n} = \frac{(1-2\beta)+2\beta\cos k\,\Delta x}{(1+2\beta)-2\beta\cos k\,\Delta x}$$

Because the cosine is periodic, the range of values $0<k\,\Delta x<\pi$ will produce all the possible values of ϕ^{n+1}/ϕ^n. This ratio is never greater than unity and its minimum value is $(1-4\beta)/(1+4\beta)$, which is attained when $k\,\Delta x=\pi$. Thus the Crank–Nicolson method is indeed unconditionally stable (von Neumann criterion).

We can go further. If $\beta<\frac{1}{4}$, ϕ^{n+1}/ϕ^n is between 0 and 1 for all values of k. In this case the waves with high values of k, that is, those that oscillate more rapidly in space, are damped. When β is greater than $\frac{1}{4}$, some of the waves will have a negative amplification factor; that is, they will change sign on each successive time step but they will still be damped. Finally, when β is very large, some waves will be multiplied by factors close to -1 at successive steps and will change sign at each time step without significant damping. This observation plays an important role in the following example and in the study of methods for elliptic equations.

Example 4.3

We now compute the solution to the problem of Example 4.1 using the Crank–Nicolson method. The subroutine used for this calculation is given below. It was used together with the program of Example 4.1, changing only the subroutine name in the CALL statement. As shown in Table 4.2, the second order Crank–Nicolson method produces a more accurate solution for a wider range of time steps than does the first order Euler method. Furthermore, the stability of the method allows moderately accurate results even for rather large time steps.

Although the solution at the midpoint seems to behave reasonably well even for β as large as 5, the solution near the edge has already deteriorated, which is

```
      SUBROUTINE CRANK (N,F,BETA,NSTEP)
C- ---THIS SUBROUTINE ADVANCES THE SOLUTION OF THE HEAT EQUATION
C-----BY THE CRANK-NICOLSON IMPLICIT METHOD.  ALL OF THE PARAMETERS
C-----HAVE THE SAME MEANING AS IN SUBROUTINE PEULER ABOVE.
      DIMENSION F(51),R(51),A(51),B(51),C(51)
C-----FIRST SET UP THE MATRIX.  IT WILL BE THE SAME AT EACH TIME
C-----STEP AND WE COULD BYPASS THIS IF THE PROGRAM IS CALLED MORE
C-----THAN ONCE.   HOWEVER, IT IS EASY AND WE DO IT AT EACH CALL.
      NM1 = N - 1
      NM2 = N - 2
      DO 2 J=1,NSTEP
      DO 1 I=1,NM1
      A(I) = -BETA
      B(I) = 2. * (1. + BETA)
      C(I) = -BETA
    1 CONTINUE
C----·NOW COMES THE MAIN LOOP.   FIRST COMPUTE THE RIGHT HAND SIDE.
      DO 3 I=2,NM2
      R(I) = BETA * (F(I-1) + F(I+1)) + 2. * (1. - BETA) * F(I)
    3 CONTINUE
      R(1) = BETA * F(2) + 2. * (1. - BETA) * F(1)
      R(NM1) = BETA * F(NM1-1) + 2. *  (1. - BETA) * F(NM1)
C-----SOLVE THE SYSTEM AND REPEAT FOR THE DESIRED NUMBER OF STEPS.
      CALL TRDIAG (NM1,A,B,C,F,R)
    2 CONTINUE
      RETURN
      END
```

Table 4.2. Crank-Nicolson Solution of the Heat Equation

Δt	β	$\phi(.5,.1)$	$\phi(.1,.1)$
0.001	0.1	.47435	0.14670
0.002	0.2	.47434	0.14669
0.005	0.5	.47428	0.14666
0.01	1.	.47403	0.14655
0.02	2.	.47375	0.14146
0.0333	3.33	.46045	0.08283
0.05	5.	.46625	0.32688
0.1	10.	.56972	−0.30495

in agreement with our expectation. The problem comes from the rapidly oscillating modes. In this case many modes contribute and the results show the oscillation observed in the Euler method.

The problem we have shown here for illustration is not a difficult case. Had we used more points, the equations would have been stiffer and the relative advantage of the Crank–Nicolson method would be greater.

In many problems of practical interest nonuniform meshes are advantageous. This approach is economical in terms of the amount of computation needed for a given resolution. It is difficult to do the analysis for problems of this kind, but the stability of the Euler method is approximately determined by using the smallest Δx in the criterion (see Eq. 4.1.14 and Example 4.2). This is very severe and the relative advantage of an implicit method is even larger when nonuniform meshes are used.

3. PARABOLIC PDEs: III. THE DUFORT–FRANKEL METHOD

The Crank–Nicolson method has two advantages over the Euler method: stability and improved accuracy. There are higher order explicit methods with reasonable stability properties. Among the second order explicit methods at our disposal are the Adams–Bashforth and leapfrog methods. The former suffers from the same stability limitation as the Euler method. The latter, which when applied to the diffusion equation Eq. 4.1.1, yields

$$\phi_j^{n+1}-\phi_j^{n-1}=\frac{2D\Delta t}{h^2}\left(\phi_{j+1}^n-2\phi_j^n+\phi_{j-1}^n\right) \tag{4.3.1}$$

and is unconditionally unstable because it is equivalent to using leapfrog on $y'=\alpha y$ with real α. It does not, therefore, seem to be much of a candidate.

Dufort and Frankel, however, have pointed out that the leapfrog method is second order accurate (for what little that is worth when it is also unstable) and the approximation

$$\phi_j^n \simeq \tfrac{1}{2}\left(\phi_j^{n+1}+\phi_j^{n-1}\right) \tag{4.3.2}$$

is also second order accurate (compare with Eq. 4.2.4). Substituting this expression into Eq. 4.3.1 for ϕ_j^n, but leaving the terms involving ϕ_{j-1}^n and ϕ_{j+1}^n alone, we have

$$(1+2\beta)\phi_j^{n+1}=(1-2\beta)\phi_j^{n-1}+2\beta\left(\phi_{j+1}^n+\phi_{j-1}^n\right) \tag{4.3.3}$$

where β is as defined in the preceding section. This method turns out to be rather surprising. To begin with, it is explicit and unconditionally stable. (The proof of this is not difficult, but it does involve a fair amount of tedious algebra, so we will not go into it. The interested reader is encouraged to try his or her hand at it.)

Another rather surprising result is discovered when one derives the modified equation for the method:

$$\frac{\partial\phi}{\partial t}-D\frac{\partial^2\phi}{\partial x^2}=-\frac{\Delta t^2}{6}\frac{\partial^3\phi}{\partial t^3}+\frac{Dh^2}{12}\frac{\partial^4\phi}{\partial x^4}-\frac{D\Delta t^2}{h^2}\frac{\partial^2\phi}{\partial t^2}-\frac{D\Delta t^4}{12h^2}\frac{\partial^4\phi}{\partial t^4}$$

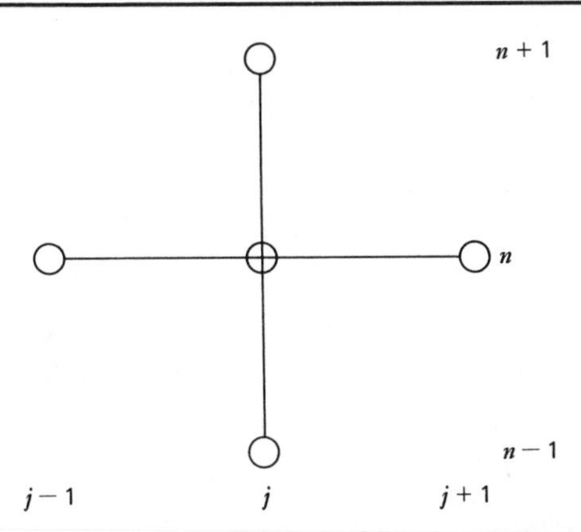

Fig. 4.5. Computational molecule for the Dufont–Frankel method.

What is unusual here is the presence of a term with the coefficient $\Delta t^2/h^2$. This means that when one takes the limit in which Δt and h go to zero, the original PDE is not recovered unless $\Delta t/h$ also goes to zero. In terms of the definition given in the preceding section, the Dufort–Frankel method is not convergent. In other words, unless Δt and h are properly chosen, the method is not even first order accurate! If $D\Delta t^2/h^2$ is large, Eq. 4.3.4 becomes in fact a wave equation. The hyperbolic character of the approximation might be anticipated by looking at the computational molecule of Fig. 4.5. These results demonstrate that caution should be exercised in choosing a numerical method. Surprises are certainly possible and are usually undesirable. We note in passing that the famous meteorologist L. F. Richardson (of extrapolation and dimensionless number fame) actually did computations of atmospheric flows using the leapfrog method on a desk calculator in the 1920s for two years before discovering that the method was unstable.

Despite these difficulties, the Dufort–Frankel method has been fairly popular, one reason being that it can be applied to problems in two or three space dimensions with little more effort than is required in one dimension. Since the Crank–Nicolson method has difficulty in this case, Dufort–Frankel was used for many years with reasonable success. In recent years it has been largely supplanted by the Alternating Direction Implicit (ADI) method described in Section 5.

Example 4.4

Since the Dufort–Frankel method is a variation of the leapfrog method, it requires another method to start it. For this example the Euler method is used for the first time step. The subroutine used is given below; the only change from the preceding subroutines is that since Dufort–Frankel requires a starting method, a parameter is needed to indicate the first call.

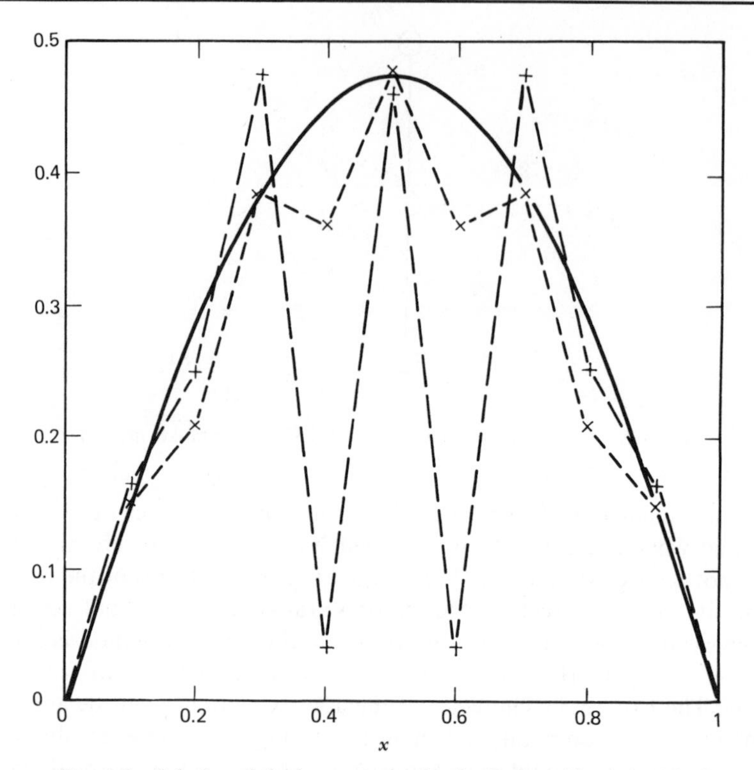

Fig. 4.6. Solution of the heat equation by the Dufont–Frankel method

$(\cdot\beta=0.01$ ———; $\times\beta=1$ ---; $+\beta=1.667$ --).

The results are shown in Fig. 4.6. When β is small, the results obtained are accurate and smooth. As β is increased, however, the results deteriorate. The oscillatory behavior for $\beta=1$ and $\beta=1.667$ in Fig. 4.6 can be regarded in at least two different ways.

First, since the method is based on leapfrog, there is a tendency for the solution to oscillate about the exact result at alternate time steps. Thus the solutions shown in Fig. 4.6 would look quite different at the next step (in fact the points at which the solution is low would become the high ones).

Another way to look at the results is that the Euler starting method for $\beta>1$ produces a negative "spike" near each boundary at the first time step. Then since the Dufort–Frankel method has a hyperbolic character, the spikes propagate inward as time increases. A "motion picture" of this solution would show this clearly.

Whichever way we choose to look at it we see that even though the Dufort–Frankel method is not unstable, it can behave very poorly. The results are dependent on the starting method and, in general, the method is not as well behaved as the implicit Crank–Nicolson method. As a result, it has been losing favor in the last few years.

```
      SUBROUTINE DUFORT (N,F,BETA,NSTEP,IDUF)
C-----THIS PROGRAM USES THE DUFORT-FRANKEL METHOD TO ADVANCE THE
C-----SOLUTION OF THE HEAT EQUATION BY NSTEP TIME STEPS.  THE METHOD
C-----IS STARTED BY USING THE EULER METHOD.  THE VARIABLES ARE THE
C-----SAME AS IN THE PREVIOUS SUBROUINES.
      DIMENSION F(51),FF(51)
C-----IF THIS IS THE FIRST CALL, WE NEED TO USE A STARTING METHOD.
      IF (IDUF.NE.1) GO TO 10
      NM1 = N - 1
      NM2 = N - 2
      DO 1 I=1,NM1
      FF(I) = F(I)
    1 CONTINUE
C-----CALL EULER TO GET THINGS GOING.
      CALL PEULER (N,FF,BETA,1)
      NSHALF = NSTEP/2
   10 IDUF = 2
      DO 2 J=1,NSHALF
      DO 3 I=2,NM2
C-----WE DO TWO STEPS IN A SWEEP; THIS AVOIDS THE NEED FOR A
C-----FOR A TEMPORARY ARRAY.
      F(I) = ((1.-2.*BETA)*F(I)+2.*BETA*(FF(I-1)+FF(I+1)))/(1.+2.*BETA)
    3 CONTINUE
      F(1) = ((1.-2.*BETA)*F(1)+2.*BETA*FF(2))/(1.+2.*BETA)
      F(NM1) = ((1.-2.*BETA)*F(NM1)+2.*BETA*FF(NM2))/(1.+2.*BETA)
      DO 4 I=2,NM2
      FF(I) = ((1.-2.*BETA)*FF(I)+2.*BETA*(F(I-1)+F(I+1)))/(1.+2.*BETA)
    4 CONTINUE
      FF(1) = ((1.-2.*BETA)*FF(1)+2.*BETA*F(2))/(1.+2.*BETA)
      FF(NM1) = ((1.-2.*BETA)*FF(NM1)+2.*BETA*F(NM2))/(1 +2.*BETA)
    2 CONTINUE
      RETURN
      END
```

4. PARABOLIC PDEs: IV. THE KELLER BOX METHOD AND HIGHER ORDER METHODS

A method that has achieved some popularity recently is the box method developed by H. B. Keller. Essentially, it is an alternative to the Crank–Nicolson method to which it is very closely related. The basic difference is that Keller chooses to write the PDE as a system of first order equations. For the diffusion equation of Eq. 4.1.1 this is easily done. Defining

$$\psi = \frac{\partial \phi}{\partial x} \tag{4.4.1}$$

Eq. 4.4.1 can be written

$$\frac{\partial \phi}{\partial t} = D\frac{\partial \psi}{\partial x} \tag{4.4.2}$$

and Eqs. 4.4.1 and 4.4.2 are a coupled set of first order equations equivalent to the diffusion equation.

To discretize these equations, Keller integrates Eq. 4.4.2 over the box shown in Fig. 4.7, obviously the origin of the name of the method. Using the

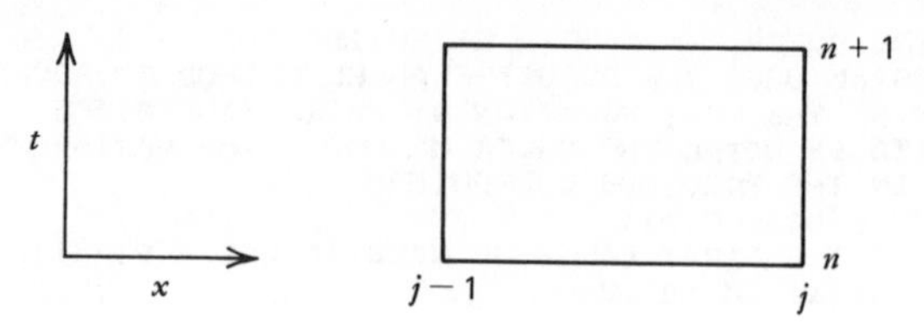

Fig. 4.7. The Keller Box.

trapezoid rule for integration whenever necessary, we have

$$\iint \frac{\partial \phi}{\partial t} dx\, dt = \int [\phi(x, t_{n+1}) - \phi(x, t_n)]\, dx$$

$$\simeq \frac{h}{2}\left[\left(\phi_j^{n+1} + \phi_{j-1}^{n+1}\right) - \left(\phi_j^n + \phi_{j-1}^n\right)\right] \tag{4.4.3a}$$

$$\iint \frac{\partial \psi}{\partial x} dx\, dt = \int [\psi(x_j, t) - \psi(x_{j-1}, t)]\, dt$$

$$= \frac{\Delta t}{2}\left[\left(\psi_j^{n+1} + \psi_j^n\right) - \left(\psi_{j-1}^{n+1} + \psi_{j-1}^n\right)\right] \tag{4.4.3b}$$

so that Eq. 4.4.2 is approximated by

$$\left(\phi_j^{n+1} + \phi_{j-1}^{n+1}\right) - \alpha\left(\psi_j^{n+1} - \psi_{j-1}^{n+1}\right) = \left(\phi_j^n + \phi_{j-1}^n\right) + \alpha\left(\psi_j^n - \psi_{j-1}^n\right) \tag{4.4.4}$$

where $\alpha = D\Delta t/h$. Equation 4.4.1, which is essentially an ODE, can be approximated by

$$\frac{h}{2}\left(\psi_j^{n+1} + \psi_{j-1}^{n+1}\right) = \left(\phi_j^{n+1} - \phi_{j-1}^{n+1}\right) \tag{4.4.5}$$

Equations 4.4.4 and 4.4.5 are a coupled set of linear algebraic equations for ϕ and ψ. Note that we now have two variables at each mesh point but that each equation references only two mesh points. The matrix corresponding to this set of equations is thus somewhat different from the ones we encountered earlier, but by taking advantage of the structure of the matrix, it is possible to construct routines that solve this system of equations almost as efficiently as the tridiagonal solvers used in the Crank–Nicolson method.

For an equation as simple as the heat equation we can actually manipulate the equations to a more recognizable form. Subtracting Eq. 4.4.5 with index j from the same equation with $j+1$, we get

$$\frac{h}{2}\left(\psi_{j+1}^{n+1} - \psi_{j-1}^{n+1}\right) = \left(\phi_{j+1}^{n+1} - 2\phi_j^{n+1} + \phi_{j-1}^{n+1}\right) \tag{4.4.6}$$

which is also valid with $n+1$ replaced by n. On the other hand, if Eq. 4.4.4

with indices j and $j+1$ are added, we get

$$\left(\phi_{j+1}^{n+1}+2\phi_j^{n+1}+\phi_{j-1}^{n+1}\right)-\alpha\left(\psi_{j+1}^{n+1}-\psi_{j-1}^{n+1}\right)=\left(\phi_{j+1}^{n}+2\phi_j^{n}+\phi_{j-1}^{n}\right)+\alpha\left(\psi_{j+1}^{n}-\psi_{j-1}^{n}\right) \tag{4.4.7}$$

If Eq. 4.4.6 with indices n and $n+1$ are substituted into Eq. 4.4.7, we have

$$\begin{aligned}\left(\phi_{j+1}^{n+1}+2\phi_j^{n+1}+\phi_{j-1}^{n+1}\right)&-2\beta\left(\phi_{j+1}^{n+1}-2\phi_j^{n+1}+\phi_{j-1}^{n+1}\right)\\ &=\left(\phi_{j+1}^{n}+2\phi_j^{n}+\phi_{j-1}^{n}\right)+2\beta\left(\phi_{j-1}^{n}-2\phi_j^{n}+\phi_{j-1}^{n}\right)\end{aligned} \tag{4.4.8}$$

These equations would become the Crank–Nicolson equations precisely if $\phi_{j+1}+2\phi_j+\phi_{j-1}$ were replaced by $4\phi_j$. Thus the results of the Keller Box method are very similar to those of Crank–Nicolson; the error terms are slightly different but both are second order.

As we noted earlier, the Keller Box method is a competitor to Crank–Nicolson. It requires slightly more computation per point, but it is a bit more accurate and it may be possible to get by with fewer mesh points. More significantly, the box method lends itself very well to the use of extrapolation methods (e.g., Richardson) and can thus be used as the basis for high order methods that do not require an inordinate amount of computer storage. It is this last property that has been responsible for its increasing popularity.

It is also possible to construct higher order methods for parabolic partial differential equations from the higher order methods that were applied to the boundary value problem for ordinary differential equations in Section 3.13. All of the remarks made about those methods apply here as well. Specifically, the regions near the boundaries require special attention, it is difficult to apply high order methods on nonuniform grids, and when these methods are used in implicit methods for parabolic equations (which is the only practical case), they require special subroutines for solving the systems of equations they produce. Since the best of these methods was found to be the compact fourth order method (see Eq. 3.13.6), we apply it to the heat equation here. One has

$$\begin{aligned}\tfrac{1}{6}\left(\phi_{j+1}^{n+1}+10\phi_j^{n+1}+\phi_{j-1}^{n+1}\right)&-\beta\left(\phi_{j+1}^{n+1}-2\phi_j^{n+1}+\phi_{j-1}^{n+1}\right)\\ &=\tfrac{1}{6}\left(\phi_{j+1}^{n}+10\phi_j^{n}+\phi_{j-1}^{n}\right)+\beta\left(\phi_{j+1}^{n}-2\phi_j^{n}+\phi_{j-1}^{n}\right)\end{aligned} \tag{4.4.9}$$

where, as earlier, $\beta=D\Delta t/\Delta x^2$. This method is no more difficult to use than Crank–Nicolson. The error due to the spatial finite differencing is reduced; however, if the time step is such that the time and space differencing errors are approximately equal, the overall error reduction may not be that large. To truly reduce the error, the time step also needs to be reduced.

5. PARABOLIC PDES: V. TWO AND THREE DIMENSIONS—ALTERNATING DIRECTION IMPLICIT (ADI) METHODS

In the case of parabolic PDEs in two and three dimensions, special attention again is given to the heat equation. There are some significant differences between one- and two-dimensional cases but most of the ideas developed in two dimensions can be generalized to three dimensions. Thus the two-dimensional case is considered first.

In two dimensions the heat equation becomes

$$\frac{\partial \phi}{\partial t} = D\left(\frac{\partial^2 \phi}{\partial x^2} + \frac{\partial^2 \phi}{\partial y^2}\right) = D\nabla^2\phi \tag{4.5.1}$$

As in the one-dimensional case, spatial derivatives are treated by means of the second order central difference formula. (Other approximations can be used, of course, but this one has been the workhorse and the others are used only occasionally.) To prevent the notation from becoming too cumbersome, we introduce the abbreviations

$$\left.\frac{\delta^2 \phi}{\delta x_1^2}\right|_{j,k}^{n} = h_1^{-2}(\phi_{j+1,k} - 2\phi_{j,k} + \phi_{j-1,k}) \tag{4.5.2a}$$

$$\left.\frac{\delta^2 \phi}{\delta x_2^2}\right|_{j,k}^{n} = h_2^{-2}(\phi_{j,k+1} - 2\phi_{j,k} + \phi_{j,k-1}) \tag{4.5.2b}$$

We denote the independent spatial variables as x_1 and x_2 in order to facilitate extension to three dimensions. The semidiscretized heat equation in two dimensions, that is,

$$\frac{\partial \phi}{\partial t} = D\left(\frac{\delta^2 \phi}{\delta x_1^2} + \frac{\delta^2 \phi}{\delta x_2^2}\right) \tag{4.5.3}$$

can be represented in linear algebra fashion by creating the vector

$$\boldsymbol{\phi} = \begin{bmatrix} \phi_{11} \\ \phi_{12} \\ \vdots \\ \phi_{1M} \\ \phi_{21} \\ \vdots \\ \phi_{2M} \\ \phi_{31} \\ \vdots \\ \phi_{NM} \end{bmatrix} \tag{4.5.4}$$

In this form we have placed the elements of the solution on the first vertical line at the top; they are followed by the elements on the next vertical line, and so on. We have assumed that M interior mesh points are used in the x_2 direction, that N is the x_1 direction, and that the region is rectangular. The elements could be be entered by horizontal rather than vertical lines—in this instance the choice is not critical—but, as we see later, it may be important in elliptic problems. The treatment of nonrectangular regions, an important and difficult problem, is taken up in Section 7.

With the vector written as in Eq. 4.5.4, the matrix form of the operator appearing in Eq. 4.5.3 becomes

$$A=\begin{bmatrix} -2(h_1^{-2}+h_2^{-2}) & h_2^{-2} & 0 & \cdot\;\cdot & 0 & h_1^{-2} & 0 & 0\ldots \\ h_2^{-2} & -2(h_1^{-2}+h_2^{-2}) & h_2^{-2} & \cdot\;\cdot & & & h_1^{-2} & \\ 0 & h_2^{-2} & \cdot & & & & & \\ 0 & \cdot & \cdot & \cdot\;\cdot & \cdot & \cdot & & \\ h_1^{-2} & 0 & \cdot & \cdot\;\cdot & \cdot & \cdot & \cdot & \\ 0 & h_1^{-2} & \cdot & \cdot\;\cdot & \cdot & \cdot & \cdot & \cdot \\ & 0 & \cdot & \cdot\;\cdot & \cdot & \cdot & \cdot & \cdot \\ & & \cdot & \cdot\;\cdot & \cdot & \cdot & \cdot & \cdot \end{bmatrix} \tag{4.5.5}$$

Only five diagonals of this $MN\times MN$ matrix contain elements other than zero: the main diagonal [all the elements of which are $-2(h_1^{-2}+h_2^{-2})$], the diagonals immediately above and below it (with elements h_2^{-2}), and the diagonals M rows above and below the main diagonal (with elements h_1^{-2}). Although this matrix has a very simple band structure, two of the diagonals of nonzero elements are far from the main diagonal. This makes the two-dimensional case much harder than the one-dimensional one, and the three-dimensional case is even harder.

From what we discovered for the one-dimensional problem, we expect the two-dimensional problem to be stiff. In order to assess the difficulty of the two-dimensional problem, let us consider the eigenvalue problem

$$A\psi=\lambda\psi \tag{4.5.6}$$

which can also be written as

$$\frac{\delta^2\psi}{\delta x_1^2}+\frac{\delta^2\psi}{\delta x_2^2}=\lambda\phi \tag{4.5.7}$$

The solution can again be obtained by guessing. The partial differential equation from which this problem was derived has as its eigenfunctions products of eigenfunctions of the corresponding one-dimensional problem.

Thus we expect the solution to Eq. 4.5.7 to have the form

$$\psi_{j,k}^{m,n} = \sin\frac{m\pi}{M}j \quad \sin\frac{n\pi}{N}k \tag{4.5.8}$$

This turns out to be correct and with some calculation we can show that, as in the PDE problem, the eigenvalues are the sum of the one-dimensional eigenvalues

$$\lambda_{m,n} = \frac{2}{h_1^2}\left(1-\cos\frac{m\pi}{M+1}\right) + \frac{2}{h_2^2}\left(1-\cos\frac{n\pi}{N+1}\right) \tag{4.5.9}$$

where m takes the values $1,2,\ldots M$ and $n=1,2,\ldots N$. From the expression 4.5.9 for the eigenvalues we find that the problem in two dimensions has about the same degree of stiffness as the one-dimensional case. The extensions of these results to the three-dimensional case is, to use a timeworn phrase, tedious but straightforward, and the stiffness problem is again about the same.

We cannot expect to use explicit methods, except possibly the Dufort–Frankel method, which has the difficulties mentioned earlier. The Crank–Nicolson method is a natural to try and leads to

$$-\left(\frac{D\Delta t}{2}\right)\left(\frac{\delta^2\phi_{j,k}^{n+1}}{\delta x_1^2} + \frac{\delta^2\phi_{j,k}^{n+1}}{\delta x_2^2}\right) + \phi_{j,k}^{n+1} = \left(\frac{D\Delta t}{2}\right)\left(\frac{\delta^2\phi_{j,k}^{n}}{\delta x_1^2} + \frac{\delta^2\phi_{j,k}^{n}}{\delta x_2^2}\right) + \phi_{j,k}^{n} \tag{4.5.10}$$

The left-hand side of this equation is a discrete version of $-(D\Delta t/2)\nabla^2\phi+\phi$ and so we need essentially to solve a two-dimensional elliptic problem. The matrix of this problem has the same form as the matrix in Eq. 4.5.5. The cost of solving a system of linear equations with a matrix of this type is much higher than the cost of solving a pentadiagonal system in which the nonzero elements are close to the main diagonal. This is the primary difficulty in solving elliptic equations and is discussed in more detail later.

A way needs to be found to get the stability benefits of implicit methods without the large penalty in computation time that comes with two-dimensional problems. One might try to compromise and treat one direction implicitly and the other explicitly. At best, such methods turn out to be conditionally stable (if one is lucky) and have time step restrictions similar to those of explicit methods. A method that treats first one direction implicitly and then the other implicitly, however, does have nice properties. This is the now-famous Alternating Direction Implicit (ADI) method originally developed by Peaceman and Rachford in the late 1950s; there are several variations of this method. Applied to the heat equation, the Peaceman–Rachford method consists of a step in which the x_1 direction is treated by backward Euler and the x_2 direction by explicit Euler:

$$\phi_{j,k}^{n+1/2}-\phi_{j,k}^{n}=\frac{D\Delta t}{2}\left(\frac{\delta^2\phi_{j,k}^{n+1/2}}{\delta x_1^2}+\frac{\delta^2\phi_{j,k}^{n}}{\delta x_2^2}\right) \tag{4.5.11a}$$

This is followed by a step in which the roles are reversed:

$$\phi_{j,k}^{n+1}-\phi_{j,k}^{n+1/2}=\frac{D\Delta t}{2}\left(\frac{\delta^2\phi_{j,k}^{n+1/2}}{\delta x_1^2}+\frac{\delta^2\phi_{j,k}^{n+1}}{\delta x_2^2}\right) \tag{4.5.11b}$$

This method has some surprising features. Considered alone, each of the steps is only first order accurate in time (as might be expected from the properties of the ODE methods to which it is related) and conditionally stable. The combined method, however, has neither of these properties. To get some idea of why this might be, we add Eqs. 4.5.11a and 4.5.11b to obtain

$$\phi_{j,k}^{n+1}-\phi_{j,k}^{n}=D\Delta t\left[\frac{\delta^2\phi_{j,k}^{n+1/2}}{\delta x_1^2}+\tfrac{1}{2}\left(\frac{\delta^2\phi_{j,k}^{n}}{\delta x_2^2}+\frac{\delta^2\phi_{j,k}^{n+1}}{\delta x_2^2}\right)\right] \tag{4.5.12}$$

Thus the overall method is equivalent to treating the x_1 direction by the midpoint rule and the x_2 direction by the trapezoid rule. It is actually second order accurate, even though both of its components are only first order. (We go more deeply into why this is so later.) First, we investigate the stability of the method in some detail, since parts of the analysis prove to be important later.

Stability is most easily investigated using the von Neumann approach so that the question of boundary conditions may be avoided. We assume that the difference equations have solutions of the form

$$\phi^n=\rho^n e^{ik_1x_1}e^{ik_2x_2}$$

$$\phi^{n+1/2}=\xi\rho^n e^{ik_1x_1}e^{ik_2x_2} \tag{4.5.13}$$

where ξ is a constant and the spatial indices are suppressed for convenience. We note that

$$\frac{\delta^2}{\delta x_i^2}e^{ik_ix_i}=2h_i^{-2}(\cos k_ih_i-1)e^{ik_ix_i} \tag{4.5.14}$$

where $i=1$ or 2. Substituting Eq. 4.5.13 into Eqs. 4.5.11a and 4.5.11b using the definition

$$\beta_i=\frac{D\Delta t}{h_i^2}(\cos k_ih_i-1) \tag{4.5.15}$$

and Eq. 4.5.14, and after cancelling out common factors, we have

$$\xi - 1 = \xi\beta_1 + \beta_2$$

$$\rho - \xi = \xi\beta_1 + \beta_2$$

Finally, solving for ρ, we have

$$\rho = \frac{1+\beta_1}{1-\beta_1}\frac{1+\beta_2}{1-\beta_2}$$

This is similar to the quantity that determines the stability of the trapezoid rule for ODEs. Since β_1 and β_2 are both negative, the ADI method is unconditionally stable.

The first step of the ADI method, represented by Eq. 4.5.11a, requires the solution of a tridiagonal system of equations on each line of constant y; that is, we have to solve N systems of M tridiagonal equations. The second step (Eq. 4.5.11b) requires solving M sets of N tridiagonal equations. Since the number of operations required to solve a tridiagonal system is proportional to the number of equations in the system, the total cost is proportional to MN, that is, to the total number of mesh points. This is the best we could hope for and this method is, therefore, hard to beat. The two-dimensional ADI method requires about twice as much computation per mesh point as the one-dimensional Crank–Nicolson method. It has all of the desirable properties of a numerical method—speed, accuracy, and stability—so it is hardly surprising that ADI is the preferred method for problems in two and three dimensions.

ADI was the first of an entire class of methods that are now called *splitting methods*. The success of ADI has prompted an extensive search for other methods with similar properties and the search has been successful to a large extent. It is, therefore, worth looking into the Peaceman–Rachford ADI and some of its relatives.

Essentially, the concept of splitting is as follows. Equation 4.5.10, the Crank–Nicolson method in two dimensions, can be written

$$\left[1 - \frac{D\Delta t}{2}(\delta_{xx} + \delta_{yy})\right]\phi^{n+1} = \left[1 + \frac{D\Delta t}{2}(\delta_{xx} + \delta_{yy})\right]\phi^n \qquad (4.5.16)$$

where δ_{xx} and δ_{yy} represent the second order central finite difference approximations to $\partial^2/\partial x^2$ and $\partial^2/\partial y^2$, respectively. On the other hand, Eqs. 4.5.11a and 4.5.11b can be written in this notation as

$$\left(1 - \frac{D\Delta t}{2}\delta_{xx}\right)\phi^{n+1/2} = \left(1 + \frac{D\Delta t}{2}\delta_{yy}\right)\phi^n \qquad (4.5.17a)$$

$$\left(1 - \frac{D\Delta t}{2}\delta_{yy}\right)\phi^{n+1} = \left(1 + \frac{D\Delta t}{2}\delta_{xx}\right)\phi^{n+1/2} \qquad (4.5.17b)$$

and these can be combined to give

$$\left(1-\frac{D\Delta t}{2}\delta_{xx}\right)\left(1-\frac{D\Delta t}{2}\delta_{yy}\right)\phi^{n+1}=\left(1+\frac{D\Delta t}{2}\delta_{yy}\right)\left(1+\frac{D\Delta t}{2}\delta_{xx}\right)\phi^{n} \tag{4.5.18}$$

We used the fact that all of the operators commute in deriving this equation. If the multiplications are performed, we find that Eq. 4.5.18 is the same as Eq. 4.5.16 except for an extra term $(D^2\Delta t^2/4)\delta_{xx}\delta_{yy}(\phi^{n+1}-\phi^n)\simeq(D^2\Delta t^3/4)\delta_{xx}\delta_{yy}\partial\phi/\partial t$. This new term is smaller than the truncation error of the Crank–Nicolson method Eq. 4.5.16 so the ADI scheme introduces no significant additional error. We call Eq. 4.5.18 an *approximate factorization* of Eq. 4.5.16. This is an alternative name for this class of methods and provides insight into finding other methods of this type. The idea is to factor an operator for which the equations are difficult to solve into the product of two (or more) operators for which the solution is computationally simpler. The savings in computational effort can be spectacular.

The Peaceman–Rachford method given above has been very popular, but other approximate factorizations of Eq. 4.5.16 have also been used. Specifically, it is difficult to extend the Peaceman–Rachford method to three dimensions; using variables at times $n+\frac{1}{3}$ and $n+\frac{2}{3}$ does not work. A scheme that does work was developed by Douglass and Gunn. One version of this method is

$$\left(1-\frac{D\Delta t}{2}\delta_{xx}\right)\phi^{n+1^*}=\left[1+\frac{D\Delta t}{2}(\delta_{xx}+2\delta_{yy}+2\delta_{zz})\right]\phi^{n} \tag{4.5.19a}$$

$$\left(1-\frac{D\Delta t}{2}\delta_{yy}\right)\phi^{n+1^{**}}=\phi^{n+1^*}-\frac{D\Delta t}{2}\delta_{yy}\phi^{n} \tag{4.5.19b}$$

$$\left(1-\frac{D\Delta t}{2}\delta_{zz}\right)\phi^{n+1}=\phi^{n+1^{**}}-\frac{D\Delta t}{2}\delta_{zz}\phi^{n} \tag{4.5.19c}$$

Showing that this is a factorization of Eq. 4.5.16 is tedious but not difficult. The two-dimensional version obtained by dropping Eq. 4.5.19c completely and δ_{zz} from Eq. 4.5.19a is essentially equivalent to the Peaceman–Rachford method. Still other approximate factorizations are possible.

Example 4.5

As a test of the ADI method, we use the two-dimensional analog of the problem treated in Example 2.1. Specifically, we solve

$$\frac{\partial\phi}{\partial t}=\frac{\partial^2\phi}{\partial x^2}+\frac{\partial^2\phi}{\partial y^2}$$

with the initial condition

$$\phi(x, y, 0) = 1$$

and the boundary conditions

$$\phi(0, y, t) = \phi(1, y, t) = \phi(x, 0, t) = \phi(x, 1, t) = 0$$

A few notes on the program used in this example (given below) are in order. Since the solution of a two-dimensional problem requires more data than does a one-dimensional problem, a number of parameters are set internally rather than being requested as input. These include the dimensions (WX and WX) and the diffusion coefficient (D). To keep the program short, CONTINUE statements are not used. Also, in order to use the tridiagonal solver TRDIAG used in Appendix B and to avoid using a zero index, it is necessary to use different indexing on the dependent variables (F and FF) and on the variables in the tridiagonal systems of equations. Finally, to avoid using new data where they should not be used, it is necessary to store the results of the first sweep in a different array (FF) than that of the result of the second sweep (F).

In the first calculation five points (including the boundaries) are used in each direction. Some of the results are shown in Table 4.3. Note that because of the symmetry in this problem it is necessary to look at only three points. Although an exact solution is not available for this problem, we can accept the solution obtained using the smallest time step as an exact solution of the semi-discretized problem. We find that the error increases quadratically with the time step as expected for a second order time method.

In order to assess the error due to spatial differencing, the calculation is repeated using half the interval size, that is, nine points in each direction. The results at the points used in the first calculation are given in the bottom line of

Table 4.3. Solution of the Heat Equation by the ADI Method

Δt	β	$\phi(0.25,0.25,0.1)$	$\phi(0.25,0.5,0.1)$	$\phi(0.5,0.5,0.1)$
Solution obtained with 5 points in each direction				
0.0005	0.008	0.1121964	0.1580812	0.2227316
0.001	0.016	0.1121948	0.1580797	0.2227304
0.002	0.032	0.1121886	0.1580733	0.2227247
0.005	0.08	0.1121446	0.1580279	0.2226841
0.01	0.16	0.1119882	0.1578652	0.2225362
0.02	0.32	0.1113787	0.1572126	0.2219077
0.05	0.8	0.1101956	0.1525546	0.2111962
Solution obtained with 9 points in each direction				
0.001	0.064	0.1125822	0.1591227	0.2249027

the Table 4.3. The difference between $\phi(0.25, 0.25, 0.1)$ obtained in this calculation and the value obtained using the same time step in the first calculation is approximately three-fourths of the spatial error in first calculation. The spatial error is thus approximately 5.10^{-4} and the time step that produces the same error due to time differencing is a bit less than 0.02, with the corresponding value of β about 0.3. Using an argument similar to the one used in Section 3, we would predict that the errors would balance with β twice as large as this. In view of the crudity of the argument this is not bad, but a little conservatism (in using a smaller time step) is not out of place.

One problem with ADI or split methods is associated with the boundary conditions at the intermediate time steps. When the boundary conditions are time independent as in the above example, there is no problem. The exact

```
      PROGRAM ADI
C-----THIS PROGRAM SOLVES THE HEAT EQUATION IN TWO DIMENSIONS BY THE
C-----ALTERNATING DIRECTION IMPLICIT (ADI) METHOD.
      DIMENSION F(21,21),FF(21,21),A(21),B(21),C(21),AA(21),BB(21),
     1 CC(21),R(21),T(21)
C-----CALL FOR THE PROBLEM DATA.
   40 WRITE (5,100)
  100 FORMAT ( ' TYPE THE NUMBER OF X INTERVALS')
      READ (5,110) NX
  110 FORMAT (I)
      WRITE (5,120)
  120 FORMAT ( ' TYPE THE NUMBER OF Y INTERVALS')
      READ (5,110) NY
      WRITE (5,130)
  130 FORMAT ( ' GIVE THE TIME STEP')
      READ (5,140) DT
  140 FORMAT (F)
      WRITE (5,150)
  150 FORMAT ( ' GIVE THE FINAL TIME')
      READ (5,140) TF
      WRITE (5,160)
  160 FORMAT ( ' RESULTS PRINTED EVERY K STEPS, GIVE K')
      READ (5,110) K
C-----NOW SET SOME OF THE BASIC PARAMETERS.
      WX = 1.
      WY = 1.
      DX = WX / NX
      DY = WY / NY
      D = 1.
      BETX = D * DT / (DX*DX)
      BETY = D * DT / (DY*DY)
      NXM1 = NX - 1
      NXP1 = NX + 1
      NYM1 = NY - 1
      NYP1 = NY + 1
      TIME = 0.
C-----SET UP THE MATRIX ELEMENTS FOR LATER USE.
      DO 1 I=1,NXP1
      A(I) = - .5 * BETX
      B(I) = 1. + BETX
    1 C(I) = - .5 * BETX
      DO 2 I=1,NYP1
      AA(I) = - .5 * BETY
      BB(I) = 1. + BETY
    2 CC(I) = - .5 * BETY
```

```
C-----SET THE INITIAL AND BOUNDARY CONDITIONS.
      DO 3 I=2,NX
      F(I,1) = 0.
      F(I,NYP1) = 0.
      DO 3 J=2,NY
    3 F(I,J) = 1.
      DO 4 J=2,NY
      F(1,J) = 0.
    4 F(NXP1,J) = 0.
C-----MAIN LOOP
   50 DO 5 L=1,K
C-----X SWEEP.  SET UP RIGHT HAND SIDE AND SOLVE.
      DO 6 J=2,NY
      DO 7 I=1,NXM1
    7 R(I) = -AA(J)*F(I+1,J-1) + (2.-BB(J))*F(I+1,J) - CC(J)*F(I+1,J+1)
      CALL TRDIAG (NXM1,A,B,C,T,R)
      DO 6 I=1,NXM1
    6 FF(I+1,J) = T(I)
C-----Y SWEEP. MIRROR IMAGE OF THE X SWEEP
      DO 5 I=2,NX
      DO 8 J=1,NYM1
    8 R(J) = -A(I)*FF(I-1,J+1) + (2.-B(I))*FF(I,J+1) - C(I)*FF(I+1,J+1)
      CALL TRDIAG (NYM1,AA,BB,CC,T,R)
      DO 5 J=1,NYM1
    5 F(I,J+1) = T(J)
C-----NOW PRINT THE OUTPUT.
      TIME = TIME + K * DT
      WRITE (5,170) TIME,BETX,BETY
  170 FORMAT( ' TIME = ',F8.4,2X,'BETA X = ',F8.4,2X,'BETA Y = ',F8.4)
      DO 9 J=2,NY
    9 WRITE (5,180) (F(I,J) , I=2,NX)
  180 FORMAT ( 2X,10( E14.7,2X))
      IF (TIME.LT.TF) GO TO 50
C-----ASK FOR ANOTHER CASE.
      WRITE (5,190)
  190 FORMAT ( ' IF YOU WANT TO RUN ANOTHER CASE, TYPE 1')
      READ (5,110) IFF
      IF (IFF.EQ.1) GO TO 40
      STOP
      END
```

boundary conditions can be used without difficulty and excellent results are obtained. In a time dependent problem, however, a problem arises from the fact that the intermediate results (for example, $\phi^{n+1/2}$ in the Peaceman–Rachford ADI method) are less accurate representations of the solution than are the values at the end of a sweep (for example, ϕ^n or ϕ^{n+1}). If accurate values of the boundary data are given, an extra error can be introduced by the fact that the boundary conditions and the approximations are "out of synch." (This situation is reminiscent of the problem with the leapfrog method analyzed in Section 3.7; better results were obtained when initial conditions were picked to match the method rather than to match the exact solution.)

To see how the boundary conditions ought to be chosen, we subtract Eq. 4.5.18b from Eq. 4.5.18c to get

$$\phi^{n+1/2} = \tfrac{1}{2}\left(1 + \frac{D\Delta t}{2}\delta_{yy}\right)\phi^n + \tfrac{1}{2}\left(1 - \frac{D\Delta t}{2}\delta_{yy}\right)\phi^{n+1} \qquad (4.5.20)$$

Then given the values of ϕ^n and ϕ^{n+1} on the boundary, the values of $\phi^{n+1/2}$,

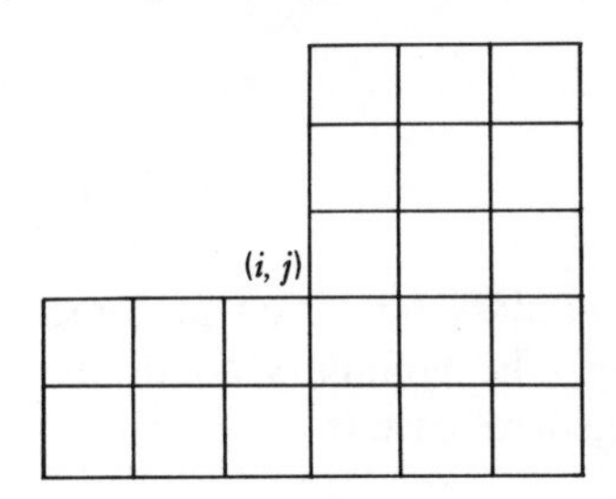

Fig. 4.8. Spatial grid for a non-rectangular region.

which are properly "synchronized" with the method, can be found. A more serious problem occurs at a corner as shown in Fig. 4.8. At the corner point (i, j), computing the right-hand side of Eq. 4.5.20 requires the value for $\phi_{i-1,j}^{n+1}$, which would not normally be computed until after all of the values of $\phi^{n+1/2}$ have been computed. To get out of this difficulty, we can compute $\phi^{n+1/2}$ on horizontal lines $1,2,\ldots j-1$. Then we have the information needed to compute ϕ^{n+1} on vertical lines $1,2,\ldots i-1$. Specifically, this provides the value of $\phi_{i-1,j}^{n+1}$ and removes the difficulty, so the calculation can proceed. Obviously, more complex geometries require still more sophisticated logic to handle the boundary conditions accurately.

6. PARABOLIC PDEs: VI. OTHER COORDINATE SYSTEMS AND NONLINEARITY

Frequently we need to solve problems in geometries that are not well suited to the use of Cartesian coordinates. Using another coordinate system in this case might make the problem simpler. This is especially true if the boundaries are coordinate surfaces in the new system. When the geometry is irregular, the problem is much more difficult. The treatment of irregular geometries is taken up later.

By far the most used coordinate system other than Cartesian is cylindrical coordinates. (Problems involving spherical geometry are less common.) In this case the heat equation takes the form

$$\frac{\partial \phi}{\partial t} = D\frac{1}{r}\frac{\partial}{\partial r}r\frac{\partial}{\partial r}\phi \tag{4.6.1}$$

where r is the radial coordinate. Typically, one is given an initial condition

$$\phi(r,0)=f(r) \tag{4.6.2a}$$

One boundary condition requires that the solution be well behaved at $r=0$. (The equation is singular at $r=0$ and there are solutions that blow up there.)

The most convenient way to do this is to apply

$$\frac{\partial \phi}{\partial r}(0,t)=0 \tag{4.6.2b}$$

as a boundary condition. Finally, one needs a boundary condition at some finite radius R representing the physical boundary. The boundary condition might specify ϕ or its derivative $\partial\phi/\partial r$. The most general case is

$$\frac{\partial \phi(R,t)}{\partial r}+k\phi(R,t)=q \tag{4.6.2c}$$

which contains the others as special cases.

To finite difference the operator on the right-hand side of Eq. 4.6.1, we can proceed in at least two different ways. The first is to note that

$$\frac{1}{r}\frac{\partial}{\partial r}r\frac{\partial \phi}{\partial r}=\frac{\partial^2\phi}{\partial r^2}+\frac{1}{r}\frac{\partial \phi}{\partial r} \tag{4.6.3}$$

and then finite difference each term using formulas developed earlier. For example,

$$\left.\frac{\partial^2\phi}{\partial r^2}\right|_{r_j,t_n}=\frac{\phi_{j+1}^n-2\phi_j^n+\phi_{j-1}^n}{\Delta r^2} \tag{4.6.4a}$$

$$\left.\frac{1}{r}\frac{\partial \phi}{\partial r}\right|_{r_j,t_n}=\frac{\phi_{j+1}^n-\phi_{j-1}^n}{2r_j\Delta r} \tag{4.6.4b}$$

where a uniform grid is used to avoid complicating the notation. These formulas are second order accurate and are satisfactory but usually not as good as the second method.

In the second approach for the derivatives at the midpoints of the intervals we write formally

$$\left.r\frac{\partial \phi}{\partial r}\right|_{r_{j+1/2},t_n}=r_{j+1/2}\frac{\phi_{j+1}^n-\phi_j^n}{\Delta r} \tag{4.6.5a}$$

$$\left.r\frac{\partial \phi}{\partial r}\right|_{r_{j-1/2},t_n}=r_{j-1/2}\frac{\phi_j^n-\phi_{j-1}^n}{\Delta r} \tag{4.6.5b}$$

These are second order accurate because they use centered differences. Then

$$\left.\frac{1}{r}\frac{\partial}{\partial r}\left(r\frac{\partial \phi}{\partial r}\right)\right|_{r_j,t_n}=\frac{1}{r_j}\left[\frac{\left(r\frac{\partial \phi}{\partial r}\right)_{r_{j+1/2},t_n}-\left(r\frac{\partial \phi}{\partial r}\right)_{r_{j-1/2},t_n}}{\Delta r}\right]$$

$$=\frac{1}{r_j\Delta r^2}\left(r_{j+1/2}\phi_{j+1}^n-2r_j\phi_j^n+r_{j-1/2}\phi_{j-1}^n\right) \tag{4.6.6}$$

The boundary condition 4.6.2b at $r=0$ can be satisfied by introducing an artificial point $r_{-1}=-\Delta r$ and letting $\phi_{-1}=\phi_1$. The PDE at $r=0$ then reduces to

$$\frac{\partial \phi_0}{\partial t}=\frac{2}{\Delta r^2}(\phi_1-\phi_0) \tag{4.6.7}$$

The boundary condition of Eq. 4.6.2c at the outer boundary is realized by introducing an artificial point at $r_{N+1}=R+\Delta r=(N+1)\Delta r$. Then

$$\frac{\phi_{N+1}^n-\phi_{N-1}^n}{2\,\Delta r}+k\phi_N^n=q \tag{4.6.8}$$

represents the boundary condition. This equation is then solved together with the $N+1$ equations that represent the approximations to the PDE at the points $r_0, r_1, \ldots r_N$. We thus have $N+2$ equations in the same number of unknowns.

The preference for the difference approximation of Eq. 4.6.6 over Eq. 4.6.4 arises from a conservation property of the differential equation. If we multiply Eq. 4.6.1 by r and integrate from 0 to R we find

$$\frac{d}{dt}\int_0^R r\phi(r,t)\,dr=Dr\frac{\partial\phi}{\partial r}\bigg|_R \tag{4.6.9}$$

Physically this equation is an overall conservation statement for the entire cylinder. In the heat transfer context it states that the rate of change of thermal energy in the cylinder is equal to the energy flux into the cylinder. It is desirable (and in some cases crucial) that the numerical approximation retain this property. The numerical analog of this property is obtained by multiplying the finite difference equation for $\phi(r_j, t)$ by r_j and summing over j. When this is done, we find that the first method does not produce anything simple but that the second one gives

$$\frac{d}{dt}\sum_{j=1}^{N} r_j\phi_j=\frac{D}{\Delta r^2}\sum_{j=1}^{N}\left(r_{j+1/2}\phi_{j+1}-2r_j\phi_j+r_{j-1/2}\phi_{j-1}\right) \tag{4.6.10a}$$

Most of the terms in the summation drop out and we have

$$\frac{d}{dt}\sum_j r_j\phi_j\Delta r=D\frac{r_{N+1/2}\phi_{N+1}+(r_{N-1/2}-2r_N)\phi_N}{\Delta r}$$

$$=Dr_{N+1/2}\left(\frac{\phi_{N+1}-\phi_N}{\Delta r}\right) \tag{4.6.10b}$$

which is a numerical approximation to the conservation property 4.6.9. Note, however, that if $k=0$ in the boundary condition 4.6.2c or 4.6.8 (which physically means that the heat flux is specified), then the right-hand side of Eq. 4.6.10b is different from the derivative approximation used in Eq. 4.6.8. Then the rate of accumulation of energy in the region will equal the heat flux only within an error of order Δr.

There are, of course, still other coordinate systems. We do not cover them here, however, because the methods are similar to what we used in the cylindrical case and space does not permit including them.

All of the material above deals only with linear problems. Nonlinearity is a difficult subject to treat with any generality because there is not just one kind of nonlinearity. We will treat a particular example, therefore, and state simply that the ideas can be applied to many other problems. The particular case is a simplification of a kind of parabolic problem that arises in fluid mechanics:

$$u\frac{\partial u}{\partial x}=\frac{\partial}{\partial x}\left(\frac{u^2}{2}\right)=\nu\frac{\partial^2 u}{\partial y^2} \tag{4.6.11}$$

This can be approximated by the Crank–Nicolson method

$$u_{i+1,j}^2-u_{i,j}^2=\frac{\nu\Delta x}{\Delta y^2}\left[\left(u_{i,j+1}-2u_{i,j}+u_{i,j-1}\right)+\left(u_{i+1,j+1}-2u_{i+1,j}+u_{i+1,j-1}\right)\right] \tag{4.6.12}$$

If the values at $x=i\Delta x$ are assumed known, we need to solve a nonlinear tridiagonal system of algebraic equations for u at $x=(i+1)\Delta x$. This can be done quite efficiently in a few iterations by a combination of the Newton–Raphson method and a tridiagonal linear system solver. (The solution at $x=i\Delta x$ is usually a good initial guess at the solution at $x=(i+1)\Delta x$ and this helps considerably.) Note also that there is no point in iterating the solution to Eq. 4.6.12 to complete convergence; the result would still be incorrect due to the truncation error in Eq. 4.6.12 itself.

An alternative applicable to this particular case is to linearize directly. Thus we can write

$$u_{i+1,j}=u_{i,j}+\left(u_{i+1,j}-u_{i,j}\right)=u_{i,j}+\Delta u_{i,j} \tag{4.6.13}$$

Then Eq. 4.6.12 becomes

$$u_{i,j}^2+2u_{i,j}\Delta u_{i,j}+\left(\Delta u_{i,j}\right)^2=\frac{\nu\Delta x}{\Delta y^2}\left[2\left(u_{i,j+1}-2u_{i,j}+u_{i,j-1}\right)\right.$$

$$\left.-\left(\Delta u_{i,j+1}-2\Delta u_{i,j}+\Delta u_{i,j-1}\right)\right. \tag{4.6.14}$$

The nonlinear term $(\Delta u_{ij})^2 \simeq \Delta x^2(\partial u/\partial x)^2_{i,j}$ is of the same order as the finite difference truncation error in Eq. 4.6.12. Neglecting it increases the error in the numerical method, but the difference is not large and the sacrifice in accuracy (or the increased number of steps need to compensate for it) may be well worth the advantage of having to solve a linear rather than a nonlinear system of equations.

This is known as the Δ *form* of a method and can be applied to linear as well as nonlinear problems. It is essentially equivalent to the first iteration of an iterative solution method. In the linear case the choice is mostly a matter of taste. In the nonlinear case it can lead to linearization with small loss of accuracy; however, this method does not work in every case and it may be necessary to use an iterative method to produce an accurate solution.

7. ELLIPTIC PDEs: I. FINITE DIFFERENCING

Elliptic partial differential equations usually arise either from the steady state cases of problems the time dependent versions of which are described by parabolic or hyperbolic PDEs or from problems in which the time dependence has an assumed form. The simplest and most common elliptic PDE is Laplace's equation, which in two dimensions is

$$\nabla^2\phi = \frac{\partial^2\phi}{\partial x_1^2} + \frac{\partial^2\phi}{\partial x_2^2} = 0 \qquad (4.7.1)$$

This equation arises in the electrostatic, heat conduction, and fluid mechanics areas, among others. Other well-known elliptic equations are Poisson's equation and Helmholtz's equation. The former is the inhomogeneous form of Laplace's equation, which occurs when there are sources (electrical charges, heat sources, or fluid sources) in the domain. The latter equation is Laplace's equation with a term proportional to ϕ included. It arises most commonly when one seeks sinusoidal solutions to the wave equation. There are other important elliptic equations, but for our purposes, the differences are less important than the similarities and the numerical methods that work for Laplace's equation can be applied to the others as well. Nonlinear elliptic equations are also common, arising typically in problems in which material properties depend on the solution (temperature dependent thermal conductivity, for example).

The simplest numerical treatment of Laplace's equation is what one might expect from what we have done up until now. One lays a rectangular mesh over the region for which the solution is desired (see Fig. 4.9). For the present we will assume that the region is rectangular so that the boundaries coincide with the mesh lines. (The more general case is discussed later, since it introduces some difficulties.) The derivatives in Laplace's equation are ap-

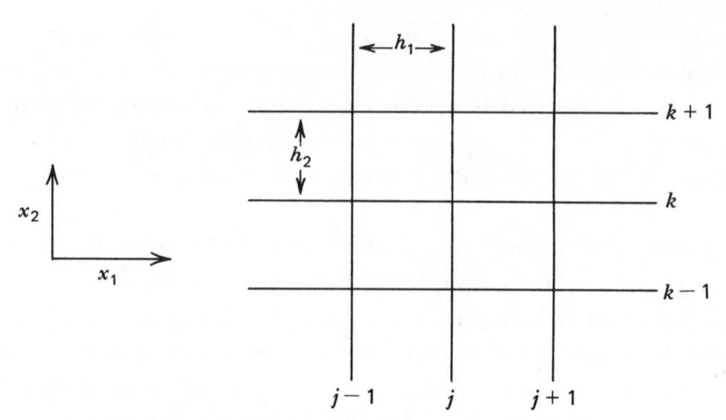

Fig. 4.9. Finite difference mark for Laplace's equation.

proximated simply by finite differences in the usual way:

$$\left.\frac{\partial^2\phi}{\partial x_1^2}\right|_{j,k} \simeq \frac{\phi_{j+1,k}-2\phi_{j,k}+\phi_{j-1,k}}{h_1^2}$$

$$\left.\frac{\partial^2\phi}{\partial x_2^2}\right|_{j,k} = \frac{\phi_{j,k+1}-2\phi_{j,k}+\phi_{j,k-1}}{h_2^2} \tag{4.7.2}$$

and the result is substituted in Eq. 4.7.1. For convenience and because the general case differs only by being more cumbersome to work with, we make the further simplifying assumption that $h_1=h_2$. Thus at any interior point of the mesh, the finite difference form of Laplace's equation is

$$\phi_{j,k}-\tfrac{1}{4}\left(\phi_{j+1,k}+\phi_{j-1,k}+\phi_{j,k+1}+\phi_{j,k-1}\right)=0 \tag{4.7.3}$$

It is interesting to note that this equation states that the value at each point is simply the average of the values at the four nearest neighboring points. The computational molecule for this method is shown in Fig. 4.10. Values of the coefficients are also shown, since this is a convenient way to display finite difference schemes for elliptic equations. This approximation is so commonly used that it is known as *the* five point different operator.

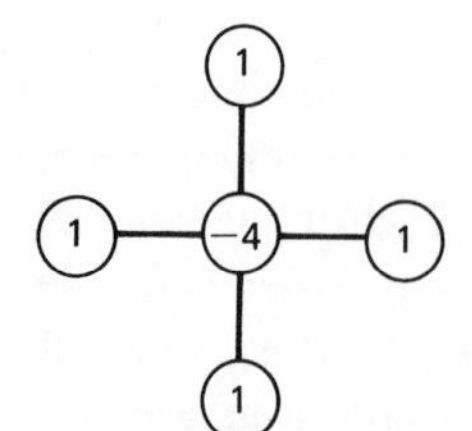

Fig. 4.10. Computational molecule for second order approximation of Laplace's equation.

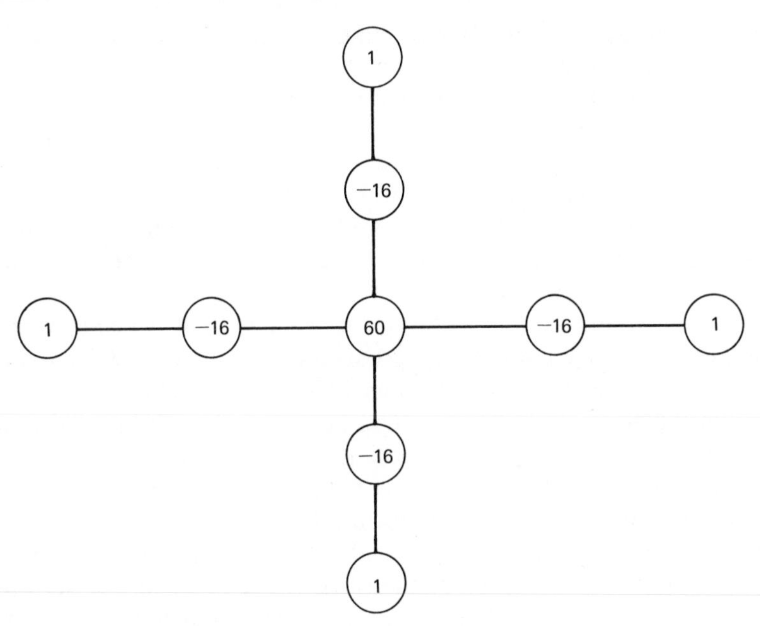

Fig. 4.11. Computational molecules for fourth order approximation of Laplace's equation.

Of course, this is not the only way in which Laplace's equation can be approximated by finite differences, but it has been the most popular choice. In cases in which higher accuracy is needed, one has to choose between using a smaller mesh or a higher order difference method, the latter being the better choice in many cases. One approach is to use the fourth order estimate of the second derivative given in Table 3.1. For Laplace's equation this leads to the difference scheme represented by Fig. 4.11. This method has obvious problems near the boundary and has been falling from favor as a result. An interesting alternative is provided by the method represented by Fig. 4.12, which is an application of the compact fourth order method of Eq. 3.13.6. This is a compact nine-point operator. Its application to the Poisson or Helmholtz

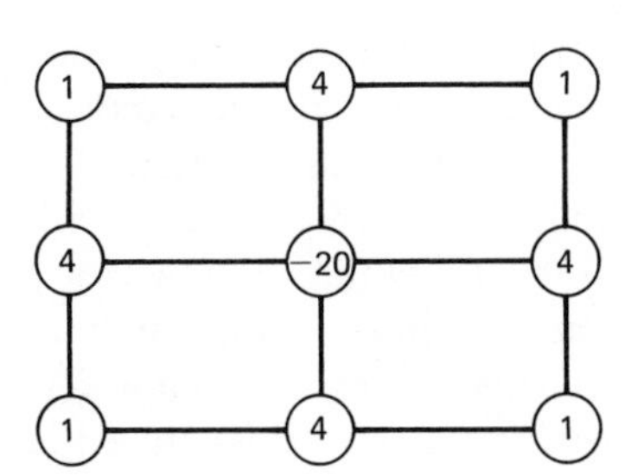

Fig. 4.12. Computational molecule for the compact fourth order method applied to Laplace's equation.

equations requires extra care. It is fourth order accurate and can be derived by the use of the approximation of Eq. 3.1.12.

In recent years the use of Fourier transforms for the estimation of derivatives and for the solution of elliptic PDEs has gained considerable popularity (see Section 15). These methods are extremely powerful for problems in which the boundaries are rectangular or nearly so.

It is also possible to derive approximations to Laplace's equation by integrating it over finite areas. If the equation is simply integrated over a small rectangular region and central difference formulas are applied to the estimation of the derivatives that arise, the results given above may be recovered. One alternative to this method is to write Laplace's equation as a system of three first order equations before integrating, which gives an elliptic version of the Keller Box method. A still more important alternative is the *finite element method*, which has gained considerable popularity in recent years because it has the ability to handle oddly-shaped regions with relatively little difficulty. This method is described briefly in Section 14.

Next, we consider the boundary conditions. Appropriate boundary conditions for elliptic problems provide the value of the function, the derivative of the function normal to the boundary, or some linear combination of the two, at each point on the boundary. For the second order method (Eq. 4.7.3) the boundary conditions may be treated similarly to the way in which they were treated in Section 3.12. When the function is given at a particular boundary point, no difference equation is written for the point and the given value is used where it occurs in the equation(s) at the neighboring point(s). When the derivative or a combination of derivative and function are given, one introduces a fictitious point outside the boundary and uses an equation analogous to Eq. 3.12.5 to represent the boundary condition and the finite difference equation for the boundary point is required. The provided boundary values cannot all be zero so some of the equations must contain inhomogeneous terms.

The method represented by Fig. 4.11 has difficulty with boundary conditions. The problem is exactly that which we encountered in Section 3.13 and can be dealt with by the methods that were used there. The best method, as was found there, is to use a third order formula at the point next to the boundary. This reduces the overall accuracy somewhat but produces reliable results.

Since we are faced with the necessity of solving a large set of linear algebraic equations, it is natural to put the problem in a standard linear algebra context. We note first that, though the dependent variable is double subscripted, it is properly regarded as a vector. As in the similar case encountered with parabolic equations, there are two natural ways of regarding the unknowns $\phi_{j,k}$ as the elements of a vector: we can list the elements in x_1 = constant lines sequentially or we can do the same for x_2 = constant lines. The choice is arbitrary, but if the number of points in the 1-direction M is less than the number of points in the 2-direction N, it is preferable to use the ordering

$$\boldsymbol{\phi}=\begin{pmatrix} \phi_{1,1} \\ \phi_{2,1} \\ \vdots \\ \phi_{M,1} \\ \phi_{1,2} \\ \vdots \\ \phi_{M,2} \\ \vdots \\ \phi_{M,N} \end{pmatrix} \tag{4.7.4}$$

The matrix corresponding to the system of equations of Eq. 4.7.3 is then

$$A=\begin{pmatrix} -4 & 1 & & & & 1 & & & \\ 1 & -4 & 1 & & & & & & \\ & 1 & -4 & 1 & & & & & \\ 1 & & & & & & & & 1 \\ & 1 & & & & & & & \\ & & & 1 & & & 1 & -4 & 1 \\ & & & & 1 & & & 1 & -4 \end{pmatrix} \tag{4.7.5}$$

where all elements not shown are zero. The lowest 1 in the first column occurs in the $M+1$st position and the last 1 in the first row is also in the $M+1$st position. The set of equations that must be solved can be written

$$A\boldsymbol{\phi}=\mathbf{b} \tag{4.7.6}$$

where $\mathbf{b}$ is a vector the elements of which are mostly zeros; the nonzero elements arise from the boundary conditions.

The method of solution of the system of equations 4.7.6 depends to a large extent on the properties of the matrix A. The size of the matrix is $MN \times MN$, which makes it quite large even for fairly small values of M and N. The matrix is very sparse (most of its elements are zero) however, and a good solution method will take advantage of this fact. The matrix A is also banded, that is, all the nonzero elements lie on diagonals not too distant from the principal diagonal. Numerical methods designed for the solution of systems involving banded matrices require a number of arithmetic operations proportional to the product of the size of the matrix (MN in the present case) and the bandwidth (M) squared. Thus to solve this system would require NM^3

operations or M^2 operations per point. This is quite a bit and it seems natural, therefore, to seek other methods. In particular, except for problems in which the geometry is very regular, the best approach to the solution of this type of problem is to use iterative methods, to which we devote the next several sections.

We end this section with a discussion of the case of irregular boundaries, for which there are several choices. One can use a coordinate transformation, which makes the solution region a rectangle in the new coordinate system. The price one pays is that equations are more complicated in the transformed coordinate system and that the generation of the coordinates will require some work as well. Generally, this method is not effective for solving simple linear elliptic problems, but it may be valuable for nonlinear equations for which the transformation does not increase the complexity of the equation.

A second choice is the finite element method mentioned above. This has become the method of choice for problems in which the equations are linear or in which the nonlinearity is not too severe. The finite element method is so effective that it now dominates the field of solid mechanics. Finite elements are discussed in Section 14, but for a complete coverage of this topic the reader is referred to texts on this specific subject.

The method that we deal with here is conceptually the simplest and is reasonably effective for simple equations. We use a rectangular grid despite the irregular boundary. At any point where the neighbors are all interior points, we use the approximation of Eq. 4.7.3. At any point that has boundary point(s) as neighbors, we use the finite difference approximations of Eq. 3.1.8 for uneven mesh spacing. Thus at point (x_i, y_j) of Fig. 4.13 we use

$$\left.\frac{\partial^2\phi}{\partial x^2}\right|_{i,j} \simeq \frac{2}{h(h+\Delta x_i)}\phi_{i-1,j} - \frac{2}{h\,\Delta x_i}\phi_{i,j} + \frac{2}{\Delta x_i(h+\Delta x_i)}\phi_{B,j} \quad (4.7.7\text{a})$$

$$\left.\frac{\partial^2\phi}{\partial y^2}\right|_{i,j} \simeq \frac{2}{h(h+\Delta y_j)}\phi_{i,j-1} - \frac{2}{h\,\Delta y_j}\phi_{i,j} + \frac{2}{\Delta y_j(h+\Delta y_j)}\phi_{i,B} \quad (4.7.7\text{b})$$

in which the values of $\phi_{B,j}$ and $\phi_{i,B}$ are determined by the boundary conditions. Here Δy_j represents the distance of the boundary point from the last regular mesh point on the line $y=y_j$, and Δx_i is the distance of the boundary from the last mesh point on the line $x=x_i$. The resulting equations are very similar to Eq. 4.7.4. The major complication is in the extra bookkeeping needed to keep track of which points are adjacent to the boundary, since each row has a different number of mesh points. Aside from this (which makes programming quite a bit more complicated), the methods of solution used in rectangular regions also work in irregular regions, although sometimes not as effectively.

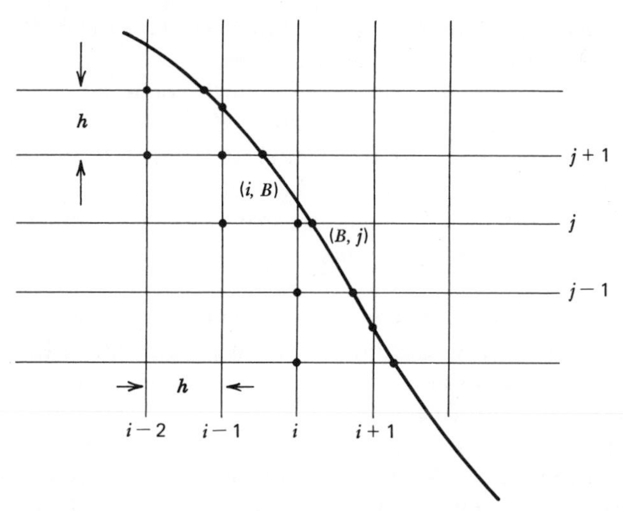

Fig. 4.13. Finite differencing of Laplace's equation for irregular boundaries.

8. ELLIPTIC PDEs: II. JACOBI ITERATION METHOD

In the preceding section we saw that direct solution of the finite difference equations for elliptical PDEs is expensive. An exception to this is the case in which the boundaries are rectangular; one can then use fast Fourier transforms (or related methods) to obtain the solution quickly. The alternative to direct solution is use of an iterative method, discussed in the next few sections. We begin with a short, general discussion of iterative methods.

Iterative methods can be applied to the solution of any set of linear (or nonlinear) algebraic equations. It is useful in what follows to use both the matrix and component forms of the equations. An iterative method applied to Eq. 4.7.6 is a procedure of the type

$$A_1\boldsymbol{\phi}^{(p+1)} = A_2\boldsymbol{\phi}^{(p)} + \mathbf{d} \tag{4.8.1}$$

where A_1 and A_2 are matrices and $\mathbf{d}$ is a vector. Given some initial guess $\phi^{(0)}$ for the solution, we use Eq. 4.8.1 to find $\phi^{(1)}$, $\phi^{(2)}$, and so on. If this is to be a satisfactory means of solving the original equation, it should have the following properties:

1. It should converge to the exact solution of the original equation. That is, we should have

$$\lim_{p\to\infty} \boldsymbol{\phi}^{(p)} = \boldsymbol{\phi} \tag{4.8.2}$$

where $\boldsymbol{\phi}$ is a solution of $A\boldsymbol{\phi} = \mathbf{b}$.

2. The convergence should be rapid if the method is to be efficient, that is, the limit (Eq. 4.8.2) should be reached quickly.
3. So that each step does not require much computation, the matrix A_1 should be easy to invert (the best possibility for this is that A_1 is diagonal; other good ones are that A_1 is tridiagonal or triangular) and the matrix A_2 should be as simple as possible to facilitate the computation of $A_2\boldsymbol{\phi}^{(p)}$. Generally, this means that A_2 should be sparse, which is possible only if A is also sparse. Iterative methods are thus best for sparse matrices.

It is not difficult to construct an iterative method, but constructing one with all of the desired properties is another story. We begin with one of the simplest possible iterative schemes and systematically improve upon it, starting with a relatively poor method for pedagogical reasons.

The iterative method for Laplace's equation that we will look at first is obtained by writing $A = -D + B$ where D is the diagonal matrix $D = 4I$ and B is the matrix obtained by zeroing out the main diagonal of the matrix 4.7.5. Then letting $A_1 = D$ and $A_2 = B$, we have the iterative scheme

$$D\boldsymbol{\phi}^{(p+1)} = B\boldsymbol{\phi}^{(p)} - \mathbf{b} \tag{4.8.3}$$

Or since $D^{-1} = (\frac{1}{4})I$, this equation is easily rewritten as

$$\boldsymbol{\phi}^{(p+1)} = D^{-1}B\boldsymbol{\phi}^{(p)} - D^{-1}\mathbf{b} \equiv M_J\boldsymbol{\phi}^{(p)} - \mathbf{d} \tag{4.8.4}$$

which is the *Jacobi iterative method*. In component form this equation is

$$\phi_{j,k}^{(p+1)} = \tfrac{1}{4}\left(\phi_{j+1,k}^{(p)} + \phi_{j-1,k}^{(p)} + \phi_{j,k+1}^{(p)} + \phi_{j,k-1}^{(p)}\right) - \tfrac{1}{4}b_{j,k} \tag{4.8.5}$$

which computes the new value of $\phi_{j,k}$ by averaging the values of its neighbors at the preceding iteration and adding the inhomogeneous (boundary condition) term. This method is easily generalized to the solution of any set of linear algebraic equations.

Next, we determine whether this method satisfies the criteria listed above, primary among which is the question of convergence. The standard method for investigating the convergence of iterative methods is to introduce an error vector that is defined as the difference between the nth iterate produced by a method and the exact solution of the system of equations

$$\boldsymbol{\varepsilon}^{(p)} = \boldsymbol{\phi}^{(p)} - \boldsymbol{\phi} \tag{4.8.6}$$

If the method converges, then for large p, $\boldsymbol{\phi}^{(p)} = \boldsymbol{\phi}^{(p+1)} = \boldsymbol{\phi}$ so that

$$\boldsymbol{\phi} = M_J\boldsymbol{\phi} - \mathbf{d} \tag{4.8.7}$$

Subtracting this equation from Eq. 4.8.4 we get a homogeneous equation for the error vector

$$\boldsymbol{\varepsilon}^{(p+1)} = M_J \boldsymbol{\varepsilon}^{(p)} \tag{4.8.8}$$

If the Jacobi method converges, we must have

$$\lim_{p \to \infty} \boldsymbol{\varepsilon}^{(p)} = \mathbf{0} \tag{4.8.9}$$

which is what we must now try to prove.

To prove convergence, it is convenient to introduce the eigenvectors and eigenvalues of the matrix M_J. Because in this particular case the matrix is simple, this is easily done. In fact in component form the eigenvectors are given by Eq. 4.5.3. The eigenvalues are easily computed and are

$$\lambda_{m,n} = \tfrac{1}{2}\left(\cos\frac{m\pi}{M+1} + \cos\frac{n\pi}{N+1}\right) \qquad \begin{matrix} m=1,2,\ldots M \\ n=1,2,\ldots N \end{matrix} \tag{4.8.10}$$

We are now in a position to give a formal analysis of the convergence properties of the Jacobi method. The method is formal because, as with some of the other proofs we have given, the computation requires more work than simple application of the method. The investigation begins by writing the initial error vector $\boldsymbol{\varepsilon}^{(0)}$ as a linear combination of the eigenvectors, which was justified in Section 1 of this chapter. We have

$$\boldsymbol{\varepsilon}^{(0)} = \sum_{m=1}^{M} \sum_{n=1}^{N} \varepsilon_{m,n} \psi^{(m,n)} \tag{4.8.11}$$

Then since $M_J \psi^{(m,n)} = \lambda_{m,n} \psi^{(m,n)}$, we find by direct computation

$$\begin{aligned} \boldsymbol{\varepsilon}^{(1)} &= M_j \boldsymbol{\varepsilon}^{(0)} = M_j \sum_m \sum_n \varepsilon_{m,n} \psi^{(m,n)} \\ &= \sum_m \sum_n \lambda_{m,n} \varepsilon_{m,n} \psi^{(m,n)} \end{aligned} \tag{4.8.12}$$

and continuing the calculation

$$\boldsymbol{\varepsilon}^{(p)} = \sum_m \sum_n \lambda_{m,n}^p \varepsilon_{m,n} \psi^{(m,n)} \tag{4.8.13}$$

from which we may obtain the information we seek.

After many iterations, in other words, for large p, the largest term in the sum of Eq. 4.8.13 will be the one containing the largest eigenvalue. It is clear that a necessary and sufficient condition for the error to go to zero for large p is that *all* of the eigenvalues of the matrix M_j must be less than unity in magnitude. This is the case for any iterative method. Since the magnitude of the largest eigenvalue of a matrix plays such a critical role in computational linear algebra, it is given a special name—the *spectral radius*.

For the particular case of the Jacobi matrix M_J associated with Laplace's equation, an explicit expression for the eigenvalues, at least for the rectangular case, is Eq. 4.8.10. By inspection it is clear that all of the $\lambda_{m,n}$ are smaller than unity and, therefore, the Jacobi method is convergent. We can learn more. The largest eigenvalue in this case is the one with $n=m=1$. The one with $n=N$ and $m=M$ is just as large and of opposite sign, but it is sufficient to deal with just one of these. Now from a straightforward Taylor series expansion, it is easy to see that for large M and N that

$$\lambda_{\max}=\tfrac{1}{2}\left(\cos\frac{\pi}{M+1}+\cos\frac{\pi}{N+1}\right)\simeq 1-\frac{\pi^2}{4}\left(\frac{1}{(M+1)^2}+\frac{1}{(N+1)^2}\right) \tag{4.8.14}$$

Thus the largest eigenvalue is indeed less than unity, but only very slightly so if M and N are large. As a result, the convergence of this method is very slow. To see just how slow it is, let us take $M=N$ and for large N we can make the approximation

$$\lambda_{\max}^p\simeq\left(1-\frac{\pi^2}{2N^2}\right)^p\simeq e^{-\pi^2 p/2N^2} \tag{4.8.15}$$

This quantity will have to be of order of the desired accuracy of the calculation. Calling δ the allowable error, setting it equal to Eq. 4.8.15, and solving for p, we see that the number of iterations required to produce the desired result is approximately

$$p\simeq-\frac{2N^2}{\pi^2}\ln\delta \tag{4.8.16}$$

Thus if the required accuracy is 10^{-3} (a typical value), the number of iterations needed to reduce the error to that value is about $1.6N^2$. Since each iteration requires four arithmetic operations per mesh point, or a total of $4N^2$ operations, the number of arithmetic operations needed to obtain the desired results is approximately $6N^4$, which is more than is required by the direct method! Thus the Jacobi method is not an attractive choice and better methods are needed.

To complete this section, we look at the Jacobi method in another way to get further insight into it in order to find improved methods. We begin with the observation made earlier that elliptic equations usually result from taking the steady state limit of a problem that is governed by a parabolic or hyperbolic equation. Thus we can find solutions of Laplace's equation by solving the heat equation with the boundary conditions given for Laplace's equation and some arbitrary initial condition. The solution should eventually settle down to the solution of Laplace's equation. Equations other than the heat equation could also be used for this purpose. All we require is that the solution settle down to a solution of Laplace's equation. There need not be any connection to a physical problem at all and purely artificial time dependent problems have been used.

Thus suppose we solve the heat equation by the Euler method. Choosing $h_1 = h_2 = h$ for convenience and letting $\beta = D\Delta t/2h^2$, we would then have

$$\phi_{j,k}^{(p+1)} = (1-2\beta)\phi_{j,k}^{(p)} + 2\beta\left(\phi_{j+1,k}^{(p)} + \phi_{j-1,k}^{(p)} + \phi_{j,k+1}^{(p)} + \phi_{j,k-1}^{(p)}\right) \quad (4.8.17)$$

The distinction between what we are trying to do now and what we were after earlier is that in the current case we have no interest in computing the time history accurately. We simple want to get to the steady state as quickly as possible. It seems logical, therefore, to take the biggest time step that stability will allow. For Eq. 4.8.17 the limit is $\beta = \frac{1}{2}$ and using it in Eq. 4.8.17 immediately reduces it to Eq. 4.8.5, the basic equation of the Jacobi method! Thus the Jacobi method for Laplace's equation is equivalent to the Euler method for the heat equation.

Iterative methods for elliptic equations are frequently known as *relaxation methods* because they cause the solution of the heat equation to relax to a steady state. The Jacobi method is also known as *simultaneous relaxation*, since the solution at each of the mesh points could be computed simultaneously from the old values. We will search for methods of speeding up the convergence of iterative methods.

Example 4.6

As our example for elliptic equation solvers, we will use Laplace's equation

$$\nabla^2\phi = 0$$

with the boundary conditions

$$\phi(0,x) = (0,y) = 0$$

$$\phi(x,1) = 100x$$

$$\phi(1,y) = 100y$$

The exact solution is $\phi = 100xy$. Since this function is linear in both variables, it is differentiated exactly by a second order finite difference "approximation" and we can follow the convergence of the method accurately.

The program given below was used to produce the solution to this problem. Note that, as with many of the methods used for parabolic equations, the Jacobi method requires the use of a second array (FF in the program) for

```
      PROGRAM JACOBI
C-----THIS PROGRAM SOLVES LAPLACE'S EQUATION IN A SQUARE.
      DIMENSION F(21,21),FF(21,21),EXACT(21,21),ERROR(21,21),X(21),
     1 Y(21)
   10 WRITE (5,110)
  110 FORMAT ( ' CIVE THE NUMBER OF X INTERVALS')
      READ (5,120) NX
  120 FORMAT (I)
      WRITE (5,130)
  130 FORMAT ( ' GIVE THE NUMBER OF Y INTERVALS')
      READ (5,120) NY
      WRITE (5,140)
  140 FORMAT ( ' RESULTS PRINTED EVERY K STEPS (K EVEN), PRINT K')
      READ (5,120) K
C-----COMPUTE THE PARAMETERS AND EXACT SOLUTION
      KH = K / 2
      NXP1 = NX + 1
      NYP1 = NY + 1
      DX = 1. / NX
      DY = 1. / NY
      EPS = 1.E-6
      DO 1 I=2,NX
      F(I,1) = 0.
      FF(I,1) = 0.
      X(I) = (I - 1) * DX
      F(I,NYP1) = 100. * X(I)
      FF(I,NYP1) = 100. * X(I)
      DO 1 J=2,NY
      F(I,J) = 1.
      Y(J) = (J - 1) * DY
    1 EXACT(I,J) = 100. * X(I) * Y(J)
      DO 2 J=2,NY
      F(1,J) = 0.
      FF(1,J) = 0.
      F(NXP1,J) = 100. * Y(J)
    2 FF(NXP1,J) = 100. * Y(J)
C-----MAIN LOOP.  DO K SWEEPS AND THEN PRINT RESULTS.
      ITNUM = 0
   20 DO 4 L=1,KH
      DO 3 I=2,NX
      DO 3 J=2,NY
    3 FF(I,J) = .25 * (F(I-1,J) + F(I+1,J) + F(I,J-1) + F(I,J+1))
      DO 4 I=2,NX
      DO 4 J=2,NY
      F(I,J) = .25 * (FF(I-1,J) + FF(I+1,J) + FF(I,J-1) + FF(I,J+1))
    4 ERROR(I,J) = EXACT(I,J) - F(I,J)
C-----WRITE THE RESULTS FOR ITERATION NUMBER ITNUM2
      ITNUM = ITNUM + K
      WRITE (5,200) ITNUM
  200 FORMAT ( ' ERROR FOR ITERATION NUMBER ',I3)
      DO 5 J=2,NY
    5 WRITE (5,210) (ERROR(I,J) , I=2,NX)
  210 FORMAT ( 21E12.6)
C-----CHECK IF RESULT IS CONVERGED.
      DO 6 I=2,NX
      DO 6 J=2,NY
    6 IF (ABS(ERROR(I,J)).CT.EPS) GO TO 20
C-----ASK FOR ANOTHER CASE.
      WRITE (5,220)
  220 FORMAT ( ' IF YOU WANT ANOTHER CASE, TYPE 1')
      READ (5,120) IFF
      IF (IFF.EQ.1) GO TO 10
      STOP
      END
```

temporary storage. This not only increases the memory requirement of the program but also increases the complexity of the program itself and a number of steps—particularly the initialization of the field and the calculation of the new values of the dependent variable—need to be done twice.

Results were obtained using $\Delta x = \Delta y = h$. The error at the centerpoint for values of h are shown in Fig. 4.14. The exponential convergence behavior anticipated by the theory is borne out in this case. Specifically, we see that with half the mesh size four times as many iterations are required to achieve the same error reduction. This is the poor behavior of the method mentioned in our discussion.

Since the method is second order accurate, in order to increase the accuracy by a factor of four we need to cut the spatial interval in half. This requires four times as many space points and thus four times as much computation per iteration. To make matters worse, it also requires four times as many iterations to achieve the *same* accuracy. A reasonable calculation, however, is one in which the method is iterated only to the point at which the error due to lack of convergence is approximately equal to the finite difference truncation error. Thus more than four times as many iterations are needed and the total cost of

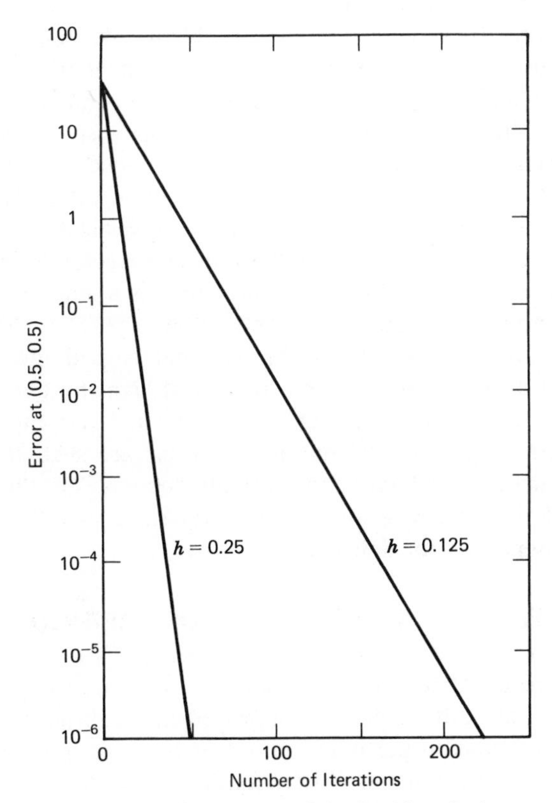

Fig. 4.14. Convergence of the Jacobi method.

the calculation is increased more than sixteen times. Clearly this is unacceptable and we need to look for other methods.

One method of reducing the cost is suggested by these results. We can begin by doing the calculation on a very coarse mesh, since it is relatively cheap. As suggested above, one should run until the convergence error is at least as small as the truncation error. Then we repeat the calculation with a half-size mesh. In this calculation one can use the results on the coarse mesh as the initial guess. The function values at the mesh points that were not part of the coarse mesh can be obtained by interpolation. Linear interpolation, which is second order accurate, is sufficient. Because the error on the coarse mesh results is approximately four times the error in the fine mesh results, we only need to iterate enough to reduce the error by a factor of four. This calculation costs much less than a fine mesh calculation starting with a poor initial guess, and it can be repeated on finer and finer meshes until the desired accuracy is achieved. This method is called *grid refinement*. A more sophisticated version of this idea—one that is based on methods better than Jacobi and both increases and decreases the mesh size—is known as the *multigrid method* and is an excellent but relatively new approach to obtaining accurate solutions to a wide variety of boundary value problems.

One final note. In solving parabolic problems, we found it valuable to expand the solution in eigenvectors of the matrix. The high index eigenvectors, the ones that change sign many times in the region of interest, were the ones that caused the stiffness of the equation and were responsible for the strict time step restriction. As we noted above, the convergence of the Jacobi method is controlled by the eigenvalues with largest absolute value. There are two of these, one positive and one negative. Further investigation shows that the largest positive eigenvalue corresponds to an eigenvector the components of which have the same sign everywhere and which is a smooth function, while the largest negative eigenvalue corresponds to an eigenvector the components of which oscillate in sign (this is the one that caused the trouble in the parabolic case). Hence the errors that remain after many iterations are composed of both very smoothly varying components and very rapidly fluctuating components. In most cases the initial conditions are such that the smooth component of the error is larger and it usually dominates in the late iterations. In Example 4.6 the error was found to be of the same sign at all points, which verifies this observation for that particular case.

9. ELLIPTIC PDEs: III. GAUSS–SEIDEL METHOD

A method is needed to reduce the amount of calculation required to achieve the desired accuracy. Although some improvement might come from reducing the amount of calculation per iteration, major improvement is more likely to come from reducing the number of iterations required. Proceeding in this latter direction, two ideas are introduced, one in this section and one in Section 11.

We have seen that each iteration of the Jacobi method for solving Laplace's equation is equivalent to a time step of the Euler method for the heat equation. All iterative methods for elliptic equations can be viewed as "relaxing" a parabolic or hyperbolic problem to a steady state. We want a method that gets to the steady state solution as quickly as possible; accurate time behavior is irrelevant. Thus we need a method that allows a bigger time step. One possibility is an implicit method. In principle, the Crank–Nicolson method allows an infinite time step and could get the solution in one "iteration," but since the problem to be solved at the new time step is essentially identical to the original problem, we must reject it. The ADI method is another possibility to be considered later, but it, too, is not free of difficulties. We begin by looking for ways to speed up the Jacobi method.

In the Jacobi method Eq. 4.8.5 shows that each new value of the function is computed entirely from old values. In fact the new values can be computed in any order but, if for no other reason than ease of programming, it makes sense to compute the new values in an orderly manner. In the program of Example 4.6 two arrays for the dependent variable and some duplicate programming was required in order to deal with this difficulty. In the program the new values are computed on one entire line, then computed on the next line up, and so on. When we come to computing $\phi_{j,k}$, we have already calculated the values of $\phi_{j-1,k}$ and $\phi_{j,k-1}$. Since we expect the new values to be better approximations than the old ones, it would seem that improvement could be achieved by using the updated values. This leads to replacing Eq. 4.8.5 by

$$\phi_{j,k}^{(p+1)} = \tfrac{1}{4}\left(\phi_{j-1,k}^{(p+1)} + \phi_{j+1,k}^{(p)} + \phi_{j,k-1}^{(p+1)} + \phi_{j,k+1}^{(p)}\right) - \tfrac{1}{4}b_{j,k} \tag{4.9.1}$$

which is the basis of the *Gauss–Seidel method*. It is also known as the *method of successive relaxation* because it computes the data in a sequential manner.

To find the matrix equivalent of this equation, we note that it is necessary to separate those parts of the matrix that represent the contributions of points $j-1, k$ and $j, k-1$ to the equation at the point j,k from the other terms in the equation. This is done easily, since the elements of the matrix A that produce these terms are those that lie above the main diagonal. We decompose the matrix, therefore, as

$$A = -D + B = -D + L + U \tag{4.9.2}$$

where $-D$ again represents the diagonal part of A, L is the (strictly) lower triangular portion, and U is the (strictly) upper triangular part. Thus the Gauss–Seidel iterative equation of Eq. 4.9.1 is equivalent to

$$(D-U)\boldsymbol{\phi}^{(p+1)} = L\boldsymbol{\phi}^{(p)} + \mathbf{b} \tag{4.9.3}$$

Now since $D-U$ is a triangular matrix, it is easily inverted (which is just another way of saying that Eq. 4.9.1 is almost trivially solved) and we can write formally

$$\boldsymbol{\phi}^{(p+1)} = (D-U)^{-1}L\boldsymbol{\phi}^{(p)} + (D-U)^{-1}\mathbf{b} = M_G\boldsymbol{\phi}^{(p)} + \mathbf{d}_G \tag{4.9.4}$$

Analysis of the rate of convergence of this method is not attempted here. We merely state that under fairly broad assumptions, the eigenvalues of the Gauss–Seidel iterative matrix M_G are precisely the squares of the corresponding eigenvalues of the Jacobi matrix M_J. This means that the Gauss–Seidel method converges whenever the Jacobi method does and, furthermore, one Gauss–Seidel iteration is roughly equivalent to two Jacobi iterations, so that the *G–S* method converges in half the number of iterations. Since this improvement is achieved at absolutely no cost in computer time and with a simpler program, there is never any reason to use the Jacobi method.

Finally, we note that the Gauss–Seidel method can be used for solving parabolic equations. For the heat equation, with $h=\Delta x=\Delta y$ and $\beta=D\Delta t/2h^2$ as before, we have

$$(1+2\beta)\phi_{j,k}^{(p+1)} = (1-2\beta)\phi_{j,k}^{(p)} + \beta\left(\phi_{j+1,k}^{(p)} + \phi_{j-1,k}^{(p+1)} + \phi_{j,k+1}^{(p)} + \phi_{j,k-1}^{(p+1)}\right) \tag{4.9.5}$$

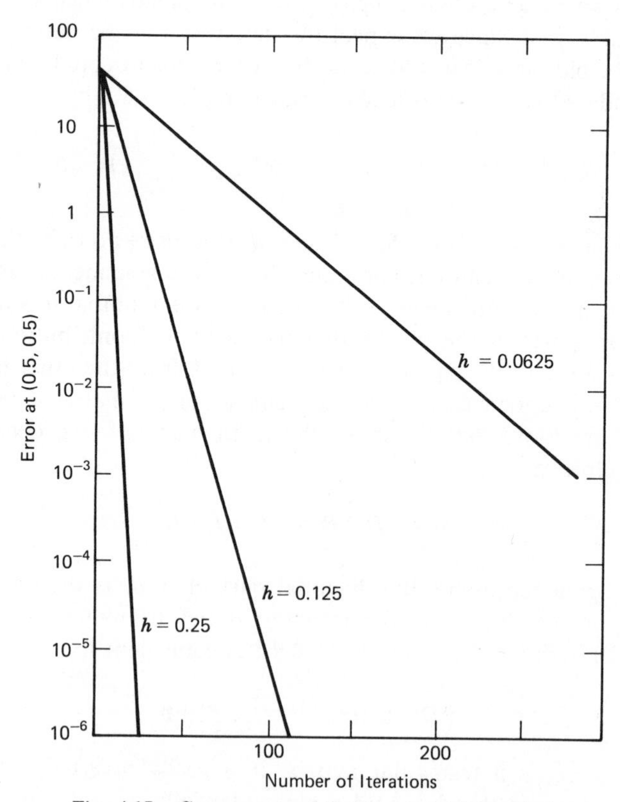

Fig. 4.15. Convergence of the Gauss–Seidel method.

This method is first order time accurate and second order space accurate (just as the Euler method) and is stable for values of β up to $\frac{1}{2}$; for $\beta=\frac{1}{2}$ it becomes Eq. 4.9.1 as might be expected.

Example 4.7

Doing the problem of Example 4.6 with Gauss–Seidel is quite simple. In fact we only need to take the program used there and eliminate parts of it. Specifically, array FF does not need to be declared in the DIMENSION statement, variable KH is not needed (and K need not be even), the statements initializing FF can be dropped, and the double DO loop (loops DO 3 and DO 4) for the main calculation can be reduced to a single loop.

Results similar to those given in Fig. 4.14 for the Jacobi method are given in Fig. 4.15 for three values of h. There are no surprises; the Gauss–Seidel method converges almost exactly twice as fast as the Jacobi method for this problem. Despite this, the method still has the principal disadvantage of the Jacobi method—increased accuracy comes only at a high price.

10. ELLIPTIC PDEs: IV. LINE RELAXATION METHOD

The Gauss–Seidel method obtains its advantage over the Jacobi method by making use of the new values of the function whenever they are available. In the Gauss–Seidel method half the data used to compute a new value are new and half are old, and half as many iterations are required. It is natural to ask whether further improvement is obtainable by including more new data in the calculation of the updated value of the function. If new data were used at all four nearest neighboring points, we would have an implicit method and the matrix would be very difficult to invert. This leaves just one further possibility—using new data at three neighboring points. Such a method is readily constructed. In Eq. 4.9.1 we can either replace $\phi_{j+1,k}^{(p)}$ by $\phi_{j+1,k}^{(p+1)}$ or $\phi_{j,k+1}^{(p)}$ by $\phi_{j,k+1}^{(p+1)}$. The choice is discussed below. For the moment let us adopt the first choice. We then have

$$\phi_{j,k}^{(p+1)}=\tfrac{1}{4}\left(\phi_{j-1,k}^{(p+1)}+\phi_{j+1,k}^{(p+1)}+\phi_{j,k-1}^{(p+1)}+\phi_{j,k+1}^{(p)}\right)-\tfrac{1}{4}b_{j,k} \qquad (4.10.1)$$

which can be written

$$-\tfrac{1}{4}\phi_{j-1,k}^{(p+1)}+\phi_{j,k}^{(p+1)}-\tfrac{1}{4}\phi_{j+1,k}^{(p+1)}=\tfrac{1}{4}\left(\phi_{j,k-1}^{(p+1)}+\phi_{j,k+1}^{(p)}\right)-\tfrac{1}{4}b_{j,k} \qquad (4.10.2)$$

In this equation the entire right-hand side is known because we already have $\phi_{j,k-1}^{(p+1)}$ when we come to the kth line. Equations 4.10.2 form a tridiagonal system of equations and can be solved by the standard algorithm. This method is known as the *line Gauss–Seidel method*, sometimes as *line relaxation*, or as the *method of lines*, and is equivalent to treating one direction implicitly. A line Jacobi method is possible but there is no reason for considering it.

We have not put this method into matrix form or written the equivalent time method; these are not difficult to do. It is important, however, to compare this method with the (point) Gauss–Seidel method. Line relaxation requires approximately half as many iterations as the Gauss–Seidel method, or only one-fourth as many as the Jacobi method; it also requires more computation per iteration. There is usually a slight advantage to the line method.

There is, however, another factor to be considered. In some problems the solution varies much more rapidly in one direction than in the other and more points are used in the rapidly varying direction. Here there is considerable advantage to using the line relaxation method, provided the lines are in the direction of the more severe variation, which is usually the direction with the larger number of points. (It is better to treat the tough direction implicitly.)

Example 4.8

The problem of Example 4.6 is now done using line relaxation. The results are shown in Fig. 4.16 and are in accord with our expectations. The method converges in roughly half as many iterations as the point Gauss–Seidel method.

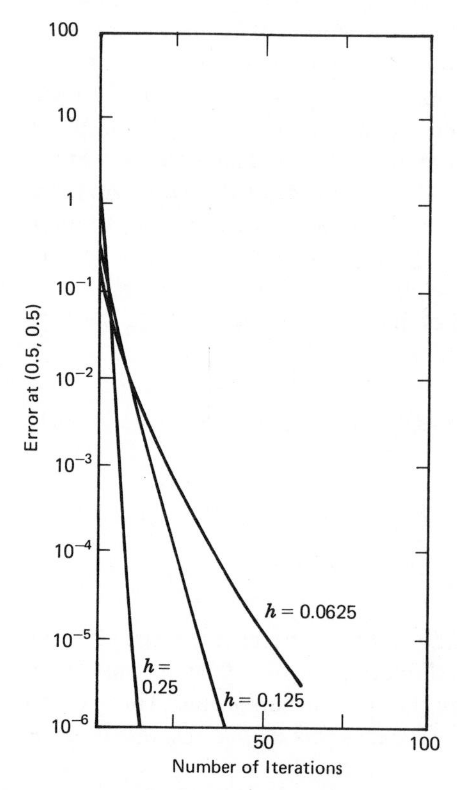

Fig. 4.16. Convergence of the line relaxation method.

11. ELLIPTIC PDEs: V. SUCCESSIVE OVERRELAXATION (SOR) METHOD

None of the methods discussed so far is enough of an improvement on direct solution methods. We still need to look for ideas to increase the rate of convergence of iterative methods. To see how improved methods might be found, let us look more closely at the convergence of the methods.

Iterative solutions of elliptic PDEs are akin to the time developing solutions of parabolic PDEs and we can use our knowledge of the latter to understand the former. We expect that the solution starts as the initial condition and relaxes in a steady smooth manner to the correct solution. The solution moves monotonically, but at a decreasing rate, toward the exact solution.* Given this behavior, we can almost guess the result of the next iteration without computing it. In other words, the past history of the calculation contains important information that can be used to aid in the convergence of the method. Instead of using $\phi_j^{(n)}$ as the input for the next iteration, we can extrapolate the preceding results. This leads to the procedure known as *extrapolation*, *acceleration*, or *overrelaxation*.

To introduce the idea, suppose that we use one of the methods previously discussed to compute $\tilde{\boldsymbol{\phi}}^{(n)}$:

$$\tilde{\boldsymbol{\phi}}^{(n)} = M\boldsymbol{\phi}^{(n-1)} + \mathbf{d} \tag{4.11.1}$$

In the preceding methods $\tilde{\boldsymbol{\phi}}^{(n)}$ would be used as the input guess for the next iteration, that is, $\boldsymbol{\phi}^{(n)} = \tilde{\boldsymbol{\phi}}^{(n)}$. Introducing the key idea of extrapolation, instead of using $\tilde{\boldsymbol{\phi}}^{(n)}$ as the input guess, we use an extrapolated value

$$\boldsymbol{\phi}^{(n)} = \tilde{\boldsymbol{\phi}}^{(n)} + \alpha(\tilde{\boldsymbol{\phi}}^{(n)} - \boldsymbol{\phi}^{(n-1)}) = (1+\alpha)\tilde{\boldsymbol{\phi}}^{(n)} - \alpha\boldsymbol{\phi}^{(n)} = \omega\tilde{\boldsymbol{\phi}}^{(n)} + (1-\omega)\boldsymbol{\phi}^{(n-1)} \tag{4.11.2}$$

For this to be an extrapolation we must have $\alpha > 0$. For safety we expect that $\alpha < 1$. Thus ω, which is called the *overrelaxation factor*, should be such that $1 < \omega < 2$.

The two steps Eqs. 4.11.1 and 4.11.2 can be combined to give

$$\boldsymbol{\phi}^{(n)} = [\omega M + (1-\omega)I]\boldsymbol{\phi}^{(n-1)} + \omega\mathbf{d} \tag{4.11.3}$$

and the equation that describes the reduction of the error vector for this method is

$$\boldsymbol{\varepsilon}^{(n)} = [\omega M + (1-\omega)I]\boldsymbol{\varepsilon}^{(n-1)} = M_\omega \boldsymbol{\varepsilon}^{(n-1)} \tag{4.11.4}$$

*Actually, as we noted earlier, the errors that control the convergence are of two types. The first type consists of smooth, slowly varying functions, while the second set contains terms that fluctuate very rapidly in space. In most cases the smooth errors are larger than the rapidly oscillating ones (this depends on the initial guess) and monotonic behavior is obtained. In other cases oscillation may occur.

Thus use of the acceleration method is equivalent to replacing the iteration matrix M by $M_\omega = \omega M + (1-\omega)I$. The simple relationship between the matrix governing the accelerated method and the original method makes it simple to compute the eigenvalues of the new matrix. In fact any eigenvector of M with eigenvalue λ is also an eigenvector of M_ω with eigenvalue

$$\lambda_\omega = \omega\lambda + (1-\omega) \tag{4.11.5}$$

This relationship is shown in Fig. 4.17 for $\omega = 1$ and $\omega = 2$, the extreme values of interest.

For the Jacobi method, the eigenvalues lie between $-\lambda_1$ and $+\lambda_1$ where λ_1 is the largest eigenvalue of the matrix M_J given by Eq. 4.8.10. From Fig. 4.17 we see that if ω is bigger than 1, the magnitude of the largest negative eigenvalue will increase and the iterative method will actually converge more slowly! This may not accord with our intuition, but the Jacobi method cannot be improved by acceleration.

The eigenvalues of the Gauss–Seidel matrix, on the other hand, are the squares of the Jacobi eigenvalues and are, therefore, real and positive. We assume that the smallest eigenvalue is zero. The optimum choice of ω is the one that makes the magnitudes of the largest positive and negative eigenvalues λ_ω equal. Any other choice would make one of them larger. Since the largest negative λ_ω corresponds to $\lambda = 0$ and the largest positive λ_ω corresponds to

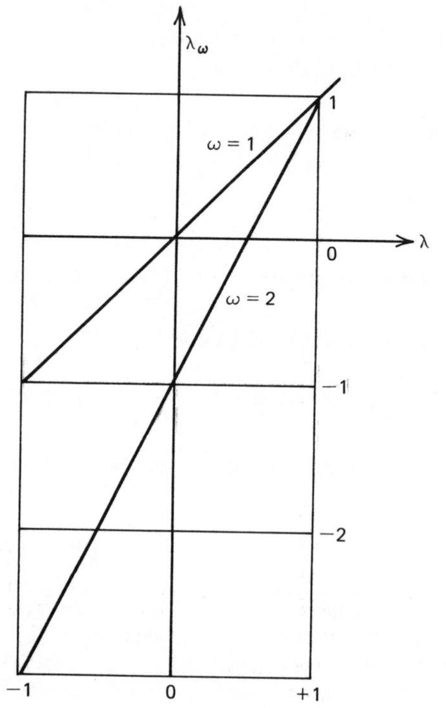

Fig. 4.17. Effect of extrapolation on the eigenvalues.

$\lambda = \lambda_1^2$, the optimum value of ω is determined by

$$(1-\omega) = -\omega\lambda_1^2 - (1-\omega) \tag{4.11.6}$$

and is

$$\omega = \frac{2}{2-\lambda_1^2} \tag{4.11.7}$$

For λ_1 near 1, say $\lambda_1 \simeq 1-\delta$, since $\lambda_1^2 \simeq 1-2\delta$ and the maximum λ_ω is $(1-2\delta)/(1+2\delta) \simeq 1-4\delta$, the accelerated method converges twice as fast as Gauss–Seidel.

This method of accelerating the solution can also be applied to the line relaxation method with a similar improvement. We are after bigger game, however, and now proceed to the Successive Overrelaxation (SOR) method.

Further improvement can be obtained by accelerating or extrapolating the solution at each point as it is calculated rather than waiting for the entire iteration to be completed. For the Jacobi method, which uses only old data, there is no difference. For the Gauss–Seidel method, however, a rather surprising improvement is possible.

The method is quite simple. At each point, a new value is computed using Eq. 4.9.1, which is then extrapolated using the point version of Eq. 4.11.2. These equations can be combined into a single operation

$$\phi_{j,k}^{(p+1)} = \frac{\omega}{4}\left(\phi_{j-1,k}^{(p+1)} + \phi_{j+1,k}^{(p)} + \phi_{j,k-1}^{(p+1)} + \phi_{j,k+1}^{(p)} - b_{j,k}\right) + (1-\omega)\phi_{j,k}^{(p)} \tag{4.11.8}$$

which can be written in matrix form

$$\begin{aligned}\boldsymbol{\phi}^{(p+1)} &= (D-\omega U)^{-1}\left[((1-\omega)D + \omega L)\boldsymbol{\phi}^{(p)} - \omega\mathbf{b}\right] \\ &= M_{\text{SOR}}\boldsymbol{\phi}^{(p)} + \mathbf{d}_{\text{SOR}}\end{aligned} \tag{4.11.9}$$

The problem is now one of finding the value of the acceleration parameter that makes the maximum eigenvalue of M_{SOR} as small as possible. This question is rather difficult and useful results are known only for particular cases. In the most important case the Jacobi matrix possesses Young's Property A. This property has to do with the placement of the zero and nonzero elements. A detailed definition of this property and a derivation of the results are beyond the scope of this book. We merely state that the five-point Laplace operator matrix has Property A but the fourth order operators do not. The importance of this property is that for matrices possessing it, it is possible to determine explicitly the eigenvalues of the SOR matrix. They turn out to be

$$\lambda_s^{1/2} = \frac{1}{2}\left\{\omega\lambda \pm \left[\omega^2\lambda^2 - 4(\omega-1)\right]^{1/2}\right\} \tag{4.11.10}$$

where λ is an eigenvalue of the Jacobi iteration matrix. Note that for $\omega=1$ (which corresponds to Gauss–Seidel iteration) $\lambda_s=\lambda^2$, which is a result shown earlier. Not surprisingly, the largest eigenvalue of the SOR matrix is the one that corresponds to the largest eigenvalue of the Jacobi matrix. One can then show without much difficulty that the optimum acceleration parameter is given by

$$\omega_{\text{opt}}=\frac{2}{1+\sqrt{1-\lambda^2_{\text{max}}}} \tag{4.11.11}$$

and the corresponding maximum eigenvalue is

$$\lambda_{\text{max}}=\omega_{\text{opt}}-1 \tag{4.11.12}$$

To get some idea of what can be achieved by the SOR method, let us assume that we have a tough problem for which the maximum Jacobi eigenvalue is $1-\delta$ where δ is small. The maximum Gauss–Seidel eigenvalue is then approximately $1-2\delta$, and the maximum SOR eigenvalue turns out to be $1-4\sqrt{\delta}$! The number of iterations required to produce accuracy ε is approximately $(N/2\pi\sqrt{2})\ln|\varepsilon|$ and is proportional to N rather than to N^2. To carry the example further, suppose that $M=N=30$. We then find that the maximum eigenvalues of the three methods and the number of iterations required for 10^{-3} accuracy are

	λ_{max}	Iterations
Jacobi	0.9945	1250
Gauss–Seidel	0.9890	625
SOR	0.7906	29

These results are extremely impressive. The SOR method in fact yields the maximum benefit in precisely those problems that are the toughest to handle by the other methods and has proven a very valuable tool since its discovery in the late 1950s. This is illustrated further in the examples below.

Example 4.9

Since the SOR method depends on the relaxation factor ω, its results are not readily displayed in the same manner as the results of the earlier methods. The reason is that, for one thing, the convergence is not always monotonic; for $\omega>\omega_{\text{opt}}$ the convergence is oscillatory. We have solved the problem of the preceding examples: $\nabla^2\phi=0$; $\phi(x,0)=\phi(0,y)=0$; $\phi(1,y)=100y$; and $\phi(x,1)=100x$, with eight intervals in each direction for various values of the relaxation factor ω. The errors at the center point after 20 iterations are given in Fig. 4.18. In this figure we have also shown a curve based on the notion that

```
      PROGRAM SOR
C-----THIS PROGRAM SOLVES LAPLACE'S EQUATION IN A SQUARE.
      DIMENSION F(21,21),EXACT(21,21),ERROR(21,21),X(21),Y(21)
   10 WRITE (5,110)
  110 FORMAT ( ' GIVE THE NUMBER OF X INTERVALS')
      READ (5,120) NX
  120 FORMAT (I)
      WRITE (5,130)
  130 FORMAT ( ' GIVE THE NUMBER OF Y INTERVALS')
      READ (5,120) NY
      WRITE (5,140)
  140 FORMAT ( ' GIVE THE RELAXATION FACTOR OMEGA')
      READ (5,150) OM
  150 FORMAT (F)
      WRITE (5,160)
  160 FORMAT ( ' RESULTS PRINTED EVERY K STEPS, PRINT K')
      READ (5,120) K
      WRITE (5,170) OM
  170 FORMAT ( ' OMEGA = ',F6.3)
C-----COMPUTE THE PARAMETERS AND EXACT SOLUTION
      NXP1 = NX + 1
      NYP1 = NY + 1
      DX = 1. / NX
      DY = 1. / NY
      EPS = 1.E-5
      DO 1 I=2,NX
      F(I,1) = 0.
      X(I) = (I - 1) * DX
      F(I,NYP1) = 100. * X(I)
      DO 1 J=2,NY
      F(I,J) = 1.
      Y(J) = (J - 1) * DY
    1 EXACT(I,J) = 100. * X(I) * Y(J)
      DO 2 J=2,NY
      F(1,J) = 0.
    2 F(NXP1,J) = 100. * Y(J)
C-----MAIN LOOP.  DO K SWEEPS AND THEN PRINT RESULTS.
      ITNUM = 0
   20 DO 3 L=1,K
      DO 3 I=2,NX
      DO 3 J=2,NY
      F(I,J) = OM * .25 * (F(I-1,J) + F(I+1,J) + F(I,J-1) + F(I,J+1))
     1 - (OM - 1.) * F(I,J)
    3 ERROR(I,J) = EXACT(I,J) - F(I,J)
C-----WRITE THE RESULTS FOR ITERATION NUMBER ITNUM
      ITNUM = ITNUM + K
      WRITE (5,200) ITNUM
  200 FORMAT ( ' ERROR FOR ITERATION NUMBER ',I3)
      DO 5 J=2,NY
    5 WRITE (5,210) (ERROR(I,J) , I=2,NX)
  210 FORMAT ( 21E12.3)
C-----CHECK IF RESULT IS CONVERGED.
      DO 6 I=2,NX
      DO 6 J=2,NY
    6 IF (ABS(ERROR(I,J)).GT.EPS) GO TO 20
C-----ASK FOR ANOTHER CASE.
      WRITE (5,220)
  220 FORMAT ( ' IF YOU WANT ANOTHER CASE, TYPE 1')
      READ (5,120) IFF
      IF (IFF.EQ.1) GO TO 10
      STOP
      END
```

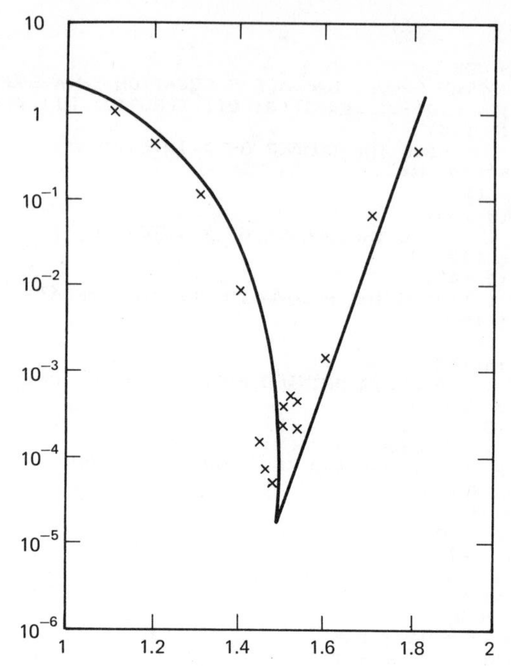

Fig. 4.18. Error in the SOR method at 20 iterations versus the overrelaxation factor (the curve is a theoretical result described in the text; points represent actual calculation).

the error at the twentieth iteration ought to be proportional to the maximum eigenvalue of the method to the twentieth power. Eq. 4.11.10, was used to estimate the largest eigenvalue of the SOR method. The results of this "theory" were normalized to the actual error at $\omega=1$. Agreement between the theory and the actual result is remarkably good and differences can be ascribed to the effects of the smaller eigenvalues. The program used in these calculations is given below. Note that the SOR method is no more difficult to program than any of the earlier methods.

A comparison between SOR and the Gauss–Seidel method is made easily by remembering that the latter can be regarded as SOR with $\omega=1$. We see from Fig. 4.18 that using the optimum relaxation factor reduces the error at iteration 20 by nearly five orders of magnitude. In the same manner, SOR obtains convergence to accuracy 10^{-6} in just over 20 iterations, while Gauss–Seidel requires approximately 100 iterations for this case. These are the kinds of improvement that were anticipated.

Using 16 intervals in each direction instead of 8 produces similar but more dramatic results. Whereas Gauss–Seidel requires nearly 400 iterations to converge, SOR does the job in approximately 50.

In a practical application the optimum relaxation factor cannot be found very easily. If a problem is to be solved only once, it is probably not very important to have the exact optimum; if the result is converging slowly, one can stop the calculation, change the acceleration factor, and try again. On the other hand, if many problems using the same geometry and same mesh are to be solved, it pays to spend some time to find the optimum value. This can be done by increasing ω until the convergence becomes oscillatory. Usually, the first value of ω for which oscillatory behavior is found is a sufficiently accurate estimate. A more accurate value can be found by a trial-and-error search, a case of which is shown (for line SOR) in Example 4.10.

The Successive Overrelaxation method is frequently applied to nonlinear elliptic problems. In these cases the nonlinearity is treated by an iterative method such as Newton–Raphson. At each iteration a linear problem must be solved and SOR can be applied to this. Usually it is best not to bother to follow SOR all the way to convergence at the early iterations as this wastes computation. A rule of thumb is that the SOR calculation should be continued until changes in the function on successive SOR iterations are an order of magnitude smaller than the changes obtained on the preceding Newton–Raphson iteration. In problems of this kind one speaks of "inner" (SOR) and "outer" (Newton–Raphson) iterations.

There is no reason that the concepts used in developing SOR cannot be applied to the line relaxation method of the preceeding section. All this method involves is using Eq. 4.10.2 to compute new estimates of the variables on the kth horizontal line and then using acceleration of Eq. 4.11.2 to give the final value. To repeat, this method solves the tridiagonal system

$$-\tfrac{1}{4}\tilde{\phi}_{j-1,k}^{(p+1)}+\tilde{\phi}_{j,k}^{(p+1)}-\tfrac{1}{4}\tilde{\phi}_{j+1,k}^{(p+1)}=\tfrac{1}{4}\left(\phi_{j,k+1}^{(p)}+\phi_{j,k-1}^{(p+1)}\right)-\tfrac{1}{4}b_{j,k} \quad (4.11.13)$$

and then computes

$$\phi_{j,k}^{(p+1)}=\omega\tilde{\phi}_{j,k}^{(p+1)}+(1-\omega)\phi_{j,k}^{(p)} \quad (4.11.14)$$

This method is no more difficult to program than the line relaxation method and is called *successive line overrelaxation* (SLOR).

Since we do not have exact eigenvalues for the line relaxation method, the theory used to compute the optimum relaxation factor for the SOR method cannot be used. Despite this, a few estimates can be given. Line relaxation converges approximately twice as fast as Gauss–Seidel, which indicates that its eigenvalues (the largest ones at least) are approximately the squares of the Gauss–Seidel eigenvalues or the fourth powers of the Jacobi eigenvalues. Since line relaxation eigenvalues are smaller than Gauss–Seidel eigenvalues, we expect that the optimum relaxation factor will be smaller for SLOR than for SOR. A first estimate can be obtained by using the SOR formula of Eq. 4.11.11 with λ^2 replaced by λ^4. Alternatively, we can use the search procedure suggested above for SOR.

Example 4.10

The problem of the preceding example is done using the SLOR method. As we have stated, the optimum overrelaxation factor is not known. Using the procedure recommended, that is, that we guess the largest eigenvalue of the line relaxation method to be the square of the largest Gauss–Seidel eigenvalue (or the fourth power of the largest Jacobi eigenvalue) and applying Eq. 4.11.11, we estimate that the optimum overrelaxation factor should be approximately 1.315.

With this in mind, we first use $\omega = 1.20$, the results of which are shown in Table 4.4. The convergence is monotonic suggesting that a larger ω should be used. Increasing ω to 1.30 produces more rapid but still monotonic convergence. Next we use $\omega = 1.40$ and, although rapid, the convergence is definitely

Table 4.4. Solution of Laplace's Equation by the SLOR Method (Error versus Number of Iterations)

Iterations	$\omega = 1.200$	$\omega = 1.300$	$\omega = 1.400$	$\omega = 1.350$	$\omega = 1.330$
1	−0.73475011	−0.67003846	−0.59530991	−0.63397267	−0.64870543
2	−0.56207564	−0.49123691	−0.41617408	−0.45426460	−0.46918312
3	−0.43125268	−0.36073786	−0.29070848	−0.32561301	−0.33964227
4	−0.33328448	−0.26727986	−0.20490835	−0.23557119	−0.24813753
5	−0.19793127	−0.10772856	−0.01197981	−0.06066309	−0.07966979
6	−0.11835144	−0.04825515	0.00889800	−0.01765680	−0.02946272
7	−0.06977715	−0.01996397	0.01090176	−0.00185897	−0.00849472
8	−0.04140361	−0.00938433	0.00221646	−0.00087584	−0.00364523
9	−0.02459643	−0.00423148	0.00094372	0.00002887	−0.00123901
10	−0.1461440	−0.00185486	0.00035583	0.00021696	−0.00033354
11	−0.00868173	−0.00080690	−0.00000741	0.00013174	−0.00007359
12	−0.00515751	−0.00035971	−0.00011892	0.00003428	−0.00002564
13	−0.00306377	−0.00015861	−0.00003657	0.00001354	−0.00000500
14	−0.00182001	−0.00007064	−0.00001501	0.00000234	−0.00000170
15	−0.00108117	−0.00003110	0.00000188	0.00000132	0.00000023
16	−0.00064227	−0.00001372	0.00000026	0.00000013	
17	−0.00038154	−0.00000607	0.00000042		
18	−0.00022665	−0.00000269			
19	−0.00013464	−0.00000119			
20	−0.00007998	−0.00000052			
21	−0.00004751				
22	−0.00002823				
23	−0.00001677				
24	−0.00000996				
25	−0.00000592				
26	−0.00000351				
27	−0.00000209				
28	−0.00000124				
29	−0.00000074				

oscillatory, so the optimum value of ω must be between 1.30 and 1.40. A further search shows that $\omega = 1.33$ is very close to the optimum; this is not far from the estimate obtained above. This procedure can be used in more complex problems.

12. ELLIPTIC PDEs: VI. ALTERNATING DIRECTION IMPLICIT (ADI) METHODS

The SOR method is very effective in a large number of problems, but there are cases in which the acceleration parameter is difficult to find and a large number of iterations may be needed, providing incentive to look for still other methods. To see what methods might be used, recall that iterative methods for elliptic problems are analogous to methods used for advancing the solutions of parabolic equations in time. The difference is that in the elliptic case we want to get the solution in the minimum number of iterations. This is equivalent to relaxing the solution of a parabolic problem to steady state as quickly as possible and it suggests that we look at methods for parabolic problems that allow large time steps, in other words, methods that are unconditionally stable for parabolic equations. Of the methods discussed earlier, the most likely candidate is the alternating direction implicit (ADI) or splitting method that was considered in Section 5 of this chapter; Dufort–Frankel is another possibility.

The basic equation that we wish to solve is again Eq. 4.7.6 with the matrix A given by Eq. 4.7.5. As we have noted, this matrix has five nonzero diagonals. The diagonals of 1's immediately above and below the main diagonal come from the finite difference representation of the operator $\partial^2/\partial x_1^2$, while the diagonals of 1's that are displaced M rows or columns from the main diagonal arise from finite differencing the operator $\partial^2/\partial x_2^2$. The main diagonal terms arise equally from both operators. If the mesh spacings were unequal, the structure of the matrix (i.e., the placement of the nonzero elements) would be the same, but the numerical values would vary. If the equation contained terms proportional to $\partial/\partial x_1$ and/or $\partial/\partial x_2$, the symmetry of the matrix would be destroyed. Finally, the presence of a term proportional to ϕ in the partial differential equation (as in the Helmholtz equation) would modify the elements on the main diagonal. None of these changes alters the applicability of any of the methods but they may make finding the optimum parameters difficult.

The splitting of the matrix A required to write the ADI method in matrix form is

$$A = H + V \tag{4.12.1}$$

where H (for horizontal) contains the terms arising from the $\partial^2/\partial x_1^2$ operator and V (for vertical) contains those coming from the $\partial^2/\partial x_2^2$ operator. These matrices are most simply defined by writing

$$H\phi_{i,j}=\phi_{i+1,j}-2\phi_{i,j}+\phi_{i-1,j} \tag{4.12.2a}$$

$$V\phi_{i,j}=\phi_{i,j+1}-2\phi_{i,j}+\phi_{i,j-1} \tag{4.12.2b}$$

Both matrices have -2 as each element on the main diagonal. H has two diagonals of $+1$s immediately adjacent to the main diagonal, while V has two such diagonals displaced by M positions from the main diagonal. In the matrix formulation of the two-dimensional Laplace operator we could have reversed the roles of the two directions so the particular structure of these matrices simply reflects a choice made earlier. Both H and V are essentially tridiagonal matrices insofar as solving systems of equations is concerned.

In terms of these matrices the ADI equations for the parabolic case, Eqs. 4.5.11a and 4.5.11b, can be written

$$(I-\beta_1 H)\phi^{n+1/2}=(I+\beta_2 V)\phi^n+\mathbf{d}_h \tag{4.12.3a}$$

$$(I-\beta_2 V)\phi^{n+1}=(I+\beta_1 H)\phi^{n+1/2}+\mathbf{d}_v \tag{4.12.3b}$$

where, as earlier, $\beta=D\Delta t/2h^2$ and $\mathbf{d}_h$ and $\mathbf{d}_v$ arise from the boundary conditions in the horizontal and vertical directions, respectively. When applied to the solution of elliptic problems, most authors write the ADI method with $\rho=1/\beta$ rather than β; the difference is a matter of taste.

Since we want to approach the steady solution as quickly as possible, it would seem that we ought to make β as large as possible, that is, use a very large time step. The matter is not that simple, however. To understand this, we can again use the method of expanding in terms of eigenvectors. The problem we want to solve is posed by Eqs. 4.12.3 in which the error satisfies the homogeneous version of these equations, that is, the same equations with the right-hand sides set to zero. The error of any iteration k can then be expanded in terms of the eigenvectors of A, which as we have seen earlier, are just the products of sine functions in the two dimensions:

$$\varepsilon^{(k)}=\sum \varepsilon_{mn}^{(k)}\psi^{(m,n)} \tag{4.12.4}$$

The analysis of Section 5 shows that

$$\varepsilon_{mn}^{(k+1)}=\rho_{mn}^{k}\,\varepsilon_{mn}^{(k)} \tag{4.12.5}$$

where

$$\rho_{mn}=\frac{(1+\beta_m)}{(1-\beta_m)}\frac{(1+\beta_n)}{(1-\beta_n)} \tag{4.12.6}$$

and where the β_j are given by Eq. 4.5.15. The behavior of ρ_{mn} as a function of

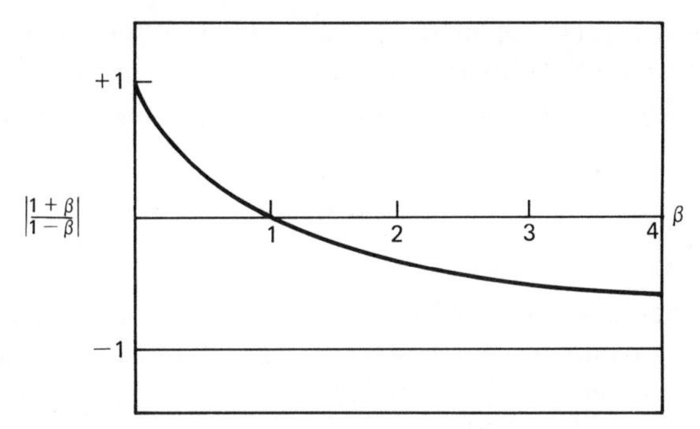

Fig. 4.19. The effective eigenvalue of the ADI method.

β_m for $\beta_n = 0$ is shown in Fig. 4.19 and the behavior for other cases is similar. The curve is asymptotic to -1 at large values of β. Thus very large time steps will produce oscillatory behavior. Since small steps will obviously produce slow convergence, it is clear that the problem now becomes one of finding an optimum time step or β for this method. From Fig. 4.19 it appears that $\beta = 1$, for which the amplification factor is zero, is ideal. Each eigenvector, however, has its own β. We can choose the time step to eliminate any one eigenvector component but there is no value of the parameter that will simultaneously knock out all of the undesired components of the solution. The procedure that is usually adopted is to use a set of iteration parameters cyclically. On the first iteration we set $\beta = \beta_1$, on the second iteration we use $\beta = \beta_2$, and so on, until on the nth iteration we use $\beta = \beta_n$. Then on the $(n+1)$st iteration we use β_1 again and repeat the cycle. The trick is to pick the set of iteration parameters such that the product of the amplification factors for n iterations is as small as possible for as many of the eigenvectors as possible. This is not an easy task and is the major reason why the ADI method is not used as often as one might expect. For geometries other than a few simple ones such as the rectangle, it is very difficult to find a good set of parameters, without which the method is not very effective.

We can give a set of parameters that is useful in the simple case. If a is a lower bound to the eigenvalues of H and V (i.e., the smallest eigenvalue of either) and b is an upper bound, then Wachspress has shown that a good set of parameters is

$$\rho_k = \frac{1}{\beta_k} = b\left(\frac{a}{b}\right)^{(k-1)/(n-1)} \qquad k = 1, 2, \ldots n \tag{4.12.7}$$

Using these parameters with a cycle of length n, one can show that the method converges in approximately $(n/8)(2N)^{1/n-1}$ iterations, which is much faster than even the SOR method. ADI is thus an extremely efficient method when the parameters can be found. In cases in which the same elliptic equation must be solved many times (for different sources or boundary conditions) it may pay the user to find a good set of iteration parameters for the ADI method by experiment. Otherwise, the method is probably not a good choice and SOR has remained popular despite the existence of a potentially much more powerful method.

Example 4.11

Laplace's equation is solved with the same boundary conditions as were used in the preceding examples. First, we use a fixed value of β and look at the behavior of the method as β is changed. The error at the centerpoint after 20 iterations is shown in Fig. 4.20.

For small values of β the method converges slowly but the rate of improvement as β is increased is quite large. The error decreases monotonically with iteration number for small values of β, indicating that the principal sources of error are the small index eigenvalues and eigenvectors.

The optimum value of β is seen to be approximately 1.75 for this case and the error behavior as a function of β is similar to the error of the SOR method as a function of ω. The major difference is that the rate of change of the error near the optimum value of β is not as great as that of the SOR method near the

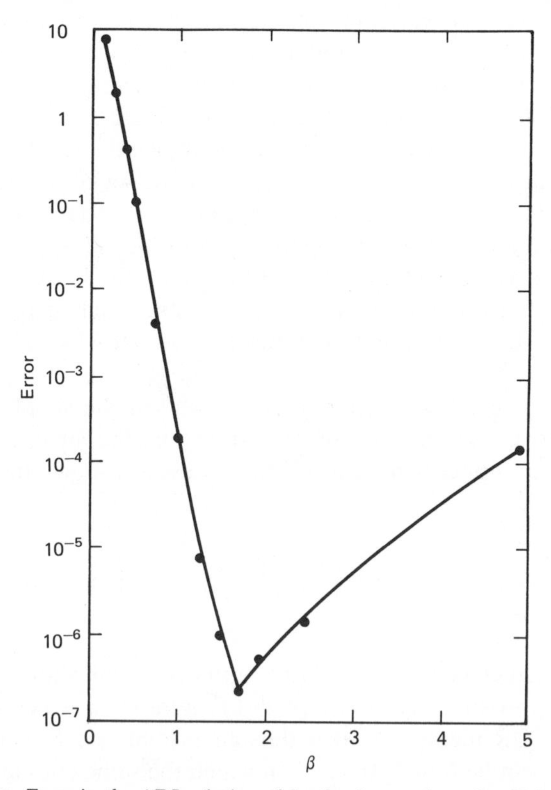

Fig. 4.20. Error in the ADI solution of Laplace's equation after 20 iterations.

optimum ω. In other words, the method is a bit less sensitive to its parameter than is SOR.

At large values of β, the convergence is oscillatory, indicating that the large index eigenvalues and eigenvectors are the principal sources of the error.

For this problem the minimum number of iterations required to reduce this error to approximately 10^{-6} is 17. This is less than SOR but slightly more than SLOR. Since the computations involved in ADI are very similar to those in SLOR, the choice between the methods is a close one and, while SLOR has been chosen more frequently, not much is lost by using ADI.

Example 4.12

Let us repeat the problem of Example 4.11 with the cyclic ADI method suggested in the discussion. With eight points in each direction we find that the smallest and largest eigenvalues of both H and V are 0.1206 and 3.8794, respectively. If we choose a cycle length of three iterations, Wachspress's

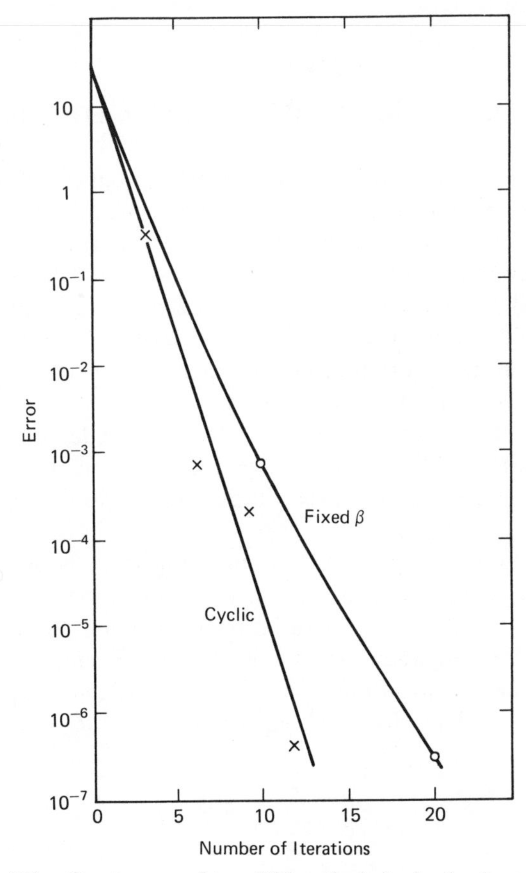

Fig. 4.21. Convergence of two ADI methods for Laplace's equation.

formula (Eq. 4.12.6) suggests that the values of β to be used are 16.5817, 2.9240, and 5.156; the order in which they are used should not be important.

The results are shown in Fig. 4.21. Both the actual results and an interpreted curve are shown for the cyclic method. We see that the cyclic method does converge more rapidly than the fixed β method. For 10^{-6} accuracy convergence is achieved in 12 iterations versus 17 for the fixed β method. This is the smallest number of iterations that we have found for any method on this problem and shows that ADI is indeed capable of extremely good performance if a good set of parameters can be found. For the square region treated here it was not difficult to find a good set of parameters. Fortunately, the method is only moderately sensitive to the parameters and it may not be too difficult to find a reasonably good set. (We ran this problem with all of the βs set to half of the values corresponding to the results shown in the figure and the method converged in 15 iterations.)

The advantage of the cyclic method would be greater in a more difficult problem, for example, in a problem with more mesh intervals. With only eight mesh intervals, the fixed β method does very well and the cyclic method cannot show all of its strength.

13. Elliptic PDEs: VII. FINITE ELEMENT METHODS

Finite element methods for ordinary differential equations were discussed in Section 3.15. The same ideas can be applied just as readily to partial differential equations in two or three dimensions. In recent years finite element methods have become dominant in some engineering fields that require the solution of partial differential equations, particularly in the analysis of stress in solids. As a result, a huge body of literature devoted to the development of these methods has been produced and separate courses on finite element methods are offered in many universities.

In a book of this kind it is impossible to provide more than a brief introduction to the subject. The reader interested in acquiring a deeper knowledge of finite element methods will find a number of well-written books devoted solely to the subject.

As was pointed out in Section 3.15, finite element methods are at their best when the problem to be solved can be stated in a variational form. The method can be applied in the absence of a variational statement but will not have the nice properties obtained when one is available. We begin, therefore, with the classic case for which a variational statement is available: Laplace's equation in two dimensions. Solving Laplace's equation

$$\nabla^2\phi = \frac{\partial^2\phi}{\partial x^2} + \frac{\partial^2\phi}{\partial y^2} = 0 \tag{4.13.1}$$

is equivalent to minimizing the integral

$$F=\int(\nabla\phi)^2\,dA=\int\left[\left(\frac{\partial\phi}{\partial x}\right)^2+\left(\frac{\partial\phi}{\partial y}\right)^2\right]dx\,dy \tag{4.13.2}$$

where the integral is over the region for which the solution is desired.

Approximate solutions to Laplace's equation are generated by creating functions that contain adjustable parameters, computing the integral of Eq. 4.13.2 as a function of the parameters, and then minimizing this result with respect to each of the parameters. We give a concrete example below.

Boundary conditions are an important issue. In order to show that a solution of Laplace's equation actually minimizes the integral, we assume that ϕ^* is the function that minimizes the integral and compute the integral using the function

$$\phi=\phi^*+\delta\phi \tag{4.13.3}$$

where $\delta\phi$ is small compared to ϕ^*. Then substituting Eq. 4.13.3 into Eq. 4.13.2 and integrating by parts, we have

$$F=F^*+\delta F=\int_A(\nabla\phi^*)^2\,dA-2\int_A\delta\phi\,\nabla^2\phi^*\,dA+\int_C\delta\phi\,\nabla\phi^*\cdot dS \tag{4.13.4}$$

where the last integral is over the bounding contour and we have neglected the term quadratic in $\delta\phi$, since it is much smaller than the terms we have kept. The first term on the right is F^*, the exact minimum according to the assumptions made. Thus the deviation in F due to the small error $\delta\phi$ in the approximating function is

$$\delta F=-2\int_A\delta\phi\,\nabla^2\phi^*\,dA+\int_C\delta\phi\,\nabla\phi^*\cdot dS \tag{4.13.5}$$

Recall that a function $f(x)$ must be a quadratic (or higher order) function of the independent variable near a minimum. In the same manner, we expect that F should be quadratic in $\delta\phi$ near its minimum. This requires that Eq. 4.13.5 must vanish and for this to be true for any $\delta\phi$, we stipulate (1) that ϕ^* must satisfy Laplace's equation 4.13.1 and (2) that the normal derivative of ϕ^* must be zero on the boundary. The latter is called the *natural boundary condition* implied by the variational statement of Eq. 4.13.2.

Suppose that we wish to solve Laplace's equation subject to some boundary condition other than the natural one implied by the statement of Eq. 4.13.2 derived above. There are two ways in which this can be done. The first is to allow only those functions that satisfy the desired boundary conditions. The conflict between the desired and natural boundary conditions can cause numerical difficulties, however. A second approach involves introducing an integral over the boundary into Eq. 4.13.2, which can be chosen so that the natural boundary condition for the modified variational statement becomes the desired boundary condition. Both things can be done simultaneously, of course, and the result is even better.

In the finite element method the region for which the equation is to be solved is broken into a number of smaller regions. The name *finite element* comes from solid mechanics in which the pieces are thought of as elements that together make up the body. The elements can be essentially any shape and it is this flexibility that gives the method one of its principal advantages: the ability to handle any kind of geometry without difficulty. A great variety of elements have been suggested and used; we discuss just a few to illustrate some of the variety that is possible.

The simplest kind of element (and the first in a historical sense) is the triangular element, an example of which is given in Fig. 4.22. Using ordinary straight-sided triangles, we are able to fill the region with elements that fit the boundary curve quite accurately. The difficulty that finite difference methods have at curved boundaries is thus eliminated. Furthermore, the elements can be of any size. If the solution is found to vary rapidly in some region, it is possible to subdivide the elements to increase the accuracy in one region without having to do so in other regions.

Within each region the function can be approximated in a number of ways. Use of constants as trial functions is not a good idea because this makes the solution discontinuous at the boundaries of the elements.

However, the linear approximation

$$\phi = c_1 + c_2 x + c_3 y \tag{4.13.6}$$

has been widely used. The constants can be expressed in terms of the values of the function at the corners of the triangular element and it is better to treat the corner values as the unknowns than to use the coefficients used in Eq. 4.13.6, the primary reason being that the same values can be used for the adjacent elements. This both reduces the number of unknowns to be computed and also

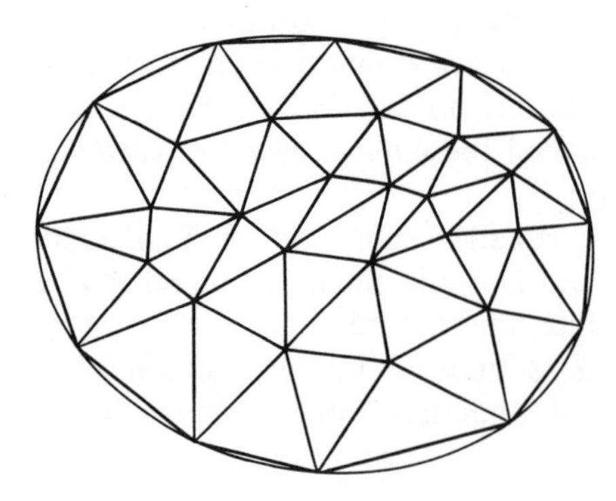

Fig. 4.22. Filling a region with triangular elements.

Fig. 4.23. Some triangular elements.

ensures that the function will be continuous across the element boundaries. This element is illustrated symbolically in Fig. 4.23*a*. Along any edge of the triangle the function is linear in the distance along that edge and is the same linear approximation for both triangles containing that edge. Thus continuity of the solution is guaranteed.

Further accuracy can be obtained with a quadratic element:

$$\phi = c_1 + c_2 x + c_3 y + c_4 x^2 + c_5 xy + c_6 y^2 \tag{4.13.7}$$

The constants can be represented in terms of the values of the solution at the corners and edge midpoints of the triangle symbolized in Fig. 4.23*b*. Still more accuracy can be obtained by using a cubic (the highest degree normally used), requiring ten coefficients in the approximation. This element is shown symbolically in Fig. 4.23*c*.

Another possibility uses the cubic approximation but uses the values of the function and its two first partial derivatives at the corners and the value of the function at the center of the triangle as parameters. This ensures that the first derivatives as well as the solution itself will be continuous across the element boundaries. It is similar to the use of Hermite interpolation in two dimensions and is useful for higher order equations.

Recently there has been a trend toward using rectangular elements with curved edges. Curved edges obviously allow one to fit the region boundaries more accurately than is possible with straight-edged triangles. To make this method computationally feasible, irregular quadrilaterals are generated by applying coordinate transformations to a standard rectangular grid on which the calculations are done. When the coordinate transformation is generated from the same kind of polynomials that are used to approximate the function the elements are said to be *isoparametric*, and the method possesses a number of advantages.

Since we do not go very deeply into the finite element method here, we adopt the use of straight-edged triangles and linear approximations for the remainder of this section. Formally, one proceeds as described above; the integral of Eq. 4.13.2 is evaluated as a sum of integrals over the elements. Within each element, the integral is evaluated by substituting the polynomial approximation for $\phi(x)$ and computing the integral analytically. For a linear problem such as the one we are dealing with, the result is bilinear in all of the parameters that are to be computed. Once the integral has been evaluated, the desired equations are obtained by differentiating it with the result with respect to each of the parameters to be computed and setting the result to zero. In practice there is an easier method, illustrated by the following example.

Example 4.9

Consider the cluster of four elements shown in Fig. 4.24. The case of four identical isosceles right triangles is chosen both for simplicity and ease of comparison with finite difference methods. When the integral described above is computed, there will be terms containing ϕ_{ij} arising from each of the four adjacent triangles. In fact it is easy to show that the contribution of the terms containing ϕ_{ij} is

$$\phi_{ij}\left[8\phi_{ij}-4\left(\phi_{i-1,j}+\phi_{i+1,j}+\phi_{i,j-1}+\phi_{i,j+1}\right)\right] \tag{4.13.8}$$

When this is differentiated with respect to ϕ_{ij}, we find

$$4\phi_{ij}-\left(\phi_{i-1,j}+\phi_{i+1,j}+\phi_{i,j-1}+\phi_{i,j+1}\right)=0 \tag{4.13.9}$$

which is the usual five-point finite difference approximation. This method of finding all the terms containing a particular variable can always be used and is the easiest way of finding the equations.

As we saw in Section 3.15, finite element methods normally do not give the same approximations as finite difference methods; it is only the simplicity of Laplace's equation that causes the two methods to yield the same approximation. If the method used above had been applied to the modified wave equation

$$\nabla^2\phi+k^2\phi=0$$

we would get

$$4\phi_{ij}-\phi_{i-1,j}+\phi_{i+1,j}+\phi_{i,j+1}+\phi_{i,j-1}$$
$$+\tfrac{1}{12}\left(8\phi_{ij}+\phi_{i-1,j}+\phi_{i+1,j}+\phi_{i,j+1}+\phi_{i,j-1}\right)=0 \tag{4.13.10}$$

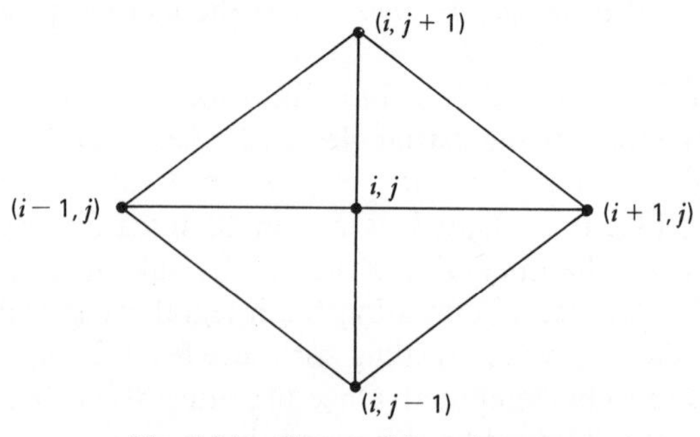

Fig. 4.24. Cluster of four elements.

which is clearly different from the usual finite difference approximation. Equation 4.13.10 can be regarded as a modified finite difference method. This leads to a useful way of comparing the methods. Almost any approximation that can be derived by the finite element method using regular arrays of points can also be derived by finite differences, and vice versa. The finite element approach however, produces approximations that might not have been thought of in the context of finite differences. Which kind of approximation is superior may depend on the problem being solved, but the fact that the finite element method has important conservation laws built into it is a big advantage. When added to the capability of dealing with difficult geometries, this has made the finite element approach the method of choice in a number of applications, particularly in solid mechanics. Whether it will be the preferred method in fluid mechanics (where there are severe nonlinearities and a variational approach is not usually available) is questionable at the present time.

One disadvantage of the finite element approach is that the structure of the matrix in the equations normally is not as simple as it is in the finite difference case; this is the price paid for the ability to deal with geometric complexity in such a flexible manner. All we can say in this short discussion is that the matrix structure depends on how the points are numbered, that routines have been developed for finding good ways of ordering the points, and that the equations are solved either by a modified Gauss elimination procedure or an iterative method. This field is still undergoing extensive development at the present time.

14. DISCRETE FOURIER TRANSFORMS

We have noted that Fourier transform methods have become popular in recent years. In this section and the next we digress to provide a brief introduction to these methods. Applications to solving PDEs will be given in Section 16.

For purposes of this discussion some familiarity with Fourier series is assumed. Any function on $0<x<L$ can be written as a Fourier series (subject to restrictions that are not usually important in physical applications)

$$f(x)=\frac{a_0}{2}+\sum_{n=1}^{\infty}\left(a_n\cos\frac{2\pi nx}{L}+b_n\sin\frac{2\pi nx}{L}\right) \tag{4.14.1}$$

the coefficients a_n and b_n can be found from the formulas

$$a_n=\frac{2}{L}\int_0^L f(x)\cos\frac{2\pi nx}{L}\,dx \tag{4.14.2a}$$

$$b_n=\frac{2}{L}\int_0^L f(x)\sin\frac{2\pi nx}{L}\,dx \tag{4.14.2b}$$

These results are generally well known. It is convenient to note that the relationship

$$e^{iy} = \cos y + i \sin y \tag{4.14.3}$$

allows Eqs. 4.14.1 and 4.14.2 to be rewritten as

$$f(x) = \sum_{n=-\infty}^{\infty} \alpha_n e^{2\pi i n x/L} \tag{4.14.4a}$$

$$\alpha_n = \frac{1}{L}\int_0^L f(x) e^{-2\pi i n x/L} \tag{4.14.4b}$$

which is just a complex form of the Fourier series; the connection is $\alpha_n = (a_n + ib_n)/2$.

It is important to note that although we have restricted the range of the independent variables to $0<x<L$, the series shown in Eqs. 4.14.1 or 4.14.4a are defined for any x. The function defined by this series is depicted in Fig. 4.25 and is called the *periodic extension* of $f(x)$. For computational applications we will, of course, need to deal with finite series. The summation in Eq. 4.14.4a is truncated after a finite number of terms and the integral in Eq. 4.14.4b is replaced by a finite sum. For "smooth" functions this causes no serious problem, but a finite series has difficulty in reproducing a function that has discontinuities. Furthermore, because the Fourier series represents the periodic extension of the function, if $f(0) \neq f(L)$, the series will "think" that there is a discontinuity in the function at $x=0$, $L, 2L$, and so on. The result is that the Fourier series of the function illustrated in Fig. 4.25 produces a result such as the one shown in Fig. 4.26. The existence of wiggles near the apparent discontinuity at the endpoints is known as *Gibbs phenomenon* and can cause considerable difficulty in applications of Fourier series.

For smooth functions Fourier series converge quite rapidly, that is, the coefficients α_n fall off rapidly with increasing n. For functions with discontinuities $\alpha_n \sim 1/n$ and truncation removes terms that are not small. The large errors are the source of the Gibbs phenomenon.

One can differentiate Fourier series for $f(x)$ term by term to produce a series representation of the derivative $f'(x)$. From Eq. 4.14.4a we find that the

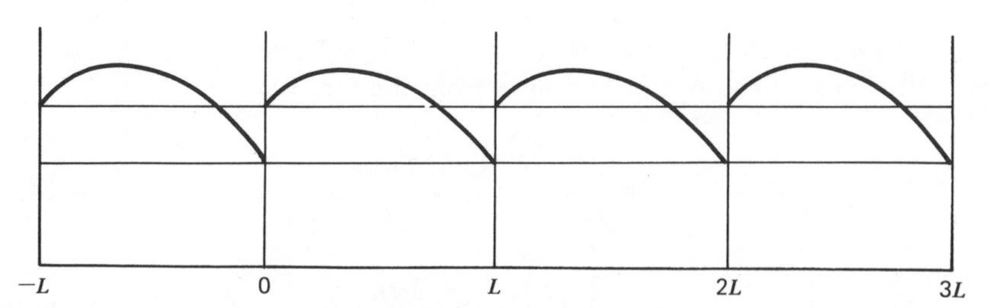

Fig. 4.25. Periodic extension of a function by Fourier series.

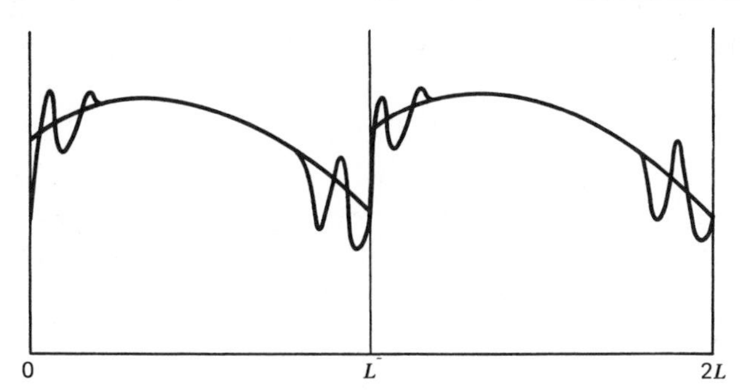

Fig. 4.26. An illustration of the Gibbs phenomenon.

coefficients of the series for the derivative $f'(x)$ are proportional to $n\alpha_n$. It is possible that this series will not converge at all (if f has a discontinuity, $\alpha_n \sim 1/n$ for large n and $n\alpha_n$ will not go to zero as $n \to \infty$). The differentiated series can be used to represent $f'(x)$ only when it converges. Practically, this means that use of Fourier series for calculating derivatives is limited to smooth, continuous functions with $f(0)=f(L)$.

Everything said about the Fourier series for the function applies to the Fourier series for the derivative. If the derivative is discontinuous at some point (including the endpoint), its series representation will exhibit the Gibbs phenomenon. This means that functions possessing discontinuous derivatives (functions with cusps) will be subject to a weaker form of the Gibbs phenomenon; the series will produce a wiggly result, but the wiggles will be milder than those shown in Figure 4.26. Still milder forms of the problem will arise for functions with discontinuous higher derivatives. Obviously, Fourier series are at their best for functions that possess continuous derivatives of all orders. This is the case for periodic functions, that is, functions such that $y(x+L)=y(x)$, which have continuous derivatives of all orders on $0<x<L$. This is rather restrictive, but a sufficient number of cases of this kind arise in applications to make the introduction of Fourier methods worthwhile.

This completes our short introduction to Fourier series. A more complete development of Fourier series is found in Bracewell (1967) (on Fourier methods in general) and Brigham (1976) (on discrete transforms and computer applications).

For the applications we have in mind, we need a discrete version of Fourier series. Suppose we have a one-dimensional mesh of grid points defined by

$$x_j = j\Delta x \qquad j=1,2,\dots N \tag{4.14.5}$$

Periodicity implies that $x=0$ and that $x=L=N\Delta x$ are equivalent points. To write a function $f(x)$ whose values we are given only at these N mesh points as a Fourier series, we need only N Fourier coefficients. For generality, $f(x)$ is allowed to be complex and we have

$$f(x_j) \equiv f_j = \sum_{l=1}^{N} \hat{f}_l e^{ik_l x_j} \tag{4.14.6}$$

(A carat ˆ is used to represent the Fourier coefficient of a function.) We now need to select the set of k_l. Some constraints on them arise from the requirements that the functions $\exp(ik_l x_j)$ be periodic and, for computational reasons, the k_l be equally spaced. A set of values that accomplishes both of these aims is

$$k_l = \frac{2\pi l}{N\Delta x} \qquad l = 1, 2, \ldots N \tag{4.14.7}$$

and the series shown in Eq. 4.14.6 becomes

$$f_j = \sum_{l=1}^{N} \hat{f}_l e^{2\pi i l j/N} \tag{4.14.8}$$

The choice of k_l made in Eq. 4.14.7 is not unique. To illustrate this, suppose we take the function $\exp(2\pi i l j/N)$ and replace l by $l+N$. We then have $\exp(2\pi i(l+N)j/N) = \exp(2\pi i l j/N)\exp(2\pi i) = \exp(2\pi i l j/N)$. Thus in terms of their values at the grid points, the set of functions given by Eq. 4.14.7 and the set with l replaced by $l+N$ are identical. Between the grid points they are different, of course, but this has no effect on Eq. 4.14.8. The existence of Fourier modes with identical values on the grid gives rise to the phenomenon known as aliasing. An illustration of how this occurs is given in Fig. 4.27. Aliasing has important consequences for computational methods, including those that do not use Fourier series explicitly. We demonstrate this later.

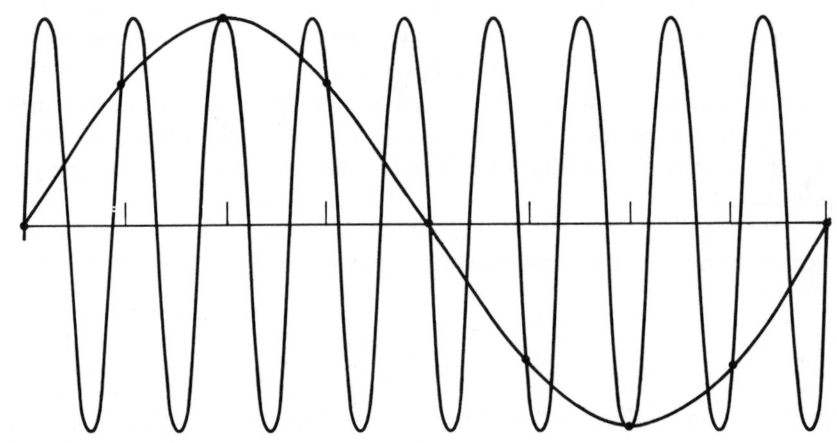

Fig. 4.27. Two Fourier waves with different wave numbers, $\sin \pi x/4$ and $\sin 5\pi x/4$, have the same values on the grid.

The inverse of Eq. 4.14.8 is given by

$$\hat{f}_l = \frac{1}{N} \sum_{j=1}^{N} f_j e^{-2\pi i l j/N} \tag{4.14.9}$$

The proof is straightforward. If we substitute Eq. 4.14.9 into Eq. 4.14.8 with j replaced by j' to avoid confusion and interchange the order of j' and l summations, we have

$$f_j = \frac{1}{N} \sum_{j'=1}^{N} f_{j'} \sum_{l=1}^{N} e^{2\pi i l(j-j')/N} \tag{4.14.10}$$

The l summation is simply a geometric series, which can be summed by a well-known formula to give

$$\sum_{l=1}^{N} e^{2\pi i l(j-j')/N} = e^{2\pi i(j-j')/N} \frac{\left(1-e^{2\pi i(j-j')N/N}\right)}{\left(1-e^{2\pi i(j-j')/N}\right)} \tag{4.14.11}$$

Now $j-j'$ is an integer that can take the values $-N+1, -N+2, \ldots, N-1$. The numerator in Eq. 4.14.11 is always zero; the denominator is not zero unless $j=j'$. Thus the right hand side of Eq. 4.14.11 is zero unless $j=j'$. When $j=j'$, the sum is easily evaluated because each term is exactly unity; the sum is simply N and the expression of Eq. 4.14.11 is thus $N\delta_{jj'}$. Applying this result to Eq. 4.14.10, we see that both sides are equal so that Eqs. 4.14.8 and 4.14.9 are inverses of each other. Some authors use a factor $N^{-1/2}$ in front of the sums in each expression rather than 1 and N^{-1}. The result is more symmetric, but the choice is a matter of taste. The only other difference between Eqs. 4.14.8 and 4.14.9 is the sign in the exponent. For this reason f_j and $\hat{f}_l$ may be regarded as discrete Fourier transforms of each other.

Although everything we have done considers only the values of the function at the mesh points, there is no reason that we cannot use Eq. 4.14.6 to define the function for any x. That is, we can use the Fourier series as an interpolation tool by replacing x_j by the continuous variable x. Having done so, we can then use Fourier series as a means of computing integrals, derivatives, and the like. A little care is needed, however. We saw earlier that there is some freedom in the choice of the set of k_l. Any set of integers that spans a range of N values, that is, any set of the type $l=j, j+1, \ldots j+N$, will reproduce the function at the mesh points. Between the mesh points, however, different sets of l give quite different results. Since we want a smooth interpolation, we should choose the set of l's that produces one. The set that gives the smoothest result is the one that contains the smallest maximum l in the absolute sense. For even N the most appropriate set is

$$l = \frac{-N}{2}, \frac{-N}{2}+1, \ldots 0, \ldots, \left(\frac{N}{2}\right)-1 \tag{4.14.12}$$

and is usually adopted ($-N/2$ could be replaced by $+N/2$).

It is now an easy matter to obtain derivatives. Differentiating Eq. 4.14.6 (with x_j replaced by the continuous variable x), we have

$$\frac{df(x)}{dx} = \sum_l ik_l \hat{f}_l e^{ik_l x} \tag{4.14.13}$$

This equation can be interpreted as saying that $ik_l \hat{f}_l$ is the Fourier transform of df/dx. This means that we can compute the derivative of a function by the following sequence of operations. First, we take the values of the function on an evenly spaced grid and calculate its discrete Fourier transform. Then we multiply the transform by ik_l and take its inverse Fourier transform. The resulting function is an accurate approximation to the derivative of the original function. One further note. For real functions, $\hat{f}_{-l}$ must be the complex conjugate of $\hat{f}_l$; for $l=-N/2$, there is no corresponding $\hat{f}_l$. The result will be a noisy derivative unless one sets $k_{-N/2}=0$. The reason for insisting that N be even is explained in the next section and some examples of this are given.

By taking the derivative of Eq. 4.14.13, we can show that $-k_l^2 \hat{f}_l$ is the Fourier transform of the second derivative d^2f/dx^2 and the second derivative can be computed in a manner very similar to that used for the first derivative. The only difference is that the transform must be multiplied by $-k_l^2$ rather than ik_l. Still higher derivatives can be computed in the same manner.

The method is effective only for functions that are smooth in the sense that they are periodic and have continuous derivatives of several orders. One can show that the approach outlined above produces derivatives with an order of accuracy (in the usual finite difference sense) that is equal to the minimum of (1) one plus the lowest order derivative that possesses a discontinuity or (2) the number of points used in the computation. For the proper functions the method is capable of extremely high accuracy. Its practical use as a numerical method requires an efficient means of computing discrete Fourier transforms, the subject of the next section.

15. THE FAST FOURIER TRANSFORM (FFT) ALGORITHM

The Fourier transforms introduced in the preceding section have the potential of providing differential equation solvers of very high accuracy. In order for them to be useful, however, we need a way to compute Fourier transforms efficiently. Direct computation of the series in Eq. 4.14.6 (or Eq. 4.14.9) requires N multiplications and N additions for each value of j (or l) and since j may have each of N values, the total number of operations is proportional to N^2 (assuming that the complex exponential factors have been precomputed). This scaling would make the method too expensive for large N. Clearly, a more efficient algorithm is necessary. Such a method was discovered by Cooley and

Tukey in 1963 and is known as the *fast Fourier transform* (FFT) or *Cooley–Tukey algorithm.*

Since the forward and backward transforms are essentially the same except for a sign change, we consider only the forward transform. Introducing the abbreviation $\omega = \exp(2\pi i/N)$ and letting the indices run from 0 to $N-1$ rather than from 1 to N, we can write Eq. 4.14.8.

$$f_j = \sum_{l=0}^{N-1} \omega^{jl} \hat{f}_l \qquad j=0,1,\ldots N-1 \tag{4.15.1}$$

The secret to the efficient computation of this transform resides in the properties $\omega^{m+n} = \omega^m \omega^n$ and $\omega^N = 1$. As a consequence of these almost trivial facts, we see immediately that many of the exponential factors in Eq. 4.13.1 are identical. Furthermore, there is a great deal of symmetry.

There are several ways to develop the concept of the FFT routine. One of the better ones is to regard ω^{jl} as a matrix in Eq. 4.15.1 and treat the operation as matrix multiplication. Thus with $N=4$ we have

$$\begin{pmatrix} f_0 \\ f_1 \\ f_2 \\ f_3 \end{pmatrix} = \begin{pmatrix} \omega^0 & \omega^0 & \omega^0 & \omega^0 \\ \omega^0 & \omega^1 & \omega^2 & \omega^3 \\ \omega^0 & \omega^2 & \omega^4 & \omega^6 \\ \omega^0 & \omega^3 & \omega^6 & \omega^9 \end{pmatrix} \begin{pmatrix} \hat{f}_0 \\ \hat{f}_1 \\ \hat{f}_2 \\ \hat{f}_3 \end{pmatrix} = \begin{pmatrix} 1 & 1 & 1 & 1 \\ 1 & \omega & -1 & -\omega \\ 1 & -1 & 1 & -1 \\ 1 & -\omega & -1 & \omega \end{pmatrix} \begin{pmatrix} \hat{f}_0 \\ \hat{f}_1 \\ \hat{f}_2 \\ \hat{f}_3 \end{pmatrix} \tag{4.15.2}$$

In this equation we can change the order in which f_i appears in the vector if we simultaneously interchange the corresponding rows of the matrix:

$$\begin{pmatrix} f_0 \\ f_2 \\ f_1 \\ f_3 \end{pmatrix} = \begin{pmatrix} 1 & 1 & 1 & 1 \\ 1 & -1 & 1 & -1 \\ 1 & \omega & -1 & -\omega \\ 1 & -\omega & -1 & \omega \end{pmatrix} \begin{pmatrix} \hat{f}_0 \\ \hat{f}_1 \\ \hat{f}_2 \\ \hat{f}_3 \end{pmatrix} \tag{4.15.3}$$

The reordered matrix can be factored into the product of two simpler matrices to give

$$\begin{pmatrix} f_0 \\ f_2 \\ f_1 \\ f_3 \end{pmatrix} = \begin{pmatrix} 1 & 1 & 0 & 0 \\ 1 & -1 & 0 & 0 \\ 0 & 0 & 1 & \omega \\ 0 & 0 & 1 & -\omega \end{pmatrix} \begin{pmatrix} 1 & 0 & 1 & 0 \\ 0 & 1 & 0 & 1 \\ 1 & 0 & -1 & 0 \\ 0 & 1 & 0 & -1 \end{pmatrix} \begin{pmatrix} \hat{f}_0 \\ \hat{f}_1 \\ \hat{f}_2 \\ \hat{f}_3 \end{pmatrix} \tag{4.15.4}$$

as can be shown by straightforward matrix multiplication. The important thing is the structure of the two matrices. The left-hand matrix is block diagonal, that is, it is a diagonal 2×2 matrix the elements of which are themselves 2×2 matrices. As a result, the first two elements of the vector produced by multiplying an arbitrary vector **b** by this matrix can be found by multiplying the 2-vector (b_1, b_2) by the upper-left 2×2 submatrix. Similarly, the last two elements of the product vector are obtained by multiplying the vector (b_3, b_4) by the lower-right 2×2 block. In other words, the process of multiplying a vector by a 4×4 matrix of this type is equivalent to two multiplications involving 2×2 matrices. It is not quite as obvious that the right-hand factor matrix in Eq. 4.15.4 has a similar structure. The difference is that now the first and third elements of the resultant vector are determined by the first and third elements of the original vector, and the second and fourth elements of the two vectors are connected to the second and fourth elements of the original vector. The net result is that the multiplication of a 4-vector by a 4×4 matrix has been replaced by 4 multiplications of 2-vectors by 2×2 matrices.

This result by itself is not of great importance, since it produces no savings in computation. The real significance lies in the generalization of this method to larger N. It turns out that as long as $N=2^n$, in other words, an integral power of 2, the matrix that represents the Fourier transform can always be represented as the product of $n=\log_2 N$ matrices, each of which is equivalent to a block diagonal matrix the elements of which are 2×2 matrices.

To illustrate the advantage in this, recall that the multiplication of a vector by a matrix requires N^2 operations. Direct application of Eq. 4.15.2 would need this many operations. The use of the factorization necessitates $N/2$ multiplications of 2-vectors by 2×2 matrices at each stage—or $(N/2)\times4=2N$ operations. Since the number of factor matrices is $\log_2 N$, the total number of numerical operations is of the order of $2N\log_2 N$. For large N, say $N\geqslant64$, the savings provided by using the factorized form can be an order of magnitude or more. It is the existence of this algorithm that makes Fourier transforms more than a special-purpose tool.

With these basic ideas as a foundation, what remains to be done is to construct a scheme for making the FFT systematic so that one subroutine can be used for any value of N. This means that we need a method of determining which of the components need to be combined at each stage of the procedure and what the multiplying factors (matrix elements) are. There is some freedom of choice in how this is done; in other words, there is more than one factorization. A simple one, which is the generalization of the method given for $N=4$ above, combines the element j of the original vector with element $j+N/2$ to produce the j and $j+N/2$ elements of the first intermediate result. At the second stage, elements j and $j+N/4$ are combined, at the third, j and $j+N/8$, and so on. The factors by which each element must be multiplied also form a systematic array. After the first stage, none of the first $N/2$ elements in the array will ever again be combined with any of the last $N/2$ elements. This means that the operations that are performed on each of these sets of $N/2$ elements from then on is precisely and $N/2$-point transform. In fact the first

```
      SUBROUTINE FFT (X,N,K)
C-----THIS SUBROUTINE COMPUTES THE FOURIER TRANSFORM OF THE VECTOR X
C-----(A COMPLEX ARRAY OF DIMENSION N).  IT IS DUE TO THE STANFORD
C---- UNIVERSITY COMPUTER SCIENCE DEPT.  THE CALLING ARGUMENTS ARE:
C-----     X IS THE DATA TO BE TRANSFORMED; ON RETURN, IT  CONTAINS
C-----        THE RESULT.
C-----     N IS THE SIZE OF THE VECTOR (MUST BE A POWER OF 2 < 32769)
C-----     K = 1  FOR FORWARD TRANSFORM
C- ---     K =-1  FOR INVERSE TRANSFORM
      IMPLICIT INTEGER (A-Z)
      REAL GAIN,PI2,ANG,RE,IM
      COMPLEX X(N),XTEMP,T,U(16),V,W
      LOGICAL NEW
      DATA PI2,GAIN,NO,KO /6.283185307,1.,0,0/
C-----TEST WHETHER THIS IS THE FIRST CALL.
      NEW = (NO .NE. N)
      IF (.NOT. NEW) GO TO 2
C-----COMPUTE LOG2 (N) IF THIS IS THE FIRST CALL.
      L2N = 0
      NO = 1
  1   L2N = L2N + 1
      NO = NO + NO
      IF (NO .LT. N) GO TO 1
      GAIN = 1./N
      ANG = PI2*GAIN
      RE  = COS (ANG)
      IM  = SIN (ANG)
C- ---COMPUTE THE COMPLEX EXPONENTIALS IF THIS IS THE FIRST CALL.
  2   IF ( .NOT. NEW .AND. K*KO .GE. 1 ) GO TO 4
      U(1) = CMPLX ( RE , -SIGN(IM,FLOAT(K)) )
      DO 3 I = 2,L2N
  3   U(I) = U(I-1)*U(I-1)
      KO = K
C-----MAIN LOOP.
  4   SBY2 = N
      DO 7 STAGE = 1,L2N
      V = U(STAGE)
      W = ( 1. ,0. )
      S = SBY2
      SBY2 = S/2
      DO 6 L = 1,SBY2
      DO 5 I = 1,N,S
      P = I+L-1
      Q = P+SBY2
      T = X(P)+X(Q)
      X(Q) = (X(P)-X(Q))*W
  5   X(P) = T
  6   W = W*V
  7   CONTINUE
C---- REORDER THE ELEMENTS USING BIT REVERSAL.
      DO 9 I = 1,N
      INDEX = I-1
      JNDEX = 0
      DO 8 J = 1,L2N
      JNDEX = JNDEX+JNDEX
      ITEMP = INDEX/2
      IF ( ITEMP+ITEMP .NE. INDEX ) JNDEX = JNDEX+1
      INDEX = ITEMP
  8   CONTINUE
      J = JNDEX+1
      IF ( J .LT. I ) GO TO 9
      XTEMP = X(J)
      X(J)  = X(I)
      X(I)  = XTEMP
  9   CONTINUE
C- ---IF THIS IS A FORWARD TRANSFORM, WE'RE ALL DONE.
      IF ( K .GT. 0 ) RETURN
C----MODIFICATION FOR INVERSE TRANSFORM.
      DO 10 I = 1,N
 10   X(I) = X(I)*GAIN
      RETURN
      END
```

stage of the transform can be regarded as the conversion of a single N-point transform into two $N/2$-point transforms. The second stage converts these to four $N/4$-point transforms, and so on, until we have $N/2$ two-point transforms, which are almost trivially done. The multipliers are also easily found.

The results come out in scrambled form; that is, they are not arranged in order of increasing l. It turns out that the relationship between the scrambled ordering and the proper one is quite simple. It is found by writing the binary representation of the number and reversing the order of the bits. Thus in a 16-point transform, the seventh element of the array (binary 7 is 0111) is the fourteenth element of the actual transform (binary 14 is 1110).

A program for computing Fourier transforms is given below. This program was written by members of the Stanford Computer Science Department.

The algorithm presented above is designed to work when $N=2^n$. For other values of N one can obtain somewhat smaller advantages. For example, a six-point transform can be reduced to two three-point transforms using the method described above and, in like manner, reductions can be made for other values of N. Newer methods do as well or better for other values of N.

The subroutine given in the program assumes that the given data are complex. If the functions to be transformed are real, it is possible to load the arrays with one function as the real part of the initial data and a second function as the imaginary part. Both transforms then can be obtained simultaneously with a little extra unscrambling.

Fourier sine and cosine transforms are also sometimes useful and N-point transforms of either type can be computed at the cost of one $2N$-point Fourier transform. The algorithm has been applied to a number of other related tasks but space does not permit covering them here.

16. ELLIPTIC PDES: VIII. FOURIER METHODS

The technique presented in the preceding section can be used to solve certain elliptic problems very efficiently. The class of treatable problems is limited, but the technique is so powerful that it is worth presenting here. Also, a technique that is as good as this one gives rise to efforts to extend the range of problems to which it may be applied. Research in this area is still continuing.

As an illustration of the method, we take one of the simplest possibilities: Poisson's equation in two dimensions

$$\frac{\partial^2\phi}{\partial x^2}+\frac{\partial^2\phi}{\partial y^2}=\rho(x, y) \tag{4.16.1}$$

We assume further that the boundary conditions are periodic in one of the dimensions, which for the sake of definiteness we choose to be the x direction. Thus

$$\phi(x+L, y)=\phi(x, y) \tag{4.16.2}$$

The boundary conditions in the y direction can be arbitrary; if they are also periodic, greater use of the Fourier method is possible.

Proceeding formally, we write the solution to Eq. 4.16.1 as a Fourier series of the type in Eq. 4.14.6 in the variable x

$$\phi(x,y)=\sum_{l=1}^{N}\hat{\phi}(k_l,y)e^{ik_lx} \tag{4.16.3}$$

where the k_l are given by Eq. 4.14.7, and substitute Eq. 4.16.3 into Eq. 4.16.1. An equation for $\hat{\phi}(k_l,y)$ is then obtained by enforcing Eq. 4.16.1 only at the mesh points x_j, $j=1,2,\dots N$, multiplying the equation by $e^{-ik_lx_j}$, and summing over j—in other words, we take the Fourier transform of Eq. 4.16.1. We then have

$$\frac{\partial^2\hat{\phi}}{\partial y^2}(k_l,y)-k_l^2\hat{\phi}(k_l,y)=\hat{\rho}(k_l,y) \tag{4.16.4}$$

where

$$\hat{\rho}(k_l,y)=\frac{1}{N}\sum_{j=1}^{N}\rho(x_j,y)e^{ik_lx_j} \tag{4.16.5}$$

Equations 4.16.4 form a set of N uncoupled ordinary differential equations—one for each value of k_l. Each ODE is to be solved together with boundary conditions that may be obtained by Fourier transforming the boundary conditions for Eq. 4.16.1. The ODE boundary values problems may be solved by the methods of Chapter 3. The usual choice is to use second order differencing and solve the resulting problems by the standard tridiagonal algorithm. If the boundary conditions in the y direction are periodic, it is possible (in fact advantageous) to use Fourier transforms to solve the problem in the y direction as well. Having solved the set of equations 4.16.4 for the $\phi(k_l,y)$, the desired solution $\phi(x,y)$ is found by another application of the fast Fourier transform algorithm.

Thus we are able to solve the Poisson equation 4.16.1 by the following sequence of steps. First, the function on the right-hand side is Fourier transformed in the x direction for each value of y (the mesh should contain $N=2^n$ evenly spaced x points for maximum efficiency). The resulting ODEs are then solved, and the solutions are Fourier transformed to produce the desired results. The most expensive parts of this procedure are the Fourier transforms. Since N transforms must be computed to find $\hat{\rho}$ and another N to produce ϕ, the number of operations required is approximately $4N^2\log_2 N$ each of additions and multiplications. The number of operations required to solve Eq. 4.16.4 is approximately $7N^2$ (total operations), which is considerably smaller than the number required by the transforms, but not negligible. The important point is that the solution is obtained by Fourier transforms with a speed that beats even the best iterative methods. Not only that, the Fourier method actually finds the "exact" solution. We put "exact" in quotation marks because the solution contains errors arising from the finite difference approximation to the ODEs produced by Fourier transforming. The error arising from the lack of iteration to convergence is absent, however. There are a number of cases in which the extra accuracy is important and for these the Fourier method is invaluable.

A rather surprising application of the method is in the exact solution of the finite difference version of the Poisson equation. Even though the Fourier method is inherently more accurate than finite differences, there are times when one wants the solution of the finite difference equations to be as accurate as possible. Provided only that the boundary conditions are periodic, this can be achieved by the use of Fourier transforms. The trick is to note that the discrete Fourier transform of the finite difference equations can be written in the form of Eq. 4.16.4 with k_l^2 replaced by the square of a "modified wavenumber," say $\bar{k}_l^2$. The modified wavenumber for a particular finite difference method can be found by applying the finite difference operator to the function $\exp(ik_l x)$. For example, the modified wavenumber for the standard second order central difference operator is $2(1-\cos k_l h)/h^2$.

There are also a number of extensions of the problem above to which the method can be applied. If the boundary conditions are not periodic but are homogeneous, the problem can be solved by expansion in an appropriate set of trigonometric functions. For example, if the boundary conditions require the function to be zero, a sine expansion is the proper choice; for zero slope conditions, a cosine expansion is the correct choice. Transforms appropriate to these expansions can be obtained at the expense of a Fourier transform with $2N$ points. The cost of solving these problems is thus about twice that of the case of periodic conditions, but is still quite advantageous. It is also worth noting that a standard way of solving problems with inhomogeneous boundary conditions both analytically and numerically is to use a transformation that introduces an inhomogeneous term into the differential equation but that renders the boundary condition homogeneous.

The method also has difficulty in dealing with regions that are not rectangular, although methods have recently been developed for cases in which the region is nearly rectangular. The problem is solved in the closest rectangular region and some of the parameters are iteratively adjusted to achieve the desired result. The method is akin to shooting, but it is not reviewed in any detail here. Finally, we note that the method can be applied to nonlinear equations by using iterative techniques.

17. ELLIPTIC PDES: IX. BOUNDARY INTEGRAL METHODS

There are a number of applications in which Laplace's equation needs to be solved in an irregular region but results are needed only on the boundary. For example, one might have a problem in which a potential is given on the boundary and the desired information is the gradient of the potential on the boundary. No interior information is necessary and if computation of it can be avoided, so much the better. Although the method we present in this section is rather limited in application, it has been used with success in fluid mechanics problems and has application in electromagnetic theory. It is also an example (along with the Fourier method) of how an analytical technique can be adapted to numerical computation, emphasizing the importance of analytical mathematics in the computer age.

The method we seek can be derived by the use of Green's formula for Laplace's equation but, in two dimensions, it is actually a little simpler to use the theory of functions of a complex variable. In three dimensions, Green's formula must be used. The theory of complex variables is not discussed in depth here, but the part of the theory we need is reviewed. If we regard $f(z)$ as a function of the complex variable $z=x+iy$, we can write

$$f(z)=\phi(z)+i\psi(z) \tag{4.17.1}$$

where ϕ and ψ can be regarded as two real valued functions of the real variables x and y. If f is to have a uniquely defined derivative as a function of z, it is necessary that ϕ and ψ satisfy the Cauchy–Riemann conditions

$$\frac{\partial \phi}{\partial x}=\frac{\partial \psi}{\partial y} \tag{4.17.2a}$$

$$\frac{\partial \phi}{\partial y}=-\frac{\partial \psi}{\partial x} \tag{4.17.2b}$$

Eliminating either of the two functions from these equations, we find that both ϕ and ψ satisfy Laplace's equation.

A function of complex variable is called *analytic* in a region if neither it nor any of its derivatives possess singularities in that region. For such a function one can show that Cauchy's theorem applies. This theorem (which is essentially a complex version of Green's formula) states that the integral on any closed contour enclosing a region of analyticity of the function is zero:

$$\oint_c f(z)\,dz=0 \tag{4.17.3}$$

From this result it is not difficult to prove Cauchy's integral formula

$$\oint \frac{f(z)}{z-z_0}\,dt=2\pi i f(z_0) \tag{4.17.4}$$

which holds under the same conditions as those for Eq. 4.17.3. These formulas are well known and appear in the standard textbooks on complex variable theory.

The next result is one that is not quite as well known but forms the basis of the method discussed in this section. Suppose that in Eq. 4.17.4 we allow the point z_0 to approach the boundary. Then the singularity due to the factor $(z-z_0)^{-1}$ is on the boundary and it is not clear how the integral is to be defined. The precise derivation of the result is difficult and can be found in Muskhelishvili's book (1957). The correct result can be derived in a simple but nonrigorous manner by realizing that what is needed is the limit of Eq. 4.18.4 as z_0 is allowed to approach the boundary from the interior. This is equivalent to computing the integral in the manner suggested by Fig. 4.28. Fig. 4.28*a* shows the original contour with z_0 approaching the contour, which is equivalent to deforming the contour slightly as shown in Fig. 4.28*b*. The integral is then composed of two parts—an integral along the contour excluding the point z_0 and an integral over the small semicircle.

For the integral around the contour it is convenient to introduce the concept of a principal value. The principal value is defined by the limiting process

$$P\int_A^B \frac{f(z)}{z-z_0}dz = \operatorname*{Lim}_{\varepsilon\to 0}\left(\int_A^{z_0-\varepsilon}\frac{f(z)}{z-z_0}dz + \int_{z_0}^B \frac{f(z)}{z-z_0}dx\right) \tag{4.17.5}$$

The integral around the semicircle is easily evaluated and is $\pi i f(z_0)$. Thus when z_0 is on the boundary, Cauchy's formula becomes

$$P\oint_C \frac{f(z)}{z-z_0}dz = \pi i f(z_0) \tag{4.17.6}$$

which is a version of Plemelj's formula. Another way of viewing it is that if $f(z)$ satisfies this equation on the boundary, it must be the limiting value of a function that is analytic within the contour. If we wish to deal with the exterior problem, that is, that of solving Laplace's equation outside the contour, and if the function vanishes at infinity, its boundary values satisfy the same equation with a negative sign on the right-hand side. If the function does not vanish at infinity, additional terms appear in the equation.

It is easy to use Eq. 4.17.6 as the basis for a numerical method. The simplest approach is to enforce the equation at a finite set of points z_i, $i=1,2\ldots N$ on the boundary. The function $f(z)$ can then be approximated by a linear function between each pair of neighboring points and the integrals computed. This requires just a bit of care and reduces Eq. 4.17.6 to a set of linear

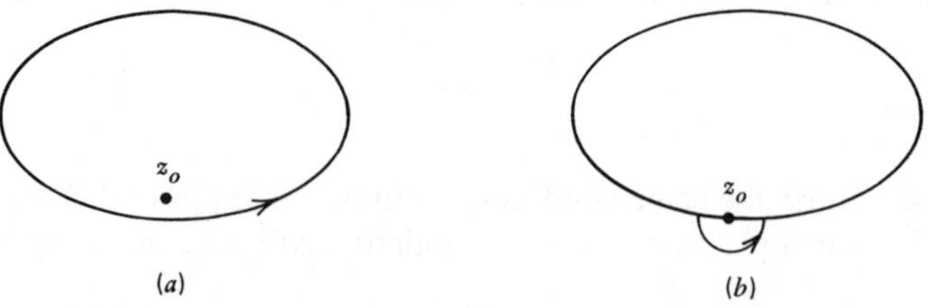

Fig. 4.28. Deformation of the integration contour used in proving Eq. 4.17.6.

algebraic equations

$$f(z_j)=\sum_{k=1}^{N} A_{jk} f(z_k) \qquad j=1,2,\dots N \tag{4.17.7}$$

where

$$A_{jk}=\frac{z_{j+1}-z_k}{z_{j+1}-z_j}\ln\left(\frac{z_{j+1}-z_k}{z_j-z_k}\right)+\frac{z_{j-1}-z_k}{z_{j-1}-z_j}\ln\left(\frac{z_j-z_k}{z_{j-1}-z_k}\right) \tag{4.17.8a}$$

for $j=k$:

$$A_{kk}=\ln\left(\frac{z_k-z_{k+1}}{z_k-z_{k-1}}\right) \tag{4.17.8b}$$

To use this method, one needs to specify either ϕ or ψ at each mesh point; the variable that is not specified is then treated as the unknown. There is one exception: If the data are all values of ϕ, one value of ψ must be given (and vice versa) or the set of equations will not have a unique solution. For computational purposes Eqs. 4.17.7 are split into $2N$ real equations by equating the real and imaginary parts of both sides of each equation. Since there are only N unknowns, only half of these equations can be used. It has been found by experience that it is usually best to use the imaginary parts of the equations because the resulting matrix is better conditioned than if the other choice were made.

For simple geometries this method turns out to be about equally as fast (for the same accuracy) as the better iterative methods. For more complicated geometries, for which the optimum parameters for the iterative methods cannot be found readily, the boundary integral method is quite advantageous.

18. HYPERBOLIC PDES: I. REVIEW OF THEORY

As was stated in the introduction to this chapter, the distinguishing feature of hyperbolic PDEs is that all of their characteristics are real and distinct. This is central to understanding the nature of the solutions of hyperbolic equations and is the basis for a number of numerical methods. Some of the salient features of hyperbolic equations and their characteristics are reviewed before proceeding to their numerical treatment.

We take as our starting point a second order partial differential equation in n independent variables that is linear in all of the second derivatives. An equation of this kind is called *quasilinear* and essentially all examples arising out of physical problems are of this type. The most general such equation is

$$\sum_{i,k=1}^{n} a_{ik}\frac{\partial^2 u}{\partial x_i \partial x_k}=F(x_i,u,\partial u/\partial x_i) \tag{4.18.1}$$

where F is given and the coefficients a_{ik} may be functions of the independent variables x_i, the dependent variable u, and the first partial derivatives $\partial u/\partial x_i$. Equation 4.18.1 is to be solved subject to initial and/or boundary conditions that we need not give explicitly at this point. Suppose that we make the transformation of independent variables

$$\xi_\alpha = \phi_\alpha(x_1, x_2, \dots x_n) \tag{4.18.2}$$

and call the dependent variable

$$\omega(\xi_1, \xi_2, \dots \xi_n) = u(x_1, x_2, \dots x_n) \tag{4.18.3}$$

It is then a straightforward and relatively simple calculation to show that under this transformation the PDE (Eq. 4.18.1) becomes

$$\sum_{\alpha,\beta=1}^{n} A_{\alpha\beta} \frac{\partial^2 \omega}{\partial \xi_\alpha \partial \xi_\beta} = G(\xi_\alpha, \omega, \partial\omega/\partial\xi_\alpha) \tag{4.18.4}$$

where the new coefficient matrix is given by

$$A_{\alpha\beta} = \sum_{i,k=1}^{n} \frac{\partial \phi_\alpha}{\partial x_i} \frac{\partial \phi_\beta}{\partial x_k} a_{ik} \tag{4.18.5}$$

We suppose further that the transformation has been chosen in such a way that the initial conditions are given on the surface $\xi_n = 0$

$$\omega(\xi_1, \xi_2, \dots \xi_{n-1}, 0) = \psi_1(\xi_1, \xi_2, \dots \xi_{n-1}) \tag{4.18.6a}$$

$$\frac{\partial \omega}{\partial \xi_n}(\xi_1, \xi_2, \dots \xi_{n-1}, 0) = \psi_2(\xi_1, \xi_2, \dots \xi_{n-1}) \tag{4.18.6b}$$

and that the domain for which the solution is desired is infinite in all dimensions other than ξ_n.

One approach to solving this problem is by use of Taylor series in n dimensions. Given the values of ω and $\partial\omega/\partial\xi_n$ on the surface represented by $\xi_n = 0$ and the PDE, we want to find their values on the surface $\xi_n = \delta\xi_n$ a short distance away. To do this we can use the series

$$\begin{aligned}
\omega(\xi_1, \xi_2, \dots \xi_{n-1}, \delta\xi_n) &= \omega(\xi_1, \xi_2, \dots \xi_{n-1}, 0) \\
&+ \sum_{j=1}^{n-1} \left(\frac{\partial \omega}{\partial \xi_j} \right)_o (\xi_j - \xi_j^o) + \left(\frac{\partial \omega}{\partial \xi_n} \right)_o \delta\xi_n \\
&+ \frac{1}{2} \sum_{j=1}^{n-1} \sum_{k=1}^{n-1} \left(\frac{\partial^2 \omega}{\partial \xi_j \partial \xi_k} \right)_o (\xi_j - \xi_j^o)(\xi_k - \xi_k^o) \\
&+ \sum_{j=1}^{n-1} \left(\frac{\partial^2 \omega}{\partial \xi_j \partial \xi_n} \right)_o (\xi_j - \xi_j^o)\delta\xi_n + \frac{1}{2} \left(\frac{\partial^2 \omega}{\partial \xi_n \partial \xi_n} \right)_o \delta\xi_n^2 + \cdots
\end{aligned} \tag{4.18.7}$$

where $\xi_1^o, \xi_2^o, \ldots \xi_{n-1}^o, 0$ is any point on the surface on which the data are given and the superscript o indicates that the function is to be evaluated at this point. The condition 4.18.6a provides the values of the function on the initial surface and, by differentiating this equation successively with respect to $\xi_1, \xi_2, \ldots \xi_{n-1}$, we can obtain the partial derivatives $\partial\omega/\partial\xi_1, \partial\omega/\partial\xi_2, \ldots \partial\omega/\partial\xi_{n-1}$ needed in the second term. Finally, $\partial\omega/\partial\xi_n$ needed in the third term is provided by Eq. 4.18.6b and we have all n first partial derivatives. The second partial derivatives that do not involve ξ_n (the fourth term) can be obtained by differentiating Eq. 4.18.6a further and the mixed second order partial derivatives $\partial^2\omega/\partial\xi_i\partial\xi_n$ (fifth term) can be obtained by differentiating Eq. 4.18.6b. This leaves us the problem of calculating $\partial^2\omega/\partial\xi_n\partial\xi_n$ and it is here that the PDE must be used. If $A_{nn} \neq 0$, we can solve Eq. 4.18.4 for this derivative in terms of information already in hand. We then proceed to the third partial derivatives, which are readily computed by further differentiation of either the initial conditions or the PDE as appropriate. Although the use of Taylor series is not recommended as a tool for obtaining solutions, it shows that the problem is well posed as long as $A_{nn} \neq 0$.

In the case in which $A_{nn}=0$ there is clearly trouble. In fact the solutions of the equation

$$A_{nn} = \sum_{i,k=1}^{n} a_{ik} \frac{\partial\phi_n}{\partial x_i}\frac{\partial\phi_n}{\partial x_k} = 0 \tag{4.18.8}$$

are the *characteristics* of the equation. (The characteristics are themselves solutions of a first order PDE and, if the coefficients a_{ik} are functions of the dependent variable, they are solution dependent and cannot, in general, be found without solving the entire problem.) Thus the problem has no solution if the initial data are given on a characteristic curve or surface. One might think that the difficulty could be cleared up by a clever choice of the conditions, but this is not possible.

An example may clarify the situation. Consider the wave equation

$$\frac{1}{c^2}\frac{\partial^2\phi}{\partial t^2} = \frac{\partial^2\phi}{\partial x^2} \tag{4.18.9}$$

with the initial conditions

$$\phi(x,0)=\alpha(x) \qquad \frac{\partial\phi}{\partial t}(x,0)=\beta(x) \tag{4.18.10}$$

This problem is one of the standards of the PDE repertoire and can be solved in a textbook manner. The general solution of the wave equation is well known:

$$\phi(x,t)=f(x-ct)+g(x+ct) \tag{4.18.11}$$

and reflects the fact that the characteristics of the wave equation are the families of straight lines $x+ct=$constant and $x-ct=$constant. In this case the coefficients are constants, so the characteristics are independent of the solution and can be found a priori. The particular solution of the wave equation that satisfies the initial conditions is found by substituting the general solution 4.18.11 into the initial conditions 4.19.10 to arrive at the equations for f and g

$$\begin{aligned} f(x)+g(x)&=\alpha(x) \\ -cf'(x)+cg'(x)&=\beta(x) \end{aligned} \tag{4.18.12}$$

where the prime denotes ordinary differentiation. If we denote the indefinite integral of $\beta(x)$ by $\gamma(x)$, we find easily that the solution is given by $f=(c\alpha-\gamma)/2c$ and $g=(c\alpha+\gamma)/2c$. Thus with the data given on the line $t=0$, which is not a characteristic of the equation, a unique solution can be found.

Now suppose that the initial data are given on one of the characteristics, for example,

$$\phi\left(x,\frac{x}{c}\right)=\alpha(x) \qquad \frac{\partial\phi}{\partial t}\left(x,\frac{x}{c}\right)=\beta(x) \tag{4.18.13}$$

Instead of Eq. 4.19.12, we would then have

$$\begin{aligned} f(0)+g(2x)&=\alpha(x) \\ -cf'(0)+cg'(2x)&=\beta(x) \end{aligned} \tag{4.18.14}$$

These equations are inconsistent unless $\beta(x)=2c\alpha(x)+$constant. Even if α and β have this property, there is no way that we can use them to determine f. Thus if the initial conditions are compatible, the solution is nonunique.

Finally, returning to the original problem, we note that if either α or β has a discontinuity, then so will f and g. Suppose that the discontinuity is at x_0. Then from Eq. 4.18.11 we see that the discontinuity will appear at time t at locations $x=x_0\pm ct$. Thus the characteristics, being the lines along which signals propagate, are also the lines (or in the more general case curves or surfaces) on which the solution may have discontinuities.

This information sets the stage for use of the knowledge of characteristics as a means of solving hyperbolic systems of equations.

19. HYPERBOLIC PDES: II. METHOD OF CHARACTERISTICS

We begin the discussion of the numerical treatment of hyperbolic PDEs by considering the case of two independent variables. A quasilinear equation can then be written in the more conventional form

$$a\frac{\partial^2 u}{\partial x^2}+2b\frac{\partial^2 u}{\partial x\,\partial y}+c\frac{\partial^2 u}{\partial y^2}=F \tag{4.19.1}$$

From the results of the preceding section the equation for the characteristics corresponding to this equation is

$$a\left(\frac{\partial\phi}{\partial x}\right)^2+2b\left(\frac{\partial\phi}{\partial x}\right)\left(\frac{\partial\phi}{\partial y}\right)+c\left(\frac{\partial\phi}{\partial y}\right)^2=0 \tag{4.19.2}$$

The characteristic lines are the level curves of the function $\phi(x, y)$, that is, the lines on which $\phi(x, y)=$constant. The equations defining these lines can be found by differentiating this relation to obtain

$$d\phi=\frac{\partial\phi}{\partial x}dx+\frac{\partial\phi}{\partial y}dy=0 \tag{4.19.3}$$

which can be rewritten as $dy/dx=-(\partial\phi/\partial x)/(\partial\phi/\partial y)$. Then dividing Eq. 4.19.2 by $(\partial\phi/\partial y)^2$, we obtain an equation for the characteristic curves

$$a\left(\frac{\partial y}{\partial x}\right)^2-2b\left(\frac{\partial y}{\partial x}\right)+c=0 \tag{4.19.4}$$

Finally, solving for dy/dx, we have

$$\frac{dy}{dx}=-\frac{b\pm\sqrt{b^2-ac}}{a}=\alpha_1,\alpha_2 \tag{4.19.5}$$

As expected, the characteristics are real and distinct if and only if $b^2-ac>0$, the criterion that determines whether or not Eq. 4.19.1 is hyperbolic. The equation for the characteristics is an ODE, which is a consequence of having only two independent variables. For an equation in more than two independent variables the characteristic equation would be a PDE.

Some of the analysis of the previous section is repeated in a slightly different form for this particular case in order to set the stage for the numerical approach. Suppose we have the solution and its normal derivative on some curve C. The data might be initial conditions or they may have been obtained by previous calculation. To construct the Taylor series, we will need the second derivatives (the first derivatives are easily computed). In contrast to what was done in the preceding section, we will use a Cartesian (x, y) coordinate system and we will need to compute the derivatives $\partial^2 u/\partial x^2$, $\partial^2 u/\partial y^2$, and $\partial^2 u/\partial x\,\partial y$ rather than the derivatives parallel and normal to the curve C. This requires a slight modification of the method. From the chain rule for differentiation we can write along any differential line element (dx, dy)

$$d\left(\frac{\partial u}{\partial x}\right)=\frac{\partial^2 u}{\partial x^2}dx+\frac{\partial^2 u}{\partial x\,\partial y}dy \tag{4.19.6a}$$

$$d\left(\frac{\partial u}{\partial y}\right)=\frac{\partial^2 u}{\partial x\,\partial y}dx+\frac{\partial^2 u}{\partial y^2}dy \tag{4.19.6b}$$

These two equations, taken together with the PDE 4.19.1, provide a set of three linear algebraic equations from which the three second order partial derivatives can be calculated. Of course this is possible only if the determinant of the system is nonzero

$$\begin{vmatrix} a & 2b & c \\ dx & dy & 0 \\ 0 & dx & dy \end{vmatrix} \neq 0 \tag{4.19.7}$$

It is no surprise that this is exactly the condition that the line element (dx, dy) *not* lie on a characteristic. In other words, the second derivatives cannot be computed solely from information given on a single characteristic. There is nothing to stop us, however, from using *both* sets of characteristics to achieve our goal. We may be able to find the second derivatives when the determinant 4.19.7 is zero, provided that the terms on the right-hand sides of these equations obey the proper conditions. One means of stating the condition that the system be solvable, which is useful for small systems of equations, is that the determinant

$$\begin{vmatrix} a & F & c \\ dx & du_x & 0 \\ 0 & du_y & dy \end{vmatrix} \tag{4.19.8}$$

be zero if dx and dy are differential elements of a characteristic curve. In the language of linear algebra Eq. 4.19.8 states that the vector (F, du_x, du_y) lies in the column space of the matrix 4.19.7. Here $u_x = \partial u/\partial x$ and $u_y = \partial u/\partial y$. Since according to Eq. 4.19.5 the characteristics are the lines for which $dy = \alpha_1 dx$ and $dy = \alpha_2 dx$, Eq. 4.19.8 is equivalent to

$$\alpha_1 a\, du_x + c\, du_y - F\, dy = 0 \tag{4.19.9a}$$

on the α_1 characteristics and

$$\alpha_2 a\, du_x + c\, du_y - F\, dy = 0 \tag{4.19.9b}$$

on the α_2 characteristics. These equations can be regarded as the PDE in the characteristic direction. Equation 4.19.9, together with the equation obtained by application of the chain rule

$$du = u_x\, dx + u_y\, dy \tag{4.19.10}$$

provide three equations that govern the way in which the unknown function u and its two first partial derivatives change with the independent variables x and y. If the PDE is nonlinear, that is, if a, b, or c are functions of u, these three equations must be solved simultaneously with Eq. 4.19.5, which defines

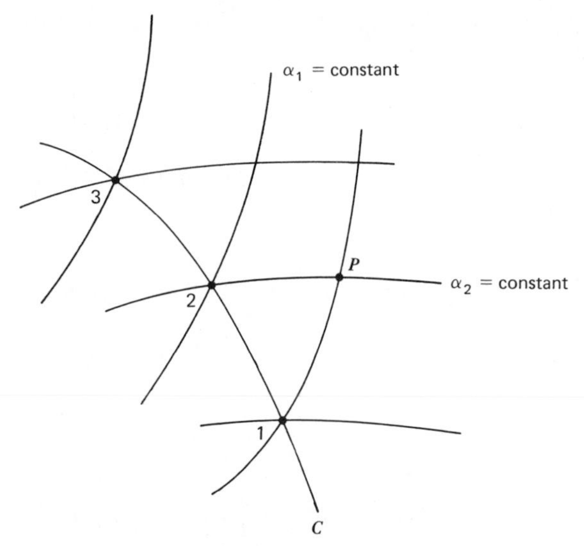

Fig. 4.29. Numerical method of characteristics.

the characteristics. In the linear case the equations for the characteristics can be solved prior to the solution of the equations for the dependent variable.

In summary, we have used the properties of characteristics to reduce the problem to a set of ordinary differential equations. Equations 4.19.5 define the characteristics and are used to locate the mesh points at the next step. Equations 4.19.9 and 4.19.10 give the values of u, u_x, and u_y at the new points in terms of the values at the old ones. We now introduce numerical approximations.

The set of equations we have derived can be solved by any of a number of methods analogous to ODE solvers. To illustrate the method, we use an implicit method that is essentially an application of the trapezoid rule (see Fig. 4.29). Suppose that we have the values of the function and its first derivatives at a set of points on curve C and we wish to advance the solution in the direction indicated. The two sets of characteristics have been drawn as if they were already known; as we have seen, in the nonlinear case they will be actually computed along with the solution. The problem is to find the coordinates of the point P and the values of the unknown function u and its derivatives at point P using the equations available. Finite differencing the equations for the characteristics (Eq. 4.19.5) provides two of the necessary equations

$$\frac{y(P)-y(1)}{x(P)-x(1)}=\tfrac{1}{2}\left[\alpha_1(P)+\alpha_1(1)\right] \tag{4.19.11a}$$

$$\frac{y(P)-y(2)}{x(P)-x(2)}=\tfrac{1}{2}\left[\alpha_2(P)+\alpha_2(2)\right] \tag{4.19.11b}$$

where α_1 and α_2 are given by Eq. 4.19.5. Differencing Eq. 4.19.9 in the same manner, we have

$$[a(P)\alpha_1(P)+a(1)\alpha_1(1)][u_x(P)-u_x(1)]+[c(P)+c(1)][u_y(P)-u_y(1)]$$
$$-[F(P)+F(1)][y(P)-y(1)]=0 \tag{4.19.11c}$$

$$[a(P)\alpha_2(P)+a(2)\alpha_2(2)][u_x(P)-u_x(2)]+[c(P)+c(2)][u_y(P)-u_y(2)]$$
$$-[F(P)+F(2)][y(P)-u(2)]=0 \tag{4.19.11d}$$

and finally, from Eq. 4.19.10

$$u(P)-u(1)=[u_x(P)+u_x(1)][x(P)-x(1)]+[u_y(P)+u_y(1)][y(P)-y(1)] \tag{4.19.11e}$$

The set of equations of Eq. 4.19.11a through 4.19.11e provides five equations for the five unknowns $x(P)$, $y(P)$, $u(P)$, $u_x(P)$, and $u_y(P)$ that need to be calculated at the new points. In general they are nonlinear, so they must be solved iteratively. In most cases the use of a predictor-corrector method is a reasonable compromise. Since the method of characteristics produces equations that behave very much like ordinary differential equations (a consequence of using characteristics as coordinates), any other ODE method could be used here as well.

For many years the method of characteristics was practically the only method used for solving hyperbolic PDEs and it is still widely used today. It has some very desirable properties: Since emphasis is placed on the importance of characteristics in determining the solution of hyperbolic PDEs, important properties of the exact solution are preserved in the numerical solution; and since discontinuities can occur only along characteristics, the method is easily adapted to the computation of problems that contain discontinuities. The most significant application of this type is in aerodynamics. In any flow that contains a region of supersonic flow, there is the possibility of shock waves (severe discontinuities in almost all of the variables) and other, milder discontinuities. For this reason the method of characteristics has been the basis for a large number of methods for treating supersonic flows. An excellent review of methods of this type, including especially methods developed in the Soviet Union, is provided by the recent monograph by Holt (1977).

The method of characteristics has some severe handicaps, however. Chief among these is the difficulty of keeping track of the locations of the characteristics and the values of the variables in three dimensions simultaneously. For this reason it has proven very difficult to write a computer program that can calculate flows around bodies of complicated shapes such as aircraft and engine inlets. Another difficulty, which also arises in aerodynamics, is that when the flow velocity is near the speed of sound (transonic flow), the characteristics tend to become nearly perpendicular to the direction of the flow. It is then necessary to take very small steps in the flow direction and the computation time becomes rather long. It was this difficulty that motivated the search for methods other than the method of characteristics for use in aerodynamic computation. These will be taken up in the sections that follow.

Example 4.10

As the example of methods for solving hyperbolic problems we solve the wave equation

$$\phi_{tt} = c^2 \phi_{xx}$$

subject to the initial conditions that the function be a "witch's hat"

$$\phi = 5(x - 0.3) \qquad 0.3 < x < 0.5$$

$$\phi = 5(0.7 - x) \qquad 0.5 < x < 0.7$$

$$\phi = 0 \text{ at all other points}$$

and that the time derivative be zero

$$\phi_t(x,0) = 0$$

For boundary conditions we insist that the function be zero at both ends

$$\phi(0,t) = \phi(1,t) = 0$$

The exact solution to this problem is easily obtained. Witch's hat-shaped waves, each half the amplitude of the initial condition, propagate to the left and right. When these waves strike the boundary, they are reflected with a change in sign. After enough time has passed for a wave to have propagated the length of the box, that is, at $t = 1/c$, the waves reinforce and the solution is identical to the initial condition except for a change in sign. One can see how well the numerical method is doing by comparing the result with the initial condition. The solution can be allowed to continue to propagate. At time $t = 2/c$, we should recover the initial condition (without change of sign), at $t = 3/c$ the solution should be the same as that at $t = 1/c$, and so on.

In this problem the characteristics are found easily. In fact $\alpha_1 = c$ and $\alpha_2 = -c$ and the two sets of characteristics are

$$x + ct = \text{constant}$$

$$x - ct = \text{constant}$$

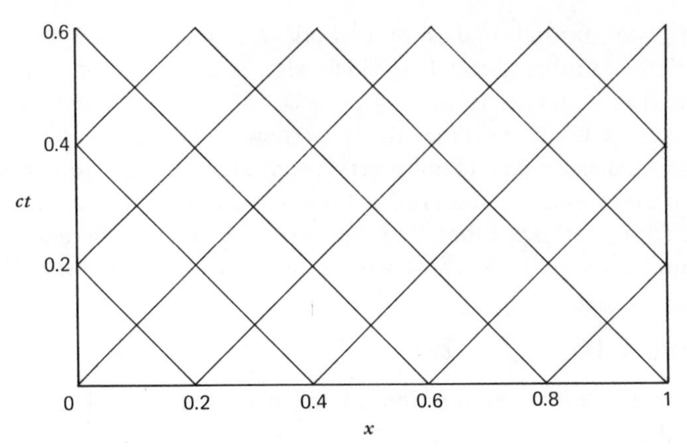

Fig. 4.30. Characteristics of the wave equation in a finite domain.

The characteristics that pass through the points $x=0$, 0.2, 0.4, 0.6, 0.8, and 1.0 at $t=0$ are shown in Fig. 4.30. Note that characteristics are also formed by reflection from the boundary and that these are shown as well. Thus the entire characteristic mesh can be computed in advance and the characteristic equation need not be treated numerically.The spatial mesh points at each time step are midway between the points at the preceding step and lie on lines of constant time. The time step is $\Delta t=\Delta x/2c$.

When finite differenced, the equations on the characteristics (Eq. 4.19.9), give

$$c\left(u_{x_j}^{n+1}-u_{x_{j+1/2}}^{n}\right)+\left(u_{t_j}^{n+1}-u_{t_{j+1/2}}^{n}\right)=0 \tag{4.19.12a}$$

$$c\left(u_{x_j}^{n+1}-u_{x_{j-1/2}}^{n}\right)-\left(u_{t_j}^{n+1}-u_{t_{j-1/2}}^{n}\right)=0 \tag{4.19.12b}$$

These equations, which are Eqs. 4.19.11c and 4.19.11d for this particular problem, are the *exact* result of integration along the characteristics. Although they have finite difference form, these equations contain no approximation. Thus u_x and u_t are propagated as they would be in the exact equations (in this linear example). The initial values of u_x were computed by central differences

$$u_{x_j}^{0}=\frac{u_{j+1}^{0}-u_{j-1}^{0}}{2\Delta x} \tag{4.19.13}$$

Equations 4.19.12a and 4.19.12b can be solved easily for $u_{x_j}^{n+1}$ and $u_{t_j}^{n+1}$. Finally, u_j^{n+1} is obtained from

$$u_j^{n+1}=\tfrac{1}{2}\left(u_{x_j}^{n+1}+u_{x_{j\pm 1/2}}^{n}\right)\Delta x+\tfrac{1}{2}\left(u_{t_j}^{n+1}+u_{j\pm 1/2}^{n}\right)\Delta t+u_{j\pm 1/2}^{n}$$

which is Eq. 4.19.11d for this problem and either sign may be used. To keep everything symmetric, we actually used the average of these in this example.

This, of course, is a second order approximation (trapezoid rule) to the exact result.

This leaves the question of handling the boundary conditions. At the odd-numbered time steps, there is no problem, since all the points are actually interior points. At the even steps, we can apply the equations resulting from solving Eqs. 4.19.12a and 4.19.12b at the boundaries. For $j=0$ the simplest thing we can do is to note that at the boundary $x=0$, we can only use the characteristics of the set $x+ct=$constant and, therefore, must use Eq. 4.20.12a. Since $u_t(0)=0$, we have

$$u_{x_0}^{n+1}=u_{x,1/2}^{n}+cu_{t,1/2}^{n}$$

At the other boundary ($j=N$), we must use the $x-ct=$constant characteristics and find

$$u_{x,N}^{n+1}=u_{x,N-1/2}^{n}-cu_{t,N-1/2}^{n}$$

The boundary conditions for u_t and u are simple; both are zero on the boundary.

The results are given in Fig. 4.31. The method has difficulty treating the corner at the top of the hat and also tends to round the corners at the bottom

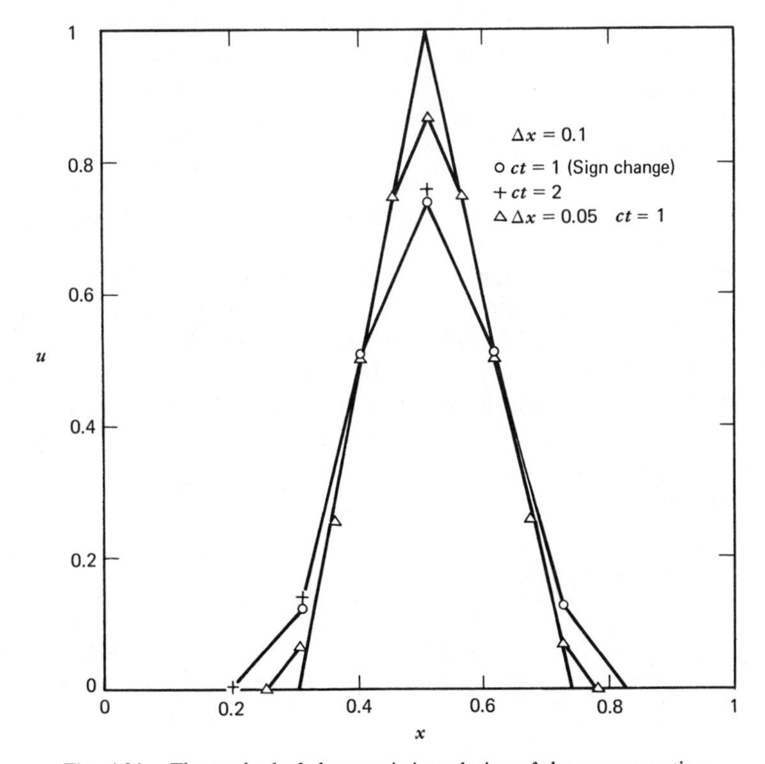

Fig. 4.31. The method of characteristics solution of the wave equation.

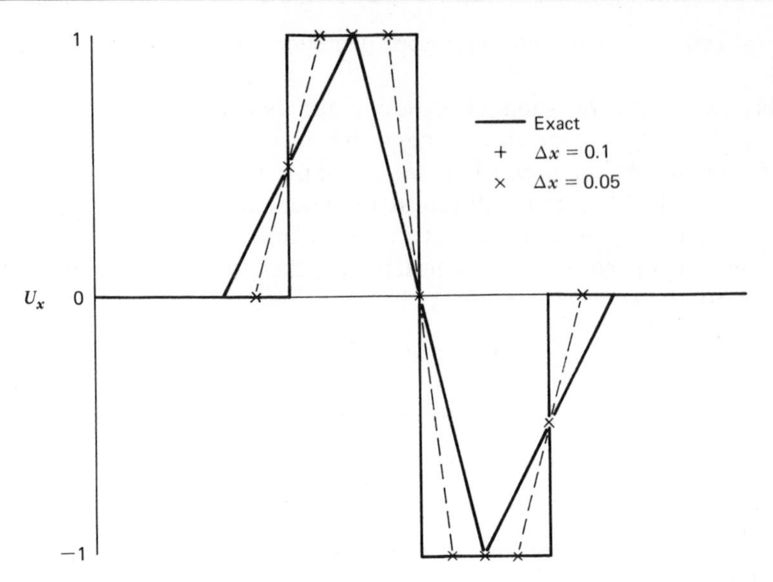

Fig. 4.32. The derivative $\partial u/\partial x$ for "witch's hat."

of the profile. What is happening is that the derivatives u_x and u_t retain their shape and are propagated as they would be by the exact wave equation. Since $u_x(x,0)$ is obtained by finite differences, however, the program "thinks" that u_x has the shape shown in Fig. 4.32 (it is the best interpolation that can be made from the limited data) and then u_t and u "lock onto" this incorrect u_x. After this happens, the profile is propagated unchanged. When the solution is carried from $t=1/c$ to $t=2/c$, no further deterioration of the solution occurs.

The effect of the finite difference approximations can be reduced by using a finer mesh. In the method of characteristics, however, reducing the spatial mesh size by half also requires that the time step be halved, which means that the cost of obtaining a solution goes up by a factor of four. As can be seen from Fig. 4.31, the error is indeed reduced by decreasing the mesh size.

Other numerical approximations can be applied to the method of characteristics and it is possible to devise methods that explicitly allow for the possibility of discontinuities. Such methods have been used to fit shock waves in gas dynamic calculations and have been quite successful.

The program used in this example is given below.

```
00100         PROGRAM CHARAC
00200   C-----THIS PROGRAM SOLVES THE WAVE EQUATION IN ONE SPACE DIMENSION
00300   C-----BY THE METHOD OF CHARACTERISTICS.  THE WAVEFORM IS A 'WITCH'S
00400   C---- HAT' BUT THE PROGRAM IS EASILY MODIFIED TO OTHER WAVEFORMS.
00500   C-----THE VELOCITY IS SET TO 1.
00600         DIMENSION U(51),U1(51),UX(51),UP(51),UTP(51),UXP(51),X(51)
00700   C-----COLLECT THE INPUT.
00800         WRITE (5,100)
```

```
00900       100 FORMAT ( ' GIVE THE NUMBER OF SPACE INTERVALS, N')
01000           READ (5,110) N
01100       110 FORMAT (I)
01200           WRITE (5,120)
01300       120 FORMAT ( ' GIVE THE FINAL TIME')
01400           READ (5,130) TF
01500       130 FORMAT (F)
01600     C-----COMPUTE THE STEP SIZES.
01700           NM1 = N - 1
01800           NP1 = N + 1
01900           DX = 1./FLOAT(N)
02000           DT = .5 * DX
02100           K = 2 * N
02200           WRITE (5,140) K
02300       140 FORMAT ( ' RESULTS WILL BE PRINTED EVERY K = ',I2,  TIME STEPS'
02400          1 ,' IF YOU WANT A DIFFERENT K, ENTER IT; OTHERWISE PUSH RETURN')
02500           READ (5,110) KK
02600           IF (KK.NE.0) K = KK
02700     C-----SET UP THE INITIAL CONDITIONS; FIRST THE FUNCTION ITSELF.
02800           DO 1 I=1,NP1
02900           X(I) = (I - 1) * DX
03000           IF (X(I).LT.0.5) U(I) = 5. * (X(I) - .3)
03100           IF (X(I).GE.0.5) U(I) = 5. * (.7 - X(I))
03200           IF (X(I).LT.0.3) U(I) = 0.
03300           IF (X(I).GT.0.7) U(I) = 0.
03400         1 CONTINUE
03500     C-----NOW, THE PARTIAL DERIVATIVES.
03600           DO 2 I=2,N
03700           UX(I) = (U(I+1) - U(I-1)) / (2. * DX)
03800           UT(I) = 0.
03900         2 CONTINUE
04000           UX(1) = 0.
04100           UX(NP1) = 0.
04200           UT(1) = 0.
04300     C-----MAIN LOOP.
04400           TIME = 0.
04500           KH = K / 2
04600        20 DO 3 L=1,KH
04700     C-----ODD SWEEP; RESULTS PUT IN UP ARRAY.
04800           DO 4 I=1,N
04900           UXP(I) = .5 * (UX(I) + UX(I+1)) + .5 * (UT(I+1) - UT(I))
05000           UTP(I) = .5 * (UT(I) + UT(I+1)) + .5 * (UX(I+1) - UX(I))
05100           UP(I) = .125 * DX * (UX(I) - UX(I+1))
05200          1        + .25 * DT * (2. * UTP(I) + UT(I) + UT(I+1))
05300          2        + .5 * (U(I) + U(I+1))
05400         4 CONTINUE
05800     C---- EVEN SWEEP; RESULTS PUT IN U ARRAY.
05900           UX(1) = UXP(1) + UTP(1)
06000           UX(NP1) = UXP(N) - UTP(N)
06100           DO 5 I=2,N
06200           UX(I) = .5 * (UXP(I-1) + UXP(I)) + .5 * (UTP(I) - UTP(I-1))
06300           UT(I) = .5 * (UTP(I-1) + UTP(I)) + .5 * (UXP(I) - UXP(I-1))
06400           U(I) = .125 * DX * (UXP(I-1) - UXP(I))
06500          1       + .25 * DT * (2. * UT(I) + UTP(I-1) + UTP(I))
06600          2       + .5 * (UP(I) + UP(I-1))
06700         5 CONTINUE
06800         3 CONTINUE
06900           TIME = TIME + K * DT
07000           WRITE (49,160) TIME
07100       160 FORMAT ( ' RESULTS AT TIME ',F10.4)
07200           WRITE (49,170) (U(I), I=2,N)
07300           WRITE (49,170) (UX(I) , I=2,N)
07400           WRITE (49,170) (UT(I) , I=2,N)
07500       170 FORMAT ( 51F12.5)
07600           IF (TIME.LT.TF) GO TO 20
07700           STOP
07800           END
```

20. HYPERBOLIC PDEs: III. EXPLICIT METHODS

The method of characteristics has some obvious advantages. Among these is the ability to compute the solution over a long span of the independent variables. The chief difficulty is that the method is very cumbersome in three dimensions; keeping the coordinates of characteristic surfaces in good order becomes very difficult, particularly when the equations are nonlinear. The method of characteristics also has difficulty in problems in which the equation is hyperbolic in one part of the region and elliptic or parabolic in another part of the region. Finally, for some problems such as slightly supersonic flow it is very slow. Consequently, the trend in recent years has been toward techniques that simply finite difference the partial differential equations. These methods are similar to the ones already described for the other types of equations, so they are not difficult to derive. There are, however, a few peculiarities associated with hyperbolic equations that require some attention.

The simplest hyperbolic equation is the first order convection equation

$$\frac{\partial \phi}{\partial t}+c\frac{\partial \phi}{\partial x}=0 \tag{4.20.1}$$

which states that the quantity ϕ is convected at constant velocity c. The characteristics of this equation are the straight lines $x-ct=$constant and reflect the physics of the problem. Although this equation is simple, it is also quite important. Almost all problems in fluid mechanics contain this operator (albeit in more than one dimension and with a nonuniform velocity) and understanding it is critical to the treatment of problems in this field. The well-known second order wave equation

$$\frac{\partial^2 \phi}{\partial t^2}=c^2\frac{\partial^2 \phi}{\partial x^2} \tag{4.20.2}$$

which allows propagation with velocity c in both the positive and negative directions, behaves very much like Eq. 4.20.1 and is equivalent to the coupled set of first order equations

$$\begin{aligned}\frac{\partial u}{\partial t}&=c\frac{\partial w}{\partial x}\\ \frac{\partial w}{\partial t}&=c\frac{\partial u}{\partial x}\end{aligned} \tag{4.20.3}$$

In this form it is evident that the wave equation (4.20.2) can be treated by the same methods as the convection equation (Eq. 4.20.1). Thus, the simpler convection equation is emphasized and a few remarks about the wave equation are made.

We begin by looking at the simplest of numerical methods, considering explicit methods first. A simple and obvious method of differencing Eq. 4.20.1

is to use the Euler forward difference method with respect to time and the second order central difference method for the spatial derivative. We then have as the difference equation

$$\frac{\phi_j^{n+1}-\phi_j^n}{\Delta t}=\frac{c}{2\,\Delta x}\left(\phi_{j+1}^n-\phi_{j-1}^n\right) \tag{4.20.4}$$

From our knowledge of the individual formulas that went into this equation, we know that it is first order accurate in time and second order accurate in space. We might expect it to be conditionally stable, but this requires investigation. To do this we use the von Neumann stability analysis, that is, we assume a solution that behaves like e^{ikx} in space and find the solutions of both the PDE and the finite difference equations that have this form. It is not difficult to see that the solution of a PDE of this kind is $\phi(x,t)=e^{ikx}e^{-ikct}$. On the other hand, the solution of Eq. 4.20.4 is

$$\phi_j^n=e^{ikj\Delta x}\left(1-\frac{ic\,\Delta t}{\Delta x}\sin k\,\Delta x\right)^n \tag{4.20.5}$$

so that the method replaces $e^{-ikc\Delta t}$ by $1-(ic\,\Delta t/\Delta x)\sin k\,\Delta x$. More important, the exact PDE propagates the solution as a traveling wave the amplitude of which is unchanged in time. This is a property that the convection equation shares with the wave equation. But the amplitude of the finite difference solution of Eq. 4.20.5 increases in time for any values of the parameters; in other words, the method is unconditionally unstable! This may seem surprising until we note that after substituting $\phi(x,t)=e^{ikx}\phi(t)$, the present problem is equivalent to $\phi'=\alpha\phi$ with α pure imaginary. Euler's method is unstable when α is imaginary, so we have reestablished an earlier result.

On the other hand, the leapfrog method, which is unstable when α is real and proved to be unsuitable for parabolic equations, turns out to be an excellent choice for the convection equation. It is not our purpose here to survey all possible methods. An important result is the one just demonstrated — that many of the methods used for hyperbolic equations are equivalent to methods for ODEs with the parameter α purely imaginary, which is the primary reason why complex α was considered in Chapter 3. We can then review the results obtained in Chapter 3 with an eye to choosing a method that behaves well when α is imaginary.

Another important idea is embedded in the results derived above. The convection equation moves the solution at constant speed without change in shape and certainly no change in amplitude. It is too much to expect that a numerical method (except perhaps a Fourier method) maintain an intact shape while moving it about, but we can require that it not destroy the amplitude of the function (this is essentially energy conservation). This is a difficult constraint. In earlier examples we asked only that a numerical method produce decay when the solution of the differential equation decays; this is the concept of stability. Now we ask for a solution that while oscillating in time and space, neither increases in time (in which case the method would be unstable, e.g., the Euler method) nor decays in time (which would eventually destroy the solution). These are stringent demands.

One consequence of these constraints on methods for solving hyperbolic equations is that it is common to discuss the accuracy of a method in terms of its amplitude and phase errors. *Stability* is the requirement that the method not increase the magnitude of the solution with increasing time. A method that produces artificial numerical damping or decay of the solution is sometimes said to be *overstable*. In time Δt, the exact solution of the PDE with e^{ikx} spatial behavior is changed by a factor $e^{i\theta}$ where $\theta = -kc\,\Delta t$; this is just a phase change. The numerical solution will produce a phase that in general differs from θ. The difference can be regarded as a phase error and is a function of the wavenumber k. Another way of looking at it is to regard the numerical solution as corresponding to a physical system in which the velocity of propagation is a function of the wavenumber, that is, $c = c(k)$. A physical system with this property is called *dispersive* and the *phase error* therefore, is sometimes known as *dispersive error*. For a similar reason amplitude error is sometimes called *dissipative error*. In most cases the spatial difference approximation introduces dispersive error but not dissipative error. The time difference approximation is a source of error of both types. It is important to note that although these concepts provide useful methods of comparing numerical methods for linear problems, they may not suffice in the nonlinear case.

In our discussion of characteristics in Section 18, we noted the importance of the domains of dependence and influence. Essentially, these concepts express the idea that there are "forbidden signals," that is, not all portions of the solution domain can influence all others. Numerical approximations ought to inherit these properties. (Similar considerations play important roles in selecting good methods for parabolic and elliptic equations.) For the convection equation (Eq. 4.20.1), the domains of influence and dependence reduce to a single straight line. For the wave equation the domains of influence and dependence are defined by a pair of intersecting straight lines in the $x-t$ plane; the regions between the lines are the domains of interest. It would seem worthwhile to seek numerical methods that have the domains illustrated in Fig. 4.33.

For the convection equation simulating the domain of dependence requires that the value of the function at the point x_j at time t_n not depend on information from points to the right of x_j at earlier times. This rules out central difference approximations even though they are desirable for accuracy reasons. The simplest approximation of this type uses backward differences in both space and time

$$\frac{\phi_j^{n+1} - \phi_j^n}{\Delta t} = -\frac{c}{\Delta x}\left(\phi_j^n - \phi_{j-1}^n\right) \tag{4.20.6}$$

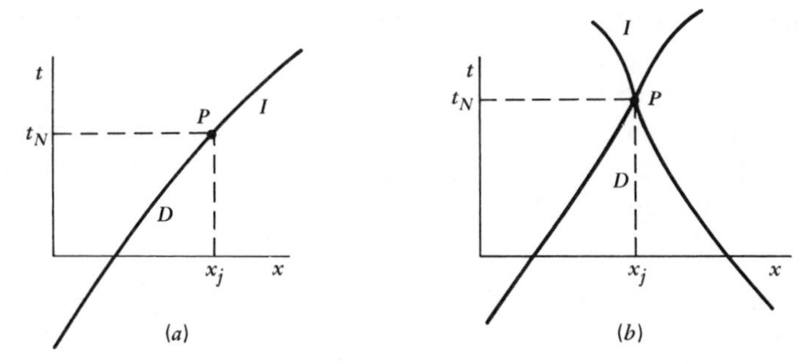

Fig. 4.33. Domains of influence (I) and dependence (D) for (a) the convection equation and (b) the wave equation.

Analyzing this scheme by the von Neumann method, we have

$$\phi_j^{n+1} = \left[1 - \frac{c\,\Delta t}{\Delta x}(1 - e^{-ik\Delta x})\right]\phi_j^n \tag{4.20.7}$$

from which it is easily shown that this method is conditionally stable. The necessary condition for stability is

$$\frac{c\,\Delta t}{\Delta x} \leqslant 1 \tag{4.20.8}$$

This result was first derived by Courant, Friedrichs, and Lewy in 1927 for a related method and, therefore, is known as the *CFL* or, sometimes, the *Courant condition*. We note that on the basis of dimensional analysis, the stability of any conditionally stable method for either of the two example equations must depend on the dimensionless parameter $c\Delta t/\Delta x$, which is sometimes called the *Courant number*.

The method defined by Eq. 4.20.6 has interesting properties. For the case in which the Courant method is unity, which happens to be the stability limit of the method, it becomes exact. This is surprising until we note that when $c\,\Delta t = \Delta x$, both the PDE and the finite difference approximation state that the solution advances a distance $c\,\Delta t$ in time Δt. In general, however, it is only first order accurate in both space and time.

Equation 4.20.8 states that the numerical method is stable as long as its domain of dependence contains the domain of dependence of the PDE. In numerical fluid dynamics the difference scheme 4.20.6 is known as *upwind differencing* and has played an important role.

For the wave equation, which propagates information equally in both directions, the use of one-sided differences does not make sense, so centered difference methods are usually chosen. The Euler scheme is unstable and cannot be used. The leapfrog method uses centered differences in both space and time and is stable for Courant number less than unity. There are other possibilities, but we will omit these in favor of going directly to methods that have more desirable properties and that receive more attention and use.

A method that is important because it is both useful in its own right and is the basis of the better methods in current use is second order accurate in both space and time. The basis for this method is the Taylor series approximation introduced in Section 3.1b. We write

$$\phi(x,t+\Delta t)=\phi(x,t)+\Delta t\frac{\partial\phi}{\partial t}(x,t)+\frac{\Delta t^2}{2}\frac{\partial^2\phi(x,t)}{\partial t^2}+\cdots \tag{4.20.9}$$

and note that the first derivative is obtained from the PDE 4.20.1 itself and the second time derivative can be found by differentiating the PDE to give

$$\frac{\partial^2\phi}{\partial t^2}=\frac{\partial}{\partial t}\left(-c\frac{\partial\phi}{\partial x}\right)=c^2\frac{\partial^2\phi}{\partial x^2} \tag{4.20.10}$$

In finite difference form this method becomes

$$\phi_j^{n+1}=\phi_j^n-c\Delta t\left(\frac{\phi_{j+1}^n-\phi_{j-1}^n}{2\Delta x}\right)+c^2\frac{\Delta t^2}{2}\left(\frac{\phi_{j+1}^n-2\phi_j^n+\phi_{j-1}^n}{\Delta x^2}\right) \tag{4.20.11}$$

and is known as the *Lax–Wendroff method*. It is applied to the wave equation in the example below.

This method is second order accurate in both space and time, explicit, generally quite easy to program, and stable for Courant numbers less than unity. For cases in which the coefficients in the PDE are not constant or the equation is nonlinear, the application of this method is not as straightforward as indicated above and a number of versions of it have been developed. Among these are a number of methods in which splitting of the equation is introduced. Some of these are discussed later.

Example 4.11

To solve the "witch's hat" problem of the preceding example using the Lax–Wendroff method, the equations are written in the first order form of Eq. 4.20.3. The finite difference equations of the Lax–Wendroff method are then

$$u_j^{n+1}=u_j^n+\frac{c\Delta t}{2\Delta x}\left(w_{j+1}^n-w_{j-1}^n\right)+\tfrac{1}{2}\left(\frac{c\Delta t}{\Delta x}\right)^2\left(u_{j+1}^n-2u_j^n+u_{j-1}^n\right) \tag{4.20.12a}$$

$$w_j^{n+1}=w_j^n+\frac{c\Delta t}{2\Delta x}\left(u_{j+1}^n-u_{j-1}^n\right)+\tfrac{1}{2}\left(\frac{c\Delta t}{\Delta x}\right)^2\left(w_{j+1}^n-2w_j^n+w_{j-1}^n\right) \tag{4.20.12b}$$

which are easily programmed. The boundary condition that requires u to be zero on the boundary is easily satisfied. The first of the equations 4.20.3 shows that $\partial w/\partial x=0$ on the boundary. This condition was satisfied by introducing an artificial point at $x_{-1}=-\Delta x$ and letting $w_{-1}=w_1$ and $u_{-1}=-u_1$. Then Eq. 4.20.12b with $j=0$ becomes

$$w_0^{n+1}=w_0^n+\frac{c\,\Delta t}{\Delta x}u_1^n+\left(\frac{c\,\Delta t}{\Delta x}\right)^2(w_1^n-w_0^n)$$

and at the right boundary

$$w_N^{n+1}=w_N^n-\frac{c\,\Delta t}{\Delta x}u_{N-1}^n-\left(\frac{c\,\Delta t}{\Delta x}\right)^2(w_N^n-w_{N-1}^n)$$

Results using ten intervals ($\Delta x=0.1$) are shown in Fig. 4.34. Note that in contrast to the method of characteristics, in the Lax–Wendroff method the time step can be chosen independently of the size of the spatial mesh. With the Courant number $c\,\Delta t/\Delta x=1$, which is the stability limit, the results are exact but this is fortuitous. In all cases other than $c\,\Delta t/\Delta x=1$ the solutions are much worse than the ones obtained using the method of characteristics. Furthermore,

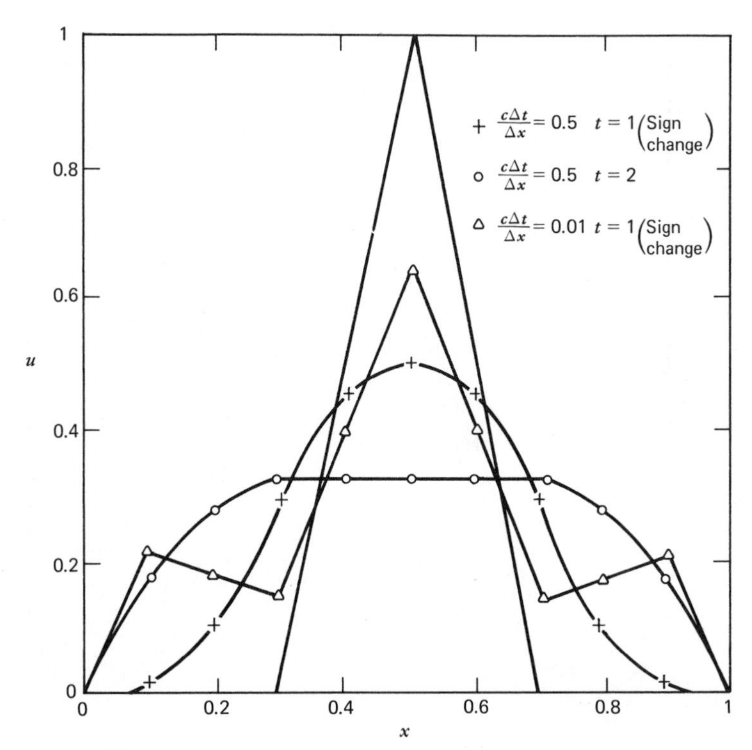

Fig. 4.34. Solution of the wave equation by the Lax–Wendroff method with 10 intervals.

they continue to deteriorate with increasing time. The results for Courant number 0.5, which approximately balances the errors due to temporal and spatial finite differences, are not very good. The broadening of the peak is due to the dispersive error in the method. When the solution is computed to $t=2$, the dispersion has broadened the original peak almost beyond recognition. Reducing the Courant number to 0.01, which eliminates the time differencing error, improves the solution near the peak but the solution develops "bumps" away from the main peak of the solution. This is due to the dispersive character of the method and can be a serious problem in computations of problems in which the solution contains discontinuities or cusps. We also note that since the wave speed is effectively reduced by the numerical method, the solution at $t=1.1$ actually shows the peak more clearly than the solution at $t=1.0$; the difference is not large, however.

Using a finer mesh improves the accuracy of the calculation and reduces the dispersion considerably as shown in Fig. 4.35 but the dispersion still increases with increasing time. Reducing the Courant number results in very little change, indicating that the major source of error is the spatial differencing approximation.

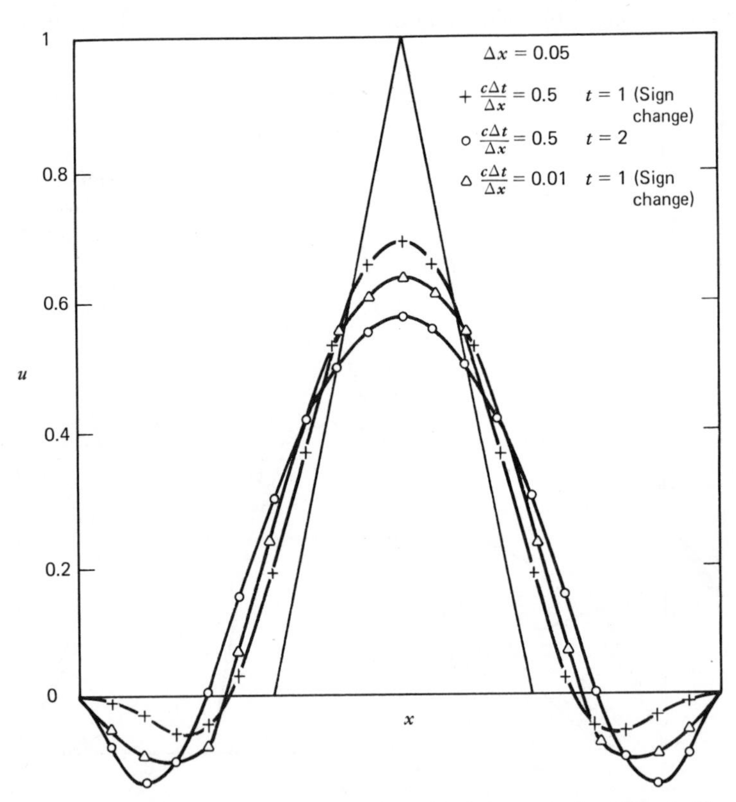

Fig. 4.35. Solution of the wave equation with 20 intervals. Lax–Wendroff.

21. HYPERBOLIC PDEs: IV. IMPLICIT METHODS

The important properties of the characteristics of hyperbolic equations can be approximated well by explicit methods. Historically, this consideration led people to stay away from implicit methods, so they have received relatively little attention. However, we have seen that an implicit method of characteristics has excellent properties.

The excellent stability properties of implicit methods strongly suggest that we see what they can do in hyperbolic problems. It turns out that they do rather well (if one is careful) and they have been increasing in popularity in recent years. There is no reason to survey many methods of this type. Their properties are very nearly what one might expect. One of the best methods is taken for further investigation. The obvious choice is the Crank–Nicolson method, which offers an excellent combination of accuracy and stability.

When applied to the convection equation 4.20.1, the Crank–Nicolson method gives

$$\frac{\phi_j^{n+1}-\phi_j^n}{\Delta t}=\frac{c}{4\Delta x}\left(\phi_{j+1}^{n+1}-\phi_{j-1}^{n+1}+\phi_{j+1}^{n}-\phi_{j-1}^{n}\right) \tag{4.21.1}$$

We have used central differences in space, but other choices are, of course, possible. As was the case with parabolic equations we are left with a tridiagonal system of equations to solve. This also means that in two or three space dimensions the method will be intractable unless a splitting (ADI) type of method is used.

The stability of this method is readily investigated using the von Neumann approach. Making the substitution $\phi(x,t)=e^{ikx}\phi^n$, we find without much effort

$$\phi^{n+1}=\left[\frac{1-\dfrac{ic\,\Delta t}{2\,\Delta x}\sin k\,\Delta x}{1+\dfrac{ic\,\Delta t}{2\,\Delta x}\sin k\,\Delta x}\right]\phi^n \tag{4.21.2}$$

so that the function in the brackets represents the approximation to $e^{-ikc\Delta t}$ produced by this method. The numerator and denominator of the function in brackets are complex conjugates of each other for any values of the parameters. Consequently, the method does not change the amplitude of the function, that is, it does not introduce any dissipation and is not unstable. We call this property *neutral stability*. There is, of course, phase error, but the second order nature of the method assures that the error will not be too serious as long as one does not take too large a time step. Thus despite the fact that implicit methods do not treat the characteristics well, they do have properties that recommend them for application to hyperbolic equations.

For the wave equation we can apply the Crank–Nicolson method just given directly to the factored form (4.20.3) of the wave equation to arrive at a method that again is unconditionally stable. An alternative is to deal directly with the second order wave equation of Eq. 4.20.2 and use a second order central time difference. For the spatial derivatives we use the average of the values at time steps $n-1$ and $n+1$

$$\frac{\phi_j^{n+1}-2\phi_j^n+\phi_j^{n-1}}{\Delta t^2}=\frac{c^2}{2\,\Delta x^2}\left[\left(\phi_{j+1}^{n+1}-2\phi_j^{n+1}+\phi_{j-1}^{n+1}\right)+\left(\phi_{j+1}^{n-1}-2\phi_j^{n-1}+\phi_{j-1}^{n-1}\right)\right] \tag{4.21.3}$$

This method is also unconditionally stable. It is slightly more accurate for the wave equation than the method applied to the factored equation. Note that although Eq. 4.21.3 represents a multistep method, there is no difficulty in starting the calculation because both the function and its time derivative are provided at the initial time. This information can be used to allow the function to be computed at the first time step without the use of a special starting method. Also note that if the spatial derivative in Eq. 4.21.3 was evaluated at the nth time step, the method would be explicit (similar to the leapfrog method) and the stability would only be conditional—the Courant number would have to be less than unity.

A final word of caution must be injected here. The Crank–Nicolson method has very attractive properties when applied to the simple hyperbolic equations chosen as examples. Specifically, the method produces no amplitude error, that is, it is neither unstable nor overstable. Unfortunately, when the method is applied to nonlinear equations, it is possible that a small amplitude error will be introduced. There is no simple analytical way to predict how a method will behave in the nonlinear case, so the method may be unstable when applied to a particular nonlinear problem or it may be overstable. A small amount of overstability can probably be tolerated, but instability is difficult to live with. This is probably the major reason why the development of implicit methods for hyperbolic problems has proceeded cautiously.

The approach to solving nonlinear hyperbolic problems using implicit methods that has been taken in recent years has been simply to try the method on a number of related problems and see how well it works. As a result there have been a number of methods suggested, all of which reduce to the Crank–Nicolson method in the linear case. Some of these are given in the next section.

Example 4.12

We proceed to treating the problem of the two preceding examples by the Crank–Nicolson method. As has been pointed out, there are at least two possible versions of the Crank–Nicolson method for the wave equation. We have used the factored form for this example. The finite difference version of

Eq. (4.20.3) is

$$u_j^{n+1} - u_j^n = \frac{c\,\Delta t}{4\,\Delta x}\left(w_{j+1}^n - w_{j-1}^n + w_{j+1}^{n+1} - w_{j-1}^{n+1}\right)$$

$$w_j^{n+1} - w_j^n = \frac{c\,\Delta t}{4\,\Delta x}\left(u_{j+1}^n - u_{j-1}^n + u_{j+1}^{n+1} - u_{j-1}^{n+1}\right)$$

Boundary conditions are treated by introducing fictitious points outside the boundary.

The set of equations, viewed as a whole, is block tridiagonal with 2×2 blocks. On closer inspection, we find that the equations break into two disjoint sets, one for the variable set

$$\left(w_0, u_1, w_2, u_2, \ldots u_{N-1}, w_N\right)$$

and another for the set

$$\left(w_1, u_2, w_3, \ldots u_{N-2}, w_{N-1}\right)$$

assuming that N, the number of intervals, is even. Since each set of equations is

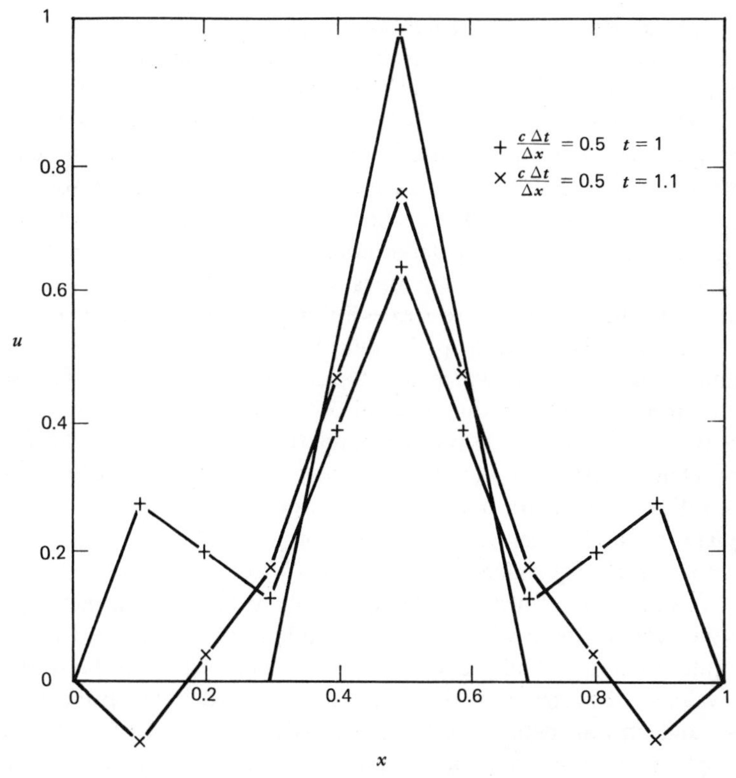

Fig. 4.36. Solution of the wave equation with the Crank–Nicolson method.

tridiagonal, it is easily solved. The matrix for the second set of variables, which does not contain the boundary values, is $(N-2)\times(N-2)$ and has constant diagonals α, 1, and $-\alpha$, where $\alpha=c\Delta t/4\Delta x$. The matrix for the first set of variables is $N\times N$ with constant diagonals α, 1, and $-\alpha$, except that the (1,2) element is -2α and the $(N, N-1)$ element is 2α.

Results obtained using $x=0.1$ are shown in Fig. 4.36. It is clear that even though the method conserves the amplitudes of all the component waves, they are phase shifted and the dispersion is quite evident. The solution is given for both $t=1.0$ and $t=1.1$, the time at which the peak in the function is best resolved. The delay in reforming the peak may be attributed to the decrease in effective wave speed inherent in the finite difference method. The dispersion is due mainly to the spatial finite difference approximation and is not cured by changing the time step. The results for smaller Courant numbers are not shown for this reason.

Reducing the mesh size does improve the results significantly. Since the results are what one might anticipate, they are not shown.

22. HYPERBOLIC PDEs: V. SPLITTING METHODS

As we might anticipate from earlier sections, implicit methods are difficult to apply to nonlinear problems. This leads to the investigation of predictor-corrector methods, and a number of these have been proposed for hyperbolic systems. These are useful methods, but since there is considerable overlap between them and splitting methods, only splitting methods are discussed.

Another problem that poses considerable difficulty is that of solving equations in two or three space dimensions. Implicit methods are impossible to apply directly without needing large computation times. The only methods with generous stability limits are ones similar to the alternating direction implicit (ADI) method that is so successful in the parabolic case. The basis of these methods is the splitting of a multidimensional problem into a set of one-dimensional problems which are then solved successively. The solution of two or three sets of one-dimensional problems is considerably easier than solving the multidimensional problem and the method retains the favorable stability characteristics of implicit methods. This suggests the possibility of using splitting methods for hyperbolic equations.

The ADI method, which is the only splitting method we have investigated so far, is not *exactly* equivalent to the Crank–Nicolson method to which it is related. A split method needs to be equivalent to the original method only to within the accuracy with which either method approximates the PDE. Thus if we are interested in second order methods, it is necessary that the split method be equivalent to the original method only to second order. The truncation errors for split and unsplit methods are not generally identical. In fact it is possible for the split version of a method to be more accurate than its unsplit ancestor. Because the splitting does not need to be exact, the method is also known as *approximate factorization*. A further consequence is that any method can be split in more than one way. To further complicate matters, methods that are identical for linear equations may behave quite differently when applied to nonlinear problems. It appears that the only reliable way to develop a method for nonlinear problems is to choose a method that works well in the linear case (this can usually be tested analytically) and to try several versions of it in the nonlinear case.

Finally, it is very difficult to attain high accuracy for nonlinear PDEs in two and three dimensions. In order to achieve more than first order time accuracy, it is necessary either to use a multistep method (with attendant problems of starting the calculation and storage), an implicit method (with expensive computation), a predictor-corrector method (not a bad choice, but usually requiring increased storage), or something like the Lax–Wendroff method (with difficulty in evaluating the second time derivative). It turns out that the use of splitting, even in an explicit method, can simplify the computation considerably and thus reduce the difficulty of programming and computation time. In this section we give split methods that are equivalent to the Lax–Wendroff method and others that are equivalent to Crank–Nicolson. By equivalence, we mean that their behavior in the linear case and their stability properties are similar. A great deal of attention has been given to split methods in the Soviet Union. See Yanenko (1970) and Marchuk (1975) for details on these developments.

It is not always obvious how the original author found the particular splitting. Looking at a split scheme, it is not clear without further analysis that the method is actually equivalent to some unsplit method. Thus let us take as an example the Richtmyer splitting of the Lax–Wendroff method. Applied to the convection equation it gives

$$\phi_j^{n+1/2} = \tfrac{1}{2}\left(\phi_{j+1}^n + \phi_{j-1}^n\right) - \frac{c\,\Delta t}{2}\left(\frac{\phi_{j+1}^n - \phi_{j-1}^n}{2\,\Delta x}\right) \tag{4.22.1a}$$

$$\phi_j^{n+1} = \phi_j^n - c\,\Delta t\left(\frac{\phi_{j+1}^{n+1/2} - \phi_{j-1}^{n+1/2}}{2\,\Delta x}\right) \tag{4.22.1b}$$

The first of these equations is simply the Euler method with central differences, except that it uses a spatial average in place of ϕ_j^n. This was proposed as a method by Lax. The use of the average stabilizes the Euler method and in this way, it is reminiscent of the Dufort–Frankel method, which followed Lax's method historically. The second equation is the leapfrog method. Thus Richtmyer's method can be regarded as a predictor-corrector method with a first order Lax predictor and a second order leapfrog corrector. The relationship between splitting methods and predictor-corrector methods is very close, and it is not always clear how a given method ought to be classified. Combining the two equations in the set 4.22.1 reveals that for linear equations the overall method is in fact exactly the Lax–Wendroff method with the spatial differences based on $2\,\Delta x$.

A different splitting, which has been very popular in aerodynamic calculations, was proposed by MacCormack. This scheme, which has many possible variations of its own, is a true split scheme in the sense that both halves of the method are of lower order accuracy than the complete method. It uses a forward spatial difference in the first half of the method and a backward difference in the second half

$$\phi_j^{n+1/2}=\phi_j^n-\frac{c\,\Delta t}{\Delta x}\left(\phi_{j+1}^n-\phi_j^n\right) \tag{4.22.2a}$$

$$\phi_j^{n+1}=\tfrac{1}{2}\left(\phi_j^n+\phi_j^{n+1/2}\right)-\frac{c\,\Delta t}{2\,\Delta x}\left(\phi_j^{n+1/2}-\phi_{j-1}^{n+1/2}\right) \tag{4.22.2b}$$

This method is exactly equivalent to the Lax–Wendroff method 4.20.11 for linear problems. It has the advantage over the Richtmyer method in that it uses only information from points $j-1$, j, and $j+1$ at the preceding time step. In light of what we have already found, it is not surprising that the Richtmyer method is more stable (Courant number up to 2) than the MacCormack method (Courant number up to 1), but the latter is more accurate and easier to program. It also has desirable nonlinear properties, which has made its use popular in fluid mechanics. In the MacCormack method the forward and backward differences can be used in the reverse order with the same accuracy, and it has been found that in the nonlinear case the best result is obtained if the order is reversed at each time step. The method also is readily extended to two and three dimensions.

It is only in the last few years that any attention has been given to implicit splitting methods for hyperbolic equations. The major hindrances seem to have been the general lack of enthusiasm for implicit methods and the problem, mentioned earlier, that since the Crank–Nicolson method is neutrally stable, it is likely to be unstable when applied to nonlinear equations or is split. Despite this, Beam and Warming have recently shown the effectiveness of an ADI method in aerodynamic applications. Of course, this method is meaningful only in two- or three-dimensional problems. When applied to the convection or wave equations, the results are very much what one might expect— the method is neutrally stable and the solution is precisely the product of factors such as those in Eq. 4.21.2, one for each dimension. There are a great many ways to produce ADI methods in nonlinear problems and the particular choice can have a great effect on the results. This is an area which is developing rapidly and it is not possible to be definitive at the present time. However, it is probable that split implicit methods will play an important role in the future.

We will satisfy ourselves with a simple example. The convective equation in two dimensions is:

$$\frac{\partial\phi}{\partial t}+u\frac{\partial\phi}{\partial x}+v\frac{\partial\phi}{\partial y}=0 \tag{4.22.3}$$

When the Crank–Nicolson method is applied to Eq. 4.22.3, we have

$$\left(1+\frac{u\Delta t}{2}\frac{\delta}{\delta x}+\frac{v\Delta t}{2}\frac{\delta}{\delta y}\right)\phi^{n+1}=\left(1-\frac{u\Delta t}{2}\frac{\delta}{\delta x}-\frac{v\Delta t}{2}\frac{\delta}{\delta y}\right)\phi^{n} \quad (4.22.4)$$

where $\delta/\delta x$ and $\delta/\delta y$ represent finite difference approximations to $\partial/\partial x$ and $\partial/\partial y$, respectively. These could be either upwind or central difference approximations. This method is second order accurate in time. We can make an approximate factorization of Eq. 4.22.4:

$$\left(1+\frac{u\Delta t}{2}\frac{\delta}{\delta x}\right)\left(1+\frac{v\Delta t}{2}\frac{\delta}{\delta y}\right)\phi^{n+1}=\left(1-\frac{u\Delta t}{2}\frac{\delta}{\delta x}\right)\left(1-\frac{v\Delta t}{2}\frac{\delta}{\delta y}\right)\phi^{n} \quad (4.22.5)$$

which one can show is equivalent to Eq. 4.22.4 within a term of order Δt^3 as required. Then the two step method

$$\left(1+\frac{u\Delta t}{2}\frac{\delta}{\delta x}\right)\phi^{*}=\left(1-\frac{u\Delta t}{2}\frac{\delta}{\delta x}\right)\phi^{n} \quad (4.22.6a)$$

$$\left(1+\frac{v\Delta t}{2}\frac{\delta}{\delta y}\right)\phi^{n+1}=\left(1-\frac{v\Delta t}{2}\frac{\delta}{\delta y}\right)\phi^{*} \quad (4.22.6b)$$

is a method of solving Eq. 4.22.3 which has all of the desired properties. It can be carried out by means of a tridiagonal system solver, it is second order accurate and it is unconditionally stable. Methods which are generalizations of this one have been playing an increasing role in computational fluid dynamics and are undergoing rapid development at the current time.

Problems

1. a. Show that the leapfrog method applied to the heat equation is in fact unstable.

   b. Compute the solution to the heat equation

   $$\frac{\partial\phi}{\partial t}=D\frac{\partial^2\phi}{\partial x^2}$$

   with $\phi(x,0)=1$, $\phi(0,t)=0$, and $\phi(1,t)=0$ using the leapfrog method with just one interior point. Use the Euler method as a starter and $\beta=D\Delta t/\Delta x^2=1$. Does reducing β to 0.5 help?

2. In many systems of interest material properties change with spatial position. For the diffusion equation an example is

$$\frac{\partial \phi}{\partial t} = \frac{\partial}{\partial x} D(x) \frac{\partial \phi}{\partial x}$$

This equation has the property that

$$\frac{d}{dt} \int_0^1 \phi(x,t)\, dx = D(1) \frac{\partial \phi(1,t)}{\partial x} - D(0) \frac{\partial \phi(0,t)}{\partial x}$$

as can be shown by integration. It states that the amount of property ϕ in the region of interest changes only due to flow through the boundaries.

a. Derive a finite difference version of this PDE.
b. Determine whether your method has the finite difference analog of the conservation property mentioned above. If it does not, find one that does.

3. Solve the heat equation with $D = 0.01$ cm^2/sec subject to the initial condition

$$\phi(x,0) = 100x(1-x)$$

with the boundary conditions $\phi(0,t) = \phi(1,t) = 0$.

a. Find the temperature distribution at $t = 1$ sec using four intervals ($\Delta x = 0.25$ cm). Use any numerical method you like, but read part (b) before proceeding.
b. Estimate (in any way you choose) the accuracy of the solution obtained in part (a). Give the result numerically.

4. The following is a method of solving the heat equation:

$$\phi^{n+1*} - \phi^n = D\Delta t \frac{\delta^2 \phi^{n+1*}}{\delta x^2} + D^2 \Delta t^2 \frac{\delta^2}{\delta x^2} \frac{\delta^2}{\delta y^2} \phi^n$$

$$\phi^{n+1} - \phi^{n+1*} = D\Delta t \frac{\delta^2 \phi^{n+1}}{\delta y^2}$$

where $\delta^2/\delta x^2$ and $\delta^2/\delta y^2$ represent finite difference approximations to $\partial^2/\partial x^2$ and $\partial^2/\partial y^2$.

a. Eliminate ϕ^{n+1*} from these equations to get an equation for ϕ^{n+1} in terms of ϕ^n.
b. Of what method is this an approximate factorization?
c. What can you say its accuracy and stability?

5. The equation

$$\frac{\partial \phi}{\partial t} + c\frac{\partial \phi}{\partial x} = D\frac{\partial^2 \phi}{\partial x^2}$$

is called the convective heat equation and with c and D constants represents a very simple model of the dispersion of pollution in the atmosphere or the ocean; it is also used as a model for fluid mechanics.

a. Classify this equation (elliptic, parabolic, hyperbolic) and give a set of simple boundary conditions that could be applied to it.
b. Suggest a numerical method for solving this problem. Be specific as to difference method, treatment of boundary conditions, spatial mesh size, and time step. Give your rationale for the choices made.

6. Solve Laplace's equation on a square using four mesh points (including boundaries) in each direction. The boundary conditions are given below.

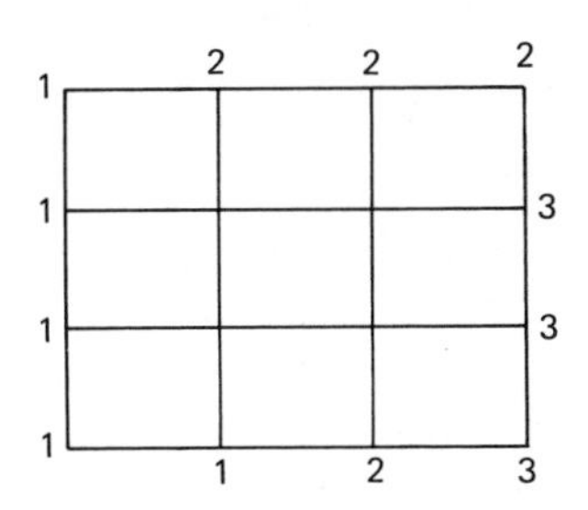

a. Write the finite difference equations.
b. Start with a guess of $\phi = 1$ at all points at which ϕ is unknown and use Gauss–Seidel iteration to find the solution accurate to two decimal places. At which point is the convergence most rapid?
c. Repeat with each iteration by doing the points in reverse order. Is the convergence any faster or slower? Explain.

7. Compute the optimum SOR relaxation factor for Problem 6 and solve the problem using this method.

8. Solve Problem 6 using the ADI method with a fixed β. Vary β and look at the convergence as a function of β. Compare the results obtained with those for the SOR method obtained in Problem 7.

9. Higher order methods (e.g., the Adams methods) could be used as the basis of methods for solving elliptic equations.

a. Derive the iterative method equivalent to applying the second order Adams–Bashforth method to Laplace's equation. For which $\beta = D\Delta t/\Delta x^2$ will the convergence be most rapid?

b. Does this method have any advantage relative to the Jacobi method?

c. Would still higher order methods be advantageous?

10. It is common to use nonuniform grids in elliptic problems. (For Laplace's equation solutions are usually smooth, so this is unnecessary except, perhaps, near corners). The problem is to determine whether or not use of nonuniform grids slows down the convergence of iterative methods. As a test case try Problem 6 with $\Delta x_1 = 0.25$, $\Delta x_2 = 0.35$, and $\Delta x_3 = 0.4$. Derive the difference equations and solve them using the Gauss–Seidel method. Compare the convergence with the uniform grid case.

11. One very common type of elliptic differential equation arises from diffusion problems in nonuniform systems. Such systems lead to equations of the type

$$\frac{\partial}{\partial x} D \frac{\partial \phi}{\partial x} + \frac{\partial}{\partial y} D \frac{\partial \phi}{\partial y} = S(x, y)$$

where D is a function of x and y.

a. Write a finite difference version of this equation.

b. Determine whether the approximation developed in part (a) satisfies the global conservation statement that the net flux of ϕ out of the region of interest is equal to the total source within the region. If your approximation does not have this property, try to find one that does.

c. Recommend a method for solving the system of equations. Pay particular attention to the effect of the nonuniform diffusion coefficient on the rate of convergence.

12. Propagation of waves through nonuniform media are governed by the wave equation

$$\frac{\partial^2 \phi}{\partial t^2} = \frac{\partial}{\partial x} c^2(x) \frac{\partial \phi}{\partial x}$$

a. What are the characteristics of this equation?

b. Give a numerical method for solving this problem.

c. Suppose we have an infinite medium in which c^2 varies as shown in (a).

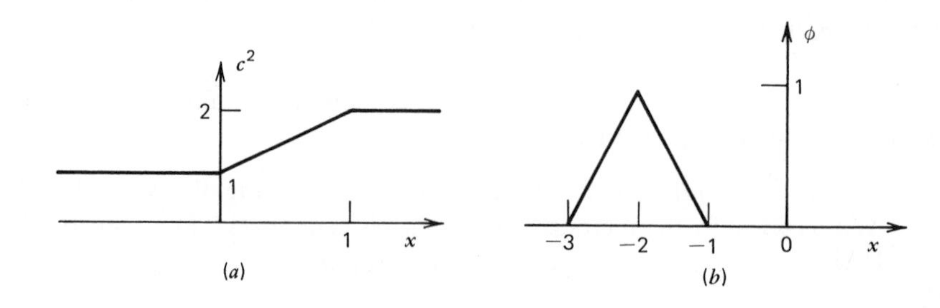

A pulse of witch's hat shape (see (b)) is introduced into the system at time $t=0$. When the position of the disturbance that propagates to the right reaches the ramp in $c^2(x)$, part of it will be transmitted and part reflected. Determine the fractions that are reflected and transmitted to an accuracy of 1%.

13. Show how the leapfrog method could be applied to solving the wave equation. Find the amplitude and phase errors of the method in a manner similar to those used in the text.

14. Consider the convection equation in two space dimensions

$$\frac{\partial \phi}{\partial t}+u\frac{\partial \phi}{\partial x}+v\frac{\partial \phi}{\partial y}=0$$

a. Apply the Euler method to this equation.
b. Write the method given in part (a) in operator form.
c. Show how the operator can be approximately factored into the product of an operator involving only x and another involving only y.
d. Consider a solution of the form $\phi=e^{ik_1x}e^{ik_2y}e^{i\omega t}$. Find the exact relation between k_1, k_2 and ω, and the analogous relations for the methods found in parts (a) and (c). How does the approximate factorization method compare with the original method in its treatment of waves of this type?

15. Repeat problem 14 for the Crank–Nicolson method.

16. An equation frequently used as a model for fluid mechanics is Burgers' equation

$$\frac{\partial u}{\partial t}+u\frac{\partial u}{\partial x}=\nu\frac{\partial^2 u}{\partial x^2}$$

a. Give an explicit method for solving this equation (there are many).
b. Given the initial condition

$$u(x,0)=\sin \pi x \qquad 0<x<1$$

and the boundary condition

$$u(0,t)=u(1,t)=0$$

solve the problem, plot the solution for various times, and explain the results.

17. Solve the Poisson equation

$$\nabla^2\phi=(1-x^2)(1-y^2)$$

subject to the periodic boundary conditions

$$\phi(x+2, y)=\phi(x, y)$$
$$\phi(x, y+2)=\phi(x, y)$$

using the Fourier transform method.

Appendix A

Solution of Tridiagonal Systems

Tridiagonal systems arise when differential equations are finite differenced. By definition, the matrices of such systems have nonzero elements only in the main diagonal and the diagonals immediately above and below it. For this reason it is best to use a special notation for these systems. We write the matrix

$$A=\begin{bmatrix} b_1 & c_1 & & & & & \\ a_2 & b_2 & c_2 & & & & \\ & a_3 & b_3 & c_3 & & & \\ & & & \ddots & & & \\ & & & & a_{n-1} & b_{n-1} & c_{n-1} \\ & & & & & a_n & b_n \end{bmatrix}$$

and the system of equations is

$$A\mathbf{x}=\mathbf{g}$$

This system is solved by standard Gauss elimination. The following is a subroutine for this purpose. Note that we have used an extra array BB in this subroutine. Although this uses a little extra storage, it is a useful feature because there are a number of applications in which this routine is called many times with the same matrix but different right-hand sides. In these cases the use of the extra array eliminates the need to redefine B before each call of the subroutine.

```
00100         SUBROUTINE TRDIAG (N,A,B,C,X,G)
00200         DIMENSION A(1000),B(1000),C(1000),X(1000),G(1000),BB(1000)
00300   C.....THIS SUBROUTINE SOLVES TRIDIAGONAL SYSTEMS OF EQUATIONS
00400   C.....BY GAUSS ELIMINATION
00500   C.....THE PROBLEM SOLVED IS MX=G WHERE M=TRI(A,B,C)
00600   C.....THIS ROUTINE DOES NOT DESTROY THE ORIGINAL MATRIX
00700   C.....AND MAY BE CALLED A NUMBER OF TIMES WITHOUT REDEFINING
00800   C.....THE MATRIX
00900   C.....N = NUMBER OF EQUATIONS SOLVED (UP TO 1000)
01000   C.....FORWARD ELIMINATION
01100   C.....BB IS A SCRATCH ARRAY NEEDED TO AVOID DESTROYING B ARRAY
01200         DO 1 I=1,N
01300         BB(I) = B(I)
01400       1 CONTINUE
01500         DO 2 I=2,N
01600         T = A(I)/BB(I-1)
01700         BB(I) = BB(I) - C(I-1)*T
01800         G(I) = G(I) - G(I-1)*T
01900       2 CONTINUE
02000   C.....BACK SUBSTITUTION
02100         X(N) = G(N)/BB(N)
02200         DO 3 I=1,N-1
02300         J = N-I
02400         X(J) = (G(J)-C(J)*X(J+1))/BB(J)
02500       3 CONTINUE
02600         RETURN
02700         END
02800
02900
03000         SUBROUTINE DTRIDG (N,A,B,C,X,G)
03100         IMPLICIT REAL*8 (A-H,O-Z)
03200         DIMENSION A(1000),B(1000),C(1000),X(1000),G(1000),BB(1000)
03300   C.....THIS SUBROUTINE SOLVES TRIDIAGONAL SYSTEMS OF EQUATIONS
03400   C.....BY GAUSS ELIMINATION
03500   C.....THE PROBLEM SOLVED IS MX=G WHERE M=TRI(A,B,C)
03600   C.....THIS ROUTINE DOES NOT DESTROY THE ORIGINAL MATRIX
03700   C.....AND MAY BE CALLED A NUMBER OF TIMES WITHOUT REDEFINING
03800   C.....THE MATRIX
03900   C.....N = NUMBER OF EQUATIONS SOLVED (UP TO 1000)
04000   C.....FORWARD ELIMINATION
04100   C.....BB IS A SCRATCH ARRAY NEEDED TO AVOID DESTROYING B ARRAY
04200         DO 1 I=1,N
04300         BB(I) = B(I)
04400       1 CONTINUE
04500         DO 2 I=2,N
04600         T = A(I)/BB(I-1)
04700         BB(I) = BB(I) - C(I-1)*T
04800         G(I) = G(I) - G(I-1)*T
04900       2 CONTINUE
05000   C.....BACK SUBSTITUTION
05100         X(N) = G(N)/BB(N)
05200         DO 3 I=1,N-1
05300         J = N-I
05400         X(J) = (G(J)-C(J)*X(J+1))/BB(J)
05500       3 CONTINUE
05600         RETURN
05700         END
```

Appendix B

The Newton–Raphson Method

At a number of places throughout this book, we have needed a method of finding the solution of nonlinear systems of equations. One of these methods is the Newton–Raphson method and there are a number of variations on the method that are also quite good.

To begin with, let us take the case of a single equation. Suppose the problem is to find the root of

$$f(x)=0 \tag{B.1}$$

Conceptually, the method is extremely simple and involves nothing more than a local linearization. Suppose, for example, we guess that the root of Eq. B.1 is x_0. In the neighborhood of x_0, we can approximate the function by the first two terms in the Taylor series

$$f(x)=f(x_0)+(x-x_0)f'(x_0) \tag{B.2}$$

Geometrically, this is equivalent to approximating the function by the tangent line as shown in Fig. B.1. More terms in the Taylor series could be used, but this is not very beneficial. The next approximation to the root, which we will call x_1, is taken to be the root of the linear approximation of Eq. B.2

$$x_1=x_0-\frac{f(x_0)}{f'(x_0)} \tag{B.3}$$

having this improved (presumably) estimate of the root, we can simply repeat the procedure using the formula

$$x_{j+1}=x_j-\frac{f(x_j)}{f'(x_j)} \tag{B.4}$$

until, it is hoped, we get a solution that is as accurate as necessary. This is all there is to the method.

The convergence of the method is extremely good. If we let $\varepsilon_j=\xi-x_j$ be the difference between the current (jth) estimate and the exact solution, we can

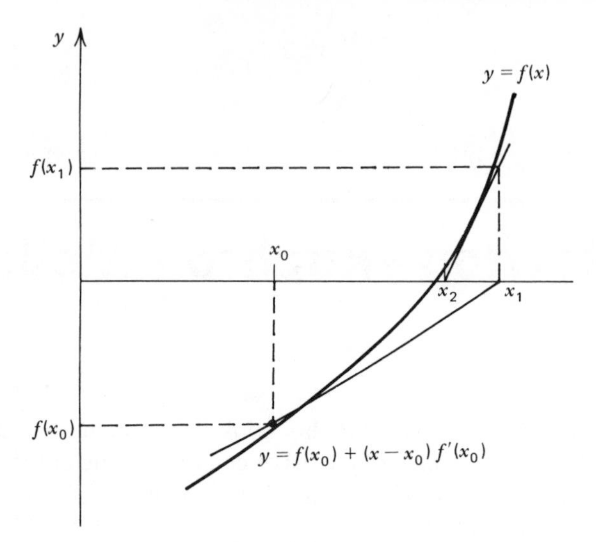

Fig. B.1.

show for small ε_j

$$\varepsilon_{j+1} \simeq -\tfrac{1}{2}\frac{f''(x_j)}{f'(x_j)}\varepsilon_j^2 \tag{B.5}$$

There are two important observations here. First, the error is reduced quadratically when ε_j is small. That is, if $\varepsilon_j = 10^{-3}$, then we expect $\varepsilon_{j+1} \simeq 10^{-6}$. The rate of convergence of the Newton–Raphson method is one of the best of any numerical method in common use. It is so good in fact that there is little reason to look for improvement on it. The second observation is that the rate of convergence depends on f'', that is, on the curvature of the function in the vicinity of the root. This accords with our intuition. If the line were in fact straight, the method would converge in one iteration; the more curvaceous the line, the slower the convergence. To reinforce this finding, redraw Fig. B.1 with different curvatures. Although the Newton–Raphson method is very good, it has two limitations. The first is that unless the initial guess is "sufficiently close" to the correct answer, the method is not guaranteed to converge. The second is that if f' is zero at the root, the method will have problems. We look at these one at a time and look for methods of overcoming them.

By far, the most common practical difficulty is that of starting too far from the root. The difficulty is best illustrated graphically. Figure B.2 shows what can happen. Starting with x_0 as the initial guess, we compute x_1 on the first iteration. On the second iteration, we find ourselves at x_2, which is further from the root, and x_3 will be worse than x_2. Before long, we will find ourselves somewhere off the graph. A related difficulty is shown in Fig. B.3. Here the problem is to find the root ξ_1, so we start to the left of it. Despite this, the method converges to ξ_2.

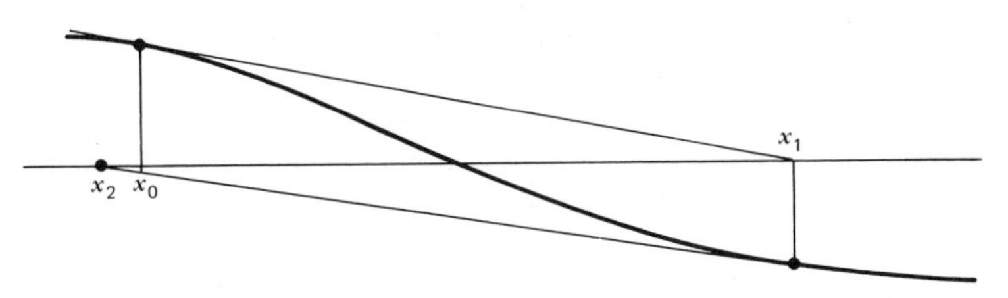

Fig. B.2.

These problems occur occasionally in practice and can be cured by using variations on the method. The reader is referred to any of a number of good numerical analysis texts for this.

We go on to systems of equations. For the case of two equations in two unknowns, the Newton–Raphson method can be given a geometric interpretation. (It is hard to think in hyperspace.) Each function can be imagined to represent a surface in the three-dimensional space in which the coordinate axes are the two independent variables (x, y) and the value of the function. In the neighborhood of the point $[x_0, y_0, f(x_0, y_0)]$ the function is approximated by the tangent plane. From the two functions two tangent planes are found and their intersection is, of course, a straight line. The intersection of this line with the plane $f=0$ gives the next estimate of the root. We then continue to iterate in the manner used for a single equation.

Mathematically the problem we seek to solve can be represented by the system of equations

$$f_i(x_1, x_2, \dots x_n)=0 \qquad i=1,2,\dots n \tag{B.5}$$

(We assume that the number of equations and unknowns are equal.) Suppose that as an initial guess we use the point $(x_1^0, x_2^0, \dots x_n^0)$. Each of the functions in the system in Eq. B.5 can be expanded in a Taylor series about this point. Keeping only the linear terms, we have

$$\begin{aligned} f_i(x_1, x_2, \dots x_n) &\simeq f_i(x_1^0, x_2^0, \dots x_n^0)+(x_1-x_1^0)\frac{\partial f_i}{\partial x_1}(x_1^0, x_2^0, \dots x_n^0) \\ &\quad +(x_2-x_2^0)\frac{\partial f_i}{\partial x_2}(x_1^0, x_2^0, \dots x_n^0)+\cdots+(x_n-x_n^0)\frac{\partial f_i}{\partial x_n}(x_1^0, x_2^0, \dots x_n^0) \\ &= f_i^0+\sum_j \frac{\partial f_i^0}{\partial x_j}(x_j-x_j^0) \qquad i=1,2,\dots n \end{aligned} \tag{B.6}$$

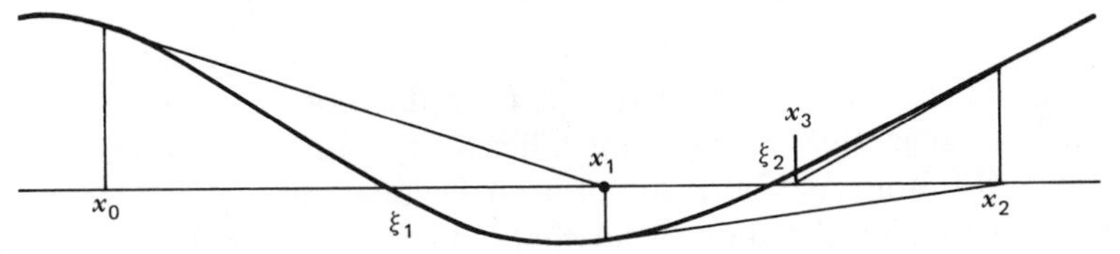

Fig. B.3.

where we have used the superscript 0 to denote a function evaluated at the point $(x_1^0, x_2^0, \ldots x_n^0)$. The equations for the next estimate of the root are found by setting *each* of the linearized equations equal to zero. This results in a set of n linear equations in n unknowns. Before writing the equations explicitly, a few definitions are needed. Let the change in x_j be denoted by h_j^1

$$h_j^1 = x_j^1 - x_j^0 \tag{B.7}$$

where x_j is the new estimate of the jth component of the root that will be obtained when the linear system is solved. We let the matrix in the system in Eq. 4.5.2 be denoted by J^0, which has as its elements

$$J_{ij}^0 = \frac{\partial f_i^0}{\partial x_j} \tag{B.8}$$

which is the *Jacobian* of the system. The system of equations obtained by setting each member of Eq. B.6 to zero is

$$\sum_{j=1}^{n} j_{ij}^0 h_j^1 = -f_i^0 \qquad i = 1, 2, \ldots n \tag{B.9}$$

This is a set of n linear equations in n unknowns and can be solved by standard methods

$$J^0 \mathbf{h}^1 = -\mathbf{f}^0 \tag{B.10}$$

The succeeding steps follow the pattern of the Newton–Raphson method for a single unknown. At the kth step we have to solve the set of equations

$$\sum_{j=1}^{n} j_{ij}^{k-1} h_j^k = -f_i\left(x_1^{k-1}, x_2^{k-1}, \ldots x_n^{k-1}\right) \qquad i = 1, 2, \ldots n \tag{B.11}$$

and the kth estimate of the root is obtained from the expression

$$x_i^k = x_i^{k-1} + h_i^k \tag{B.12}$$

The Newton–Raphson method for systems of equations converges quadratically when the estimate is near the root.

The method also has all of the other characteristics of the Newton–Raphson method for a single equation. It may converge slowly in the initial stages and may diverge if the initial guess is too far from the root. If an attempt to find a root fails, the suggested remedies are either to try a different initial guess or to use underrelaxation. As in the case of a single equation, it is difficult to give a recipe for the underrelaxation factor that will always work. Unfortunately, at the present time, the solution of systems of nonlinear equations is as much art as science. This is largely a reflection of the diversity of types of nonlinear equations. Experience with equations similar to the ones that one has to deal with is always valuable and, in many cases, is more useful than any advice that can be given.

One method of reducing the cost of the method can be found by noting that as the method approaches convergence, the derivatives tend to change very slowly. This means that it is probably not necessary to recompute the Jacobian at every iteration and the cost of calculating it can be saved. This will increase the number of iterations but the cost of each iteration is reduced so sufficiently that the overall cost is decreased.

References and Annotated Bibliography

Acton, F. S., *Numerical Methods that Work*, Harper and Row, New York, 1970.

A good place for finding a method, but not really a textbook.

Ames, W. F., *Numerical Methods for Partial Differential Equations*, 2nd ed., Academic, New York, 1977.

Similar to Mitchell and Ames; I have a slight preference for this one because of its greater clarity. Though a new edition, it does not cover some of the newer methods.

Bracewell, R., *Fourier Methods*, McGraw-Hill, New York, 1967.

A good basic book in Fourier methods and their applications; written from an electrical engineering viewpoint.

Brigham, *The Fast Fourier Transform*, Prentice-Hall, Englewood Cliffs, N.J. 1976.

The only book on the fast Fourier transform, but a bit lengthier than need be.

Carnahan, B. H., Luther, H. A., and Wilkes, J. O., *Applied Numerical Methods*, Wiley, New York, 1969.

Courant, R. and Hilbert, D., *Methods of Mathematical Physics*, Vol. II, Wiley-Interscience, New York, 1962.

Dahlquist, G., Bjorck, A., and Anderson, N., *Numerical Methods*, Prentice-Hall, Englewood Cliffs, N.J., 1974.

A good compendium of numerical methods but weak on PDEs. More analytical than this book.

Dodes, I. A., *Numerical Analysis for Computer Science*, North Holland, New York, 1978.

A good introductory book, which includes linear algebra, interpolation, quadrature, and ODEs; very little on boundary value problems.

Forsythe, G. E., Malcolm, M. A., and Moler, C. B., *Computer Methods for Mathematical Computations*, Prentice-Hall, Englewood Cliffs, N.J., 1977.

An excellent source of computer subroutines for the topics covered, which do not include PDEs. Not really a text, but well worth reading.

Garabedian, P., *Partial Differential Equations*, Wiley, New York, 1964.

Gear, C. W., *Numerical Initial Value Problems in Ordinary Differential Equations*, Prentice-Hall, Englewood Cliffs, N.J., 1971.

A standard in the particular topic covered; very complete.

Gerald, C. F., *Applied Numerical Analysis*, Addison-Wesley, Reading, Mass., 1970.

A good text that has a somewhat wider scope but less depth than this one; more analytical and less intuitive.

Holt, M., *Numerical Methods in Fluid Dynamics*, Springer, New York, 1977.

Covers methods for hyperbolic equations, especially those developed in the U.S.S.R., in some detail.

Hornbeck, R. W., *Numerical Methods*, Quantum, New York, 1975.

A good textbook covering numerical linear algebra, interpolation, quadrature, and initial value problems for ODEs.

Isaacson, E. and Keller, H. B., *Analysis of Numerical Methods*, Wiley, New York, 1966.

One of the standards in the field; very much oriented toward mathematicians, but relatively weak on PDEs.

Jain, M. K., *Numerical Solution of Differential Equations*, Wiley, New York, 1979.

Covers both ODEs and PDEs on a more analytical level than this book.

Lambert, J. D., *Computational Methods in Ordinary Differential Equations*, Wiley, New York, 1973.

An alternative to Gear's book; quite good but I prefer Gear.

Marchuk, G., *Methods of Numerical Mathematics*, Springer, New York, 1975.

Translated from Russian, this book has a different flavor than do most English language books. Analytical in approach and therefore on a more advanced level than this book.

Mitchell, A. R., *Computational Methods in Partial Differential Equations*, Wiley, New York, 1969.

Similar in style and scope to Ames and Smith. More analytical than this book but a bit dated.

Mushkelishvili, N. I., *Singular Integral Equations*, Noordhoff, Gronigen, Netherlands, 1957.

A classic on boundary value problems in complex variable theory, but not easy to read.

Shampine, L. P. and Gordon, M. K., *Computer Solution of Ordinary Differential Equations*, Freeman, San Francisco, 1975.

A very complete development of the program "ODE," which was used in some of the examples and is probably the best available code for the initial value problem. Not a particularly good text otherwise.

Smith, G. D., *Numerical Solution of Partial Differential Equations*, Oxford University Press, 1965.

Similar to Ames and Mitchell, but the treatment, especially of hyperbolic equations, is dated.

Strang, G. and Fix, G. J., *An Analysis of the Finite Element Method*, Prentice-Hall, Englewood Cliffs, N.J., 1973.

One of the best introductions to the finite element method.

There are a number of other relevant texts not listed here; this bibliography reflects the author's preferences.

Index

Abscissas:
  for Gauss quadrature, 41-47
    tables, 43, 45, 46
  for integration, 24
Acceleration, 189
Accuracy, *see* Truncation error
Adams-Bashforth methods, 89-92
  coefficients, table, 89
  in PDEs, 152, 249
  as predictor, 80, 92
  stability, 89-90
Adams methods, *see* Adams-Bashforth methods; Adams-Moulton methods
Adams-Moulton methods, 89-92
  coefficients, table, 90
  as corrector, 92
  stability, 91
Adaptive integration, 37-41
  of Gaussian function, 38-41
ADI methods, *see* Alternating direction, implicit method
Aerodynamics, applications in, 228
Algebraic systems of equations, in boundary value problems, 110
Aliasing, 210
Alternating direction, implicit method, 153, 158-167, 197-202
  boundary conditions for, 166-167
  Douglass-Gunn, 163
  and elliptic equations, 185, 197-202
  FORTRAN program, 165-166
  for hyperbolic PDEs, 241, 244, 249
  Peaceman-Rachford, 160-167
  stability, 162
  vs. successive overrelaxation, 199-200
Amplitude error:
  of explicit Euler method, 67
  for hyperbolic PDEs, 236
  of implicit Euler method, 71-72
  of trapezoid rule, 76
Analytic functions, 219
Approximate factorization:
  for hyperbolic equations, 244-247, 248
  for parabolic PDEs, 163
  see also Splitting methods
A-stability, 62. *See also* Stability
Automatic error control, 92-95

Backward difference formula:
  first derivative, 53
  second derivative, 54
Backward Euler, *see* Euler's method, implicit
Banded matrices, 175-176
Beam-Warming method, 246
Block diagonal matrix, 214
Block tridiagonal matrix, 243
Boundary conditions:
  in finite element methods, 203-204
  in fourth order methods, 116
  in hyperbolic PDEs, 231
  natural, 203
  for PDEs, 135-138, 174-177
  in second order methods, 111
Boundary integral method, 218-221
Boundary value problems, 50, 105-130
  direct methods, 110-124
    FORTRAN program, 113
    nonuniform grids, 120-124
    stiff problems, 114-115
  eigenvalue problems, 126-130
  finite element methods, 124-126
  shooting methods, 105-110
Box method, *see* Keller Box method
Bulirsch-Stoer method, 92-93
Burgers' equation, 251

Calculus of variations, 125, 203
Cartesian coordinates, 167, 225
Cauchy-Riemann conditions, 219

Cauchy's integral formula, 219
Cauchy's theorem, 219
Central difference formula:
  first derivative, 53
  second derivative, 54
CFL condition, 237
Chain rule, 226
Characteristics, 136-138, 221-224
  method of, 224-233
  PDE on, 226
  of second order quasilinear PDE, 223
  in two dimensions, 225
Compact fourth order finite differences, 56, 119-120
  in elliptic PDEs, 173
  first derivative, 56
  in parabolic PDEs, 157
  second derivative, 56
Compact nine point operator, 173-174
Complex variables, 138, 218-221
Compound interest grid, 122-124
Computational molecule, 143
  for Crank-Nicolson method, 148
  for Dufort-Frankel method, 153
  for Euler's method, 143
  for Laplace's equation, 172-173
Computational root, multistep methods, 86
Conditional stability, definition, 62
  see also Stability
Conservation laws:
  in cylindrical coordinates, 169
  and finite element methods, 124, 207
Convection equation, 234-238, 246-247, 251
Convective heat equation, 249
Convergence:
  of Gauss-Seidel, 188
  of iterative methods, 177-178
  of Jacobi method, 178-181
  of numerical methods for PDEs, 149
  of shooting method, 108-110
Cooley-Tukey algorithm, 212-216
Cosine expansions, 218
Courant condition, 237
Courant number, 237-240
Crank-Nicolson method, 147-152, 154, 156-157, 160, 185, 251
  computational molecule, 148
  for hyperbolic PDEs, 241-244
  for parabolic PDEs, 147-152
    FORTRAN program, 151
    modified equation, 149
Cubic finite elements, 205
Cyclic ADI method, 199, 201-202
Cylindrical coordinates, 167-170

Delta (Δ) form, 171
Determinants, 226
Diagonalization, 96-97
Difference equations, exact solution of:
  explicit Euler method, 61
  implicit Euler method, 69-70
  leapfrog method, 85-87
  Runge-Kutta method, 78
  trapezoid rule, 73-74
Diffusion equation, 137, 247, 248, 250
  alternating direction implicit, 158-167
  Crank-Nicolson solution of, 150-152
  in cylindrical coordinates, 167-170
  Dufort-Frankel solution of, 153-155
  explicit Euler solution, 143-145
  Keller Box method solution of, 155-157
Direct methods for boundary value problems, 110-115
Discrete Fourier transforms, 207-212
  differentiation with, 212
  interpolation with, 211
  inversion, 211
Dispersive error, 236, 244
Dissipative error, 236
Divided differences, 11
Domain of dependence, 136-137, 236
Domain of influence, 136-137
Double precision, 102
  in higher order methods, 120
Douglass-Gunn method, 163
Dufort-Frankel method, 152-155, 160, 245
  computational molecule, 153
  for elliptic equations, 197
  FORTRAN program, 155

Effective eigenvalue, 199
Effective exponent:
  definition, 81
  for various methods, 82
Eigenfunctions, 127
Eigenvalue problems, 126-130
  in parabolic problems, 139-141
  stiffness in, 129
Eigenvalues:
  of alternating direction implicit, 198-199
  of Gauss-Seidel, 188
  of Jacobi method, 179
  of matrices, 96
  in parabolic problems, 139
  of successive line overrelaxation, 194-197

of successive overrelaxation, 191-192
of two-dimensional problem, 159-160
Eigenvectors, 96, 139
of Jacobi method, 179
of two-dimensional problem, 159-160
Elliptic PDEs, 136
alternating direction implicit method, 197-202
boundary integral methods, 218-221
and complex variables, 138
finite differencing, 171-177
finite element methods, 202-207
Fourier methods, 216-218
Gauss-Seidel method, 184-187
Jacobi method, 177-184
line relaxation, 187-188
successive overrelaxation, 189-197
End conditions in splines, 14-15
Euler's method, 57-60, 130, 251
in alternating direction implicit method, 160-161
for diffusion equation, 138-147
explicit, 57-60, 81, 130
accuracy, 58-60, 64, 82, 83
for nonlinear equations, 68-69
stability, 60-69
in stiff systems, 102-104
FORTRAN program, 68-69
forward, 57-60
global behavior of, 58-59
hyperbolic PDEs, 235, 245
implicit, 69-72, 73, 130
accuracy, 70, 82, 83
as corrector, 79
stability, 70
in stiff systems, 102-104
and Jacobi method, 180-181
for parabolic PDEs, 138-147, 152
FORTRAN program, 146-147
as predictor, 77-79
and relaxation methods, 181
as starter method for PDEs, 153
for stiff systems, 102-104
Explicit methods:
Adams-Bashforth, 89-92
definition, 58
Euler, 57-60
hyperbolic PDEs, 234-240
parabolic PDEs, 138-147, 152-155
shooting with, 105-110
split, 246
stability of, 60-69
Exponential function:
Gauss quadrature of, 44-45
Lagrange interpolation, 4-7
Newton-Cotes integration, 28-29
Extrapolation:
and eigenvalues, 190
of iterative methods, 189-192
in Keller Box method, 157
Lagrange, 6
polynomial, 92
rational, 92
Richardson, 30-33

Fast Fourier transform, 177, 212-216
FORTRAN program, 215
Fehlberg Runge-Kutta method, 80
FFT, *see* Fast Fourier transform
Filtering, 105
Finite differences, 51-57
backward, 53
in boundary value problems, 110-115
central, 53
on characteristics, 227
forward, 53
Lagrange interpolation, use of, 23, 52-54
table, 154
Finite element methods:
for elliptic PDEs, 174, 176, 202-207
and finite differences, 206-207
for ODEs, 124-126
Finite volume methods, for ODEs, 124
First order difference formula:
first derivative, 53
second derivative, 54
Five point operator, 172, 191
Forbidden signals, 236
FORTRAN programs:
alternating direction implicit, 165-166
Crank-Nicolson, 157
direct method for boundary value problems, 113
Euler's method:
for ODEs, 68-69
for parabolic PDEs, 146-147
fast Fourier transform, 215
Jacobi method, 182
method of characteristics, 233
Romberg integration, 36-37
Runge-Kutta (fourth order), 84

spline interpolation, 17-18
successive overrelation, 195
tridiagonal systems, 254
Forward difference formula:
first derivative, 53
second derivative, 54
Fourier coefficients, 140, 209
Fourier series, 207
derivatives of, 209
Fourier transforms, 174
discrete, 207-212
elliptic equation solution, 216-218
hyperbolic PDEs, 235
Fourth order finite differences, 54-56
in boundary value problems, 115
Fourth order methods for ODEs:
Adams-Bashforth, 89-92
Adams-Moulton, 89-92
boundary condition treatment, 116-120
for boundary value problems, 115-120
compact, 56, 119-120
Runge-Kutta, 79-84
Fourth order methods for PDEs:
for elliptic PDEs, 173-174
for parabolic PDEs, 143

Gauss elimination, 207
Gaussian function:
Gauss quadrature of, 46-47
Romberg integration of, 35-37
Gauss quadrature, 12, 41, 47
Gauss-Hermite quadrature, 44-45
table, 46
Gauss-Laguerre, 44-45
table, 45
Gauss-Legendre quadrature, 42-44
error estimate, 43
of exponential function, 44-45
table, 43
and Hermite interpolation, 42
Gauss-Seidel method, 184-187, 248, 250
convergence, 186
extrapolation of, 190-191
and Jacobi method, 186
line, 187-188
and parabolic PDEs, 186
vs. successive overrelaxation, 193
GEAR (program), 103-104
Gear's method, 98-100
coefficients of, 98
stability, 98

Gibbs phenomenon, 208
Global error:
Euler explicit, 59-60
ODE methods, 72
Green's formula, 219

Heat equation, *see* Diffusion equation
Helmholtz equation, 171, 173, 197
Hermite interpolation:
expressions for, 12
and finite differences, 53, 120
and Gauss quadrature, 41-42
Heun's method, 77-78
Homogeneous ODEs, 50
Hyperbolic PDEs, 136, 221-247
explicit methods, 234-240
method of characteristics, 224-233
splitting methods, 244-247
theory, 221-224

Ill conditioning:
in boundary value problem, 114
in Lagrange interpolation, 2
and stiffness, 97
Implicit methods:
Adams-Moulton, 90-92
alternating directions implicit, 158-167, 197-202
definition, 69
Euler, 69-72
for hyperbolic PDEs, 241-244
for parabolic PDEs, 147-152, 155-167
split, 246
in stiff systems, 98
trapezoid rule, 73-74
Inherent instability, 104-105
Inhomogeneous ODEs, 50
Initial-boundary value problems, 135
Initial conditions for PDEs, 137-138, 222-223
Initial value problems, 50
explicit Euler method, 57-60
implicit Euler method, 69-72
multistep methods, 84-92
predictor-corrector methods, 76-78
Runge-Kutta methods, 78-82
for systems of ODEs, 95-105
trapezoid rule, 73-76
Inner iterations, 194
Instability, *see* Inherent instability; Stability
Integration, 24-49

adaptive, 37-41
Newton-Cotes, 25-30
and ODEs, 56-57
Romberg, 33-37
singularities, 47-48
Integration by parts, in quadrature, 47
Interpolation, 1-23
and finite differences, 51-54
Fourier, 211-212
Hermite, 12
Lagrange, 2-11. *See also* Lagrange interpolation
least squares, 1
and mesh refinement, 184
multidimensional, 22-23
and numerical differentiation, 51-54
parametric, 21-23
polynomial, *see* Lagrange interpolation
spline, 12-21
cubic, 12-20
tension, 20-21
see also Spline interpolation
use in integration, 24-25
Irregular boundaries, 159, 166-167, 176-177, 228-229
Isoparametric elements, 205
Iterative methods:
for elliptic PDEs, 177-178
extrapolation of, 189-192
Gauss-Seidel, 184-187
Jacobi, 177-184
line relaxation, 187-188
successive line overrelaxation, 194-197
successive overrelaxation, 189-197

Jacobian of systems of equations, 258
Jacobi method, 177-184, 186-187, 250
convergence of, 178-181
eigenvalues and eigenvectors, 179
extrapolation of, 190
FORTRAN program, 182
and Gauss Seidel, 186
for Laplace's equation, 178
and line relaxation, 187
and parabolic PDEs, 181

Keller Box method, 155-157
for elliptic PDEs, 174
for parabolic PDEs, 155-157

Lagrange interpolation, 2-11, 23, 131
error estimate, 3-4
examples of, 4-11
exponential function, 4-7
extrapolation by, 6
and finite differences, 52-54, 56
and Hermite interpolation, 12
and integration, 25-27
multidimensional, 22-23
and numerical differentiation, 52-54
parametric, 22, 23
piecewise, 11
polynomials for, 2-4
quadratic, 23
roundoff error in, 5-6
sine function, 7-8
and spline, comparison, 17-20
superellipse, 8-11
tension, 20-21
Langrangian approach to classical mechanics, 124
Laplace's equation, 138, 248
alternating direction implicit method, 200-202
boundary integral method, 218-221
finite differencing of, 171-177
finite element methods, 202-207
Gauss Seidel, 187
Jacobi method, 178-181
line relaxation, 188
successive line overrelaxation, 194-197
successive overrelaxation, 192-197
Lax' method, 245
Lax-Wendroff method, 238-240, 245
Leapfrog method, 57, 85-88, 92
analysis of, 85-87
in PDEs, 152, 237, 245, 247
stability, 86-88
starting, 85-87
Legendre polynomials, and Gauss quadrature, 42
Linear finite elements, 204
Linearization of non-linear PDEs, 170
Line relaxation, 187-188
convergence, 188
extrapolation of, 191
and Jacobi, 187

MacCormack's method, 246
Matrices:
diagonalization, 96-97
in systems of ODEs, 96
Merson Runge-Kutta method, 80-81
Mesh refinement, 184

Mesh spacing, in boundary value problems, 120-124
Method of characteristics, 224-233
Method of lines, 188
Midpoint rule, 28
  as corrector, 78
  and leapfrog method, 57
  as predictor, 79
  and trapezoid rule, 57
Modified equations:
  for Crank-Nicolson, 149
  definition, 142
Modified wavenumber, 218
Multidimensional interpolation, 22-23
Multigrid method, 184
Multipoint methods, *see* Multistep methods
Multistep methods, 75, 84-92
  Adams-Bashforth methods, 89-92
  Adams-Moulton methods, 89-92
  for hyperbolic PDEs, 245
  leapfrog method, 85-88
  starting, 86-87

Natural boundary conditions, 203
Neutral stability:
  for hyperbolic PDEs, 241
  leapfrog method, 88
Newton Cotes integration, 25-30, 48
  in adaptive quadrature, 41
  closed, 25-28
  coefficients, table, 26
  for exponential function, 29
  formula, 26
  open, 28
  for superellipse, 29-30
Newton-Raphson method, 255-259
  convergence, 255-256
  for implicit methods, 72
  for nonlinear PDEs, 170
  in shooting, 107-108
  and successive overrelaxation, 194
Nonlinear ODEs, 50
  by explicit Euler method, 68-69
Nonlinear PDEs:
  elliptic, 171, 194
  hyperbolic, 238
  parabolic, 170
Nonuniform grids, 120-124, 133
  in Crank-Nicolson method, 151-152
  in ODE boundary value problems, 120-124
  in parabolic PDEs, 145-147
  in stiff problem, 123-124

Numerical differentiation, 51-57
  and interpolation, 52-54
  and numerical integration, 56-57
  Taylor Series, 54-56

ODE (computer program), comparison with other methods, 93-95
Ordinary differential equations (ODEs), 50-134
  boundary value problems, 105-130
  classification, 50-51
  eigenvalue problems, 126-130
  initial value problems, 57-105
  linearization of, 60
  stability, 60-69
Orthogonality of eigenvectors, 140
Outer iterations, 194
Overrelaxation, 189
Overrelaxation factor, 189
  optimum:
    for SLOR, 194-197
    for SOR, 192
Overstability, 236, 242

Parabolic PDEs, 136, 138-171
  in cylindrical coordinates, 167-170
  explicit methods, 138-147
  implicit methods, 147-152
  nonlinear, 170-171
  one-dimensional, 138-157
  two- and three-dimensional, 158-167
Parametric interpolation, 21-23
Parasitic root (multistep methods), 86
Partial differential equations (PDEs), 135-247
  classification, 135-138
  elliptic, 171-221
  hyperbolic, 221-251
  parabolic, 138-171
PDE, *see* Partial differential equations (PDEs)
Peaceman-Rachford method, 160-167
Pentadiagonal systems of equations, 116
Periodic extension, 208
Phase error:
  of explicit Euler method, 67
  for hyperbolic PDEs, 236
  of implicit Euler method, 71-72
  of trapezoid rule, 76
Plemelj's formula, 220
Poisson equation, 171, 173, 216-218
Predictor-corrector methods, 76-84, 131

Adams methods, 92
Heun's method, 77-78
stability of, 77
Principal value, 220
Property A, 191

QR algorithm, 128
Quadratic finite elements, 205
Quadrature, *see* Integration
QUANC8, 41
Quasilinear equations, 221

Relaxation, 181
simultaneous, 181
Richardson extrapolation, 30-33, 77, 92
calculation of π, 32-33
and finite differences, 55
in Keller Box method, 157
for ODE methods, 72-73
and Romberg integration, 33-37
Richtmyer method, 245, 246
RKF45, comparison with other methods, 93-95
Romberg integration, 33-37, 38
of exponential function, 34-35
FORTRAN subroutine for, 36-37
of Gaussian function, 35-37
Roundoff error:
explicit method, 59-60
interpolation, 6
in multistep, 86, 88
Runge-Kutta methods, 73, 76-84, 93-95, 130
accuracy, 79
Fehlberg, 80
FORTRAN program, 84
fourth order, 79
Merson, 80-81
Scraton, 80
second order, 78

Scraton Runge-Kutta method, 80
Secant method, 108
Second order difference formula:
first derivative, 53
second derivative, 54
Second order methods for ODEs:
Adams-Bashforth, 89-92
Adams-Moulton, 89-92
for boundary value problems, 110-115
for eigenvalue problems, 126-130
predictor-corrector, 77
Richardson extrapolation, 72-73
Runge-Kutta, 78
trapezoid rule, 73-74
Second order methods for PDEs:
elliptic PDEs, 171-172
hyperbolic PDEs, 238, 241-242, 244-247
parabolic PDEs, one dimensional, 138-147
Semi-discrete methods, 139
Shampine-Gordon method, 93. *See also* ODE (computer program)
Shooting, 105-110, 218
Signal propagation, 136
Simpson's rule, 27, 131
as corrector, 79, 80
ODE method, 57
Simultaneous relaxation, 181
Sine expansions, 218
Sine function, Lagrange interpolation, 7-8
Singularity subtraction, 48
Singular perturbation problems, 109
SLOR, *see* Successive line overrelaxation
SOR, *see* Successive overrelaxation method
Spectral radius, 180
Spherical coordinates, 167
Spline interpolation, 12-21, 23
cubic, 12-20
end conditions, 14-15, 23
error estimate, 15
exponential function, 15-19
FORTRAN program, 17-18
sine function, 19
superellipse, 19
theory, 12-15
Splitting methods:
for elliptic equations, 197-202
for hyperbolic PDEs, 241, 244-247
for parabolic PDEs, 162-167
for stiff systems, 100-101
Stability:
of Adams methods, 89-92
definition, 61-62
of Euler's method, 62
in hyperbolic PDEs, 236
in parabolic PDEs, 141, 145-147
of implicit Euler method, 70
overstability, 236, 242
of predictor-corrector method, 77-78
of Runge-Kutta methods, 79-80

von Neumann method, 141-142
Stiffness, 95-104
in direct methods, 114-115
in eigenvalue problems, 129
explicit methods, 102-104
and parabolic PDEs, 141
in shooting methods, 108
splitting methods for, 100-101
stability of numerical methods, 96-98
Stiffness ratio, 97
Stiff systems of equations, *see* Stiffness
Successive line overrelaxation, 194-197
vs. alternating direction implicit, 201
Successive overrelaxation method, 189-197, 249
vs. alternating direction implicit, 199-200
convergence, 191-194
FORTRAN program, 195
for nonlinear problems, 194
Successive relaxation, 185
Superellipse:
Lagrange interpolation of, 8-11
Newton-Cotes integration, 29-30
Superposition, in shooting methods, 106
SYMEIG (computer program), 128
Systems of ordinary differential equations, 50, 95-105
stiffness of, 95-104

Taylor series:
in convergence of iterative methods, 180
for explicit Euler method, 58
and finite differences, 54-56
for hyperbolic PDEs, 222, 225, 238
for implicit Euler method, 70
linearization of ODEs, 60
in Newton-Raphson, 255
for nonuniform grids, 121
in parabolic PDEs, 142
for predictor-corrector methods, 78
and Richardson extrapolation, 31
for trapezoid rule, 74
Transformation of variable, for boundary value problems, 121-122
Trapezoid rule:
as corrector, 77, 78, 80
and Crank-Nicolson method, 148
for exponential function, 28-29
for integration, 27
and Keller Box method, 156
for ODEs, 57, 73-76
accuracy, 74, 82, 83
stability, 74
TRDIAG (computer program), 164
Triangular finite elements, 126, 204-207
Tridiagonal systems, 253-254
in alternating direction implicit method, 164
in boundary value problems, 111-112
in compact fourth order method, 120
in Crank-Nicolson method, 148, 243-244
in eigenvalue problems, 127-128
FORTRAN program, 254
and numerical differentiation, 56
in parabolic PDEs, 139, 170
QR algorithm, 128
in spline interpolation, 14
Truncation error, 54-56
and approximate factorization, 163, 244-245
in boundary value problems, 116
in elliptic PDEs, 171-177, 183
in hyperbolic PDEs, 244
in parabolic PDEs, 142

Unconditional stability:
Crank-Nicolson, 150, 242
definition, 62
Dufort-Frankel, 152
implicit Euler, 70
see also Stability
Upwind differencing, 237

Variable grid spacing, *see* Nonuniform grids
von Neumann method, 141-142
for ADI method, 161
for Crank-Nicolson, 150
for explicit Euler, 141-142, 235-236
for implicit methods, 241

Wachspress parameters, 199, 201-202
Wave equation, 136, 234, 250
characteristics of, 223-224
FORTRAN program, 233
Lax-Wendroff solution, 238-240
method of characteristics solution of, 229-233
Weights:
in finite element methods, 124-126
for Gauss quadrature, 41-47
tables, 43, 45, 46
for integration, 24

Young's Property A, 191